Smart Polymer Hydrogels: Synthesis, Properties and Applications - Volume I

Smart Polymer Hydrogels: Synthesis, Properties and Applications - Volume I

Editor

Wei Ji

MDPI • Basel • Beijing • Wuhan • Barcelona • Belgrade • Manchester • Tokyo • Cluj • Tianjin

Editor
Wei Ji
College of Bioengineering
Chongqing University
Chongqing
China

Editorial Office
MDPI
St. Alban-Anlage 66
4052 Basel, Switzerland

This is a reprint of articles from the Special Issue published online in the open access journal *Gels* (ISSN 2310-2861) (available at: www.mdpi.com/journal/gels/special_issues/smart_polymer_hydrogels).

For citation purposes, cite each article independently as indicated on the article page online and as indicated below:

LastName, A.A.; LastName, B.B.; LastName, C.C. Article Title. *Journal Name* **Year**, *Volume Number*, Page Range.

ISBN 978-3-0365-6977-2 (Hbk)
ISBN 978-3-0365-6976-5 (PDF)

© 2023 by the authors. Articles in this book are Open Access and distributed under the Creative Commons Attribution (CC BY) license, which allows users to download, copy and build upon published articles, as long as the author and publisher are properly credited, which ensures maximum dissemination and a wider impact of our publications.
The book as a whole is distributed by MDPI under the terms and conditions of the Creative Commons license CC BY-NC-ND.

Contents

About the Editor vii

Wei Ji
Editorial on Special Issue: "Smart Polymer Hydrogels: Synthesis, Properties and Applications—Volume I"
Reprinted from: *Gels* **2023**, *9*, 84, doi:10.3390/gels9020084 1

Shengjia Ye, Bin Wei and Li Zeng
Advances on Hydrogels for Oral Science Research
Reprinted from: *Gels* **2022**, *8*, 302, doi:10.3390/gels8050302 3

Airi Harui and Michael D. Roth
Hyaluronidase Enhances Targeting of Hydrogel-Encapsulated Anti-CTLA-4 to Tumor Draining Lymph Nodes and Improves Anti-Tumor Efficacy
Reprinted from: *Gels* **2022**, *8*, 284, doi:10.3390/gels8050284 19

Seunghyun Lee, Shiyi Liu, Ruth E. Bristol, Mark C. Preul and Jennifer Blain Christen
Hydrogel Check-Valves for the Treatment of Hydrocephalic Fluid Retention with Wireless Fully-Passive Sensor for the Intracranial Pressure Measurement
Reprinted from: *Gels* **2022**, *8*, 276, doi:10.3390/gels8050276 29

Jianxiong Xu, Yuecong Luo, Yin Chen, Ziyu Guo, Yutong Zhang and Shaowen Xie et al.
Tough, Self-Recoverable, Spiropyran (SP3) Bearing Polymer Beads Incorporated PAM Hydrogels with Sole Mechanochromic Behavior
Reprinted from: *Gels* **2022**, *8*, 208, doi:10.3390/gels8040208 43

Muhammad Suhail, Quoc Lam Vu and Pao-Chu Wu
Formulation, Characterization, and In Vitro Drug Release Study of β-Cyclodextrin-Based Smart Hydrogels
Reprinted from: *Gels* **2022**, *8*, 207, doi:10.3390/gels8040207 59

Arshad Mahmood, Alia Erum, Sophia Mumtaz, Ume Ruqia Tulain, Nadia Shamshad Malik and Mohammed S. Alqahtani
Preliminary Investigation of *Linum usitatissimum* Mucilage-Based Hydrogel as Possible Substitute to Synthetic Polymer-Based Hydrogels for Sustained Release Oral Drug Delivery
Reprinted from: *Gels* **2022**, *8*, 170, doi:10.3390/gels8030170 79

Salah Hamri, Tewfik Bouchaour, Djahida Lerari, Zohra Bouberka, Philippe Supiot and Ulrich Maschke
Cleaning of Wastewater Using Crosslinked Poly(Acrylamide-*co*-Acrylic Acid) Hydrogels: Analysis of Rotatable Bonds, Binding Energy and Hydrogen Bonding
Reprinted from: *Gels* **2022**, *8*, 156, doi:10.3390/gels8030156 95

Dominique Larrea-Wachtendorff, Vittoria Del Grosso and Giovanna Ferrari
Evaluation of the Physical Stability of Starch-Based Hydrogels Produced by High-Pressure Processing (HPP)
Reprinted from: *Gels* **2022**, *8*, 152, doi:10.3390/gels8030152 113

Tim B. Mrohs and Oliver Weichold
Multivalent Allylammonium-Based Cross-Linkers for the Synthesis of Homogeneous, Highly Swelling Diallyldimethylammonium Chloride Hydrogels
Reprinted from: *Gels* **2022**, *8*, 100, doi:10.3390/gels8020100 129

Valeria Stagno, Alessandro Ciccola, Roberta Curini, Paolo Postorino, Gabriele Favero and Silvia Capuani
Non-Invasive Assessment of PVA-Borax Hydrogel Effectiveness in Removing Metal Corrosion Products on Stones by Portable NMR
Reprinted from: *Gels* **2021**, *7*, 265, doi:10.3390/gels7040265 . . . . **143**

Alina Elena Sandu, Loredana Elena Nita, Aurica P. Chiriac, Nita Tudorachi, Alina Gabriela Rusu and Daniela Pamfil
New Hydrogel Network Based on Alginate and a Spiroaceta Copolymer
Reprinted from: *Gels* **2021**, *7*, 241, doi:10.3390/gels7040241 . . . . **161**

About the Editor

Wei Ji

Wei Ji is a professor at the college of bioengineering, Chongqing University. He obtained his Ph.D. degree in materials science and engineering from Shanghai Jiao Tong University. He has been a faculty member at Chongqing University since 2021, following the completion of postdoctoral research at Okinawa Institute of Science and Technology and Tel Aviv University. His research interests include molecular self-assembly, bio-inspired optoelectronic materials, and biomedical engineering.

Editorial

Editorial on Special Issue: "Smart Polymer Hydrogels: Synthesis, Properties and Applications—Volume I"

Wei Ji

Key Laboratory of Biorheological Science and Technology, Ministry of Education, College of Bioengineering, Chongqing University, Chongqing 400044, China; weiji@cqu.edu.cn

Smart polymer hydrogels are soft materials formed by crosslinking with various covalent and non-covalent interactions. They can respond to different chemical and physical external stimuli, such as pH, temperature, light, redox agents, electric or magnetic field, and so on. Due to the advantages of biocompatibility, stimuli responsiveness and low cost, smart polymer hydrogels have gained increasing interest as promising soft materials for their tremendous potential applications in biomedical and nanotechnological fields. Fortunately, in this Special Issue, we obtained 11 papers from researchers working in the hydrogel related fields.

The papers in this Special Issue present the research results on the synthesis, properties and applications of smart polymer hydrogels. To obtain a new biohybrid gel structure, Sandu et al. present a strategy by coupling synthetic polymers with natural compounds using a spiroacetal polymer and alginate at different ratios. The release of carvacrol, an encapsulated bioactive compound, could be controlled by changing the ratio of the synthetic polymer [1]. Mrohs et al. reported the effect of multivalent allylammonium-based crosslinkers on the synthesis of homogeneous, highly swelling diallyldimethylammonium chloride hydrogels, which indicated that homogeneity plays important role for the formation of coherent gels with low cross-linking densities [2]. To evaluate the physical stability of starch-based hydrogels via high-pressure processing, Larrea-Wachtendorff et al. performed the accelerated methods based on temperature sweep tests and oscillatory rheological measurements, as well as temperature cycling tests, which could be used in predicting the physical stability of starch-based hydrogels [3].

As promising drug delivery and tissue engineering scaffolds, hydrogels have attracted great interest from researchers for the potential applications in tumour and disease treatment. Suhail et al. reported novel pH-responsive polymeric β-cyclodextrin-based hydrogels using the free radical polymerization technique, which could be utilized for the controlled delivery of theophylline [4]. Mahmood et al. synthesized *Linum usitatissimum* mucilage polymer-based hydrogel for sustained release of oral drugs. Compared to the conventional synthetic polymers, the developed above hydrogel showed many-fold benefits in the in vitro experiments for oral delivery of drugs [5]. Harui et al. reported a hyaluronidase and anti-CTLA-4 containing hydrogel to target tumour draining lymph nodes, which further improved the anti-tumour efficacy examined by employing live and ex vivo imaging [6]. Lee et al. reported a hydrogel-based fully realized and passive implantable valve for the treatment of hydrocephalic fluid retention. The reproducibility of the hydrogel-based valve and sensor functions was demonstrated, indicating the system's potential applications as a chronic implant [7]. Ye et al. summarized the recent advances of hydrogels as soft materials in the tissue engineering area for the treatment of oral disease [8].

Hydrogels have three-dimensional pore networks showing the application in cleaning the wastewater; Hamri et al. reported a theoretical model based on docking simulations and applied the model to analyze the adsorption process of a polymer-based hydrogel in cleaning of wastewater [9]. Stagno et al. presented a method for non-invasive assessment of a PVA hydrogel in removing metal corrosion products on stones by portable Nuclear

Citation: Ji, W. Editorial on Special Issue: "Smart Polymer Hydrogels: Synthesis, Properties and Applications—Volume I". *Gels* **2023**, *9*, 84. https://doi.org/10.3390/gels9020084

Received: 13 January 2023
Accepted: 15 January 2023
Published: 19 January 2023

Copyright: © 2023 by the author. Licensee MDPI, Basel, Switzerland. This article is an open access article distributed under the terms and conditions of the Creative Commons Attribution (CC BY) license (https://creativecommons.org/licenses/by/4.0/).

Magnetic Resonance (NMR), indicating that the NMR protocol cold be used for in situ analysis of cleaning efficacy and the effect of different hydrogels on different materials [10]. Xu et al. synthesized a new polymer hydrogel in the presence of spiropyran dimethacrylate mechanophore crosslinker. The developed polymer hydrogel could display the sole mechanochromic behavior, holding great potential in outdoor strain sensors [11].

We hope that the papers published in this Special Issue will inspire the design and synthesis of more smart polymer hydrogels for biomedical and nanotechnological applications. The Editorial Team thanks all the authors who contributed to the Special Issue of "Smart Polymer Hydrogels: Synthesis, Properties and Applications—Volume I".

Conflicts of Interest: The author declares no conflict of interest.

References

1. Sandu, A.E.; Nita, L.E.; Chiriac, A.P.; Tudorachi, N.; Rusu, A.G.; Pamfil, D. New Hydrogel Network Based on Alginate and a Spiroacetal Copolymer. *Gels* **2021**, *7*, 241. [CrossRef] [PubMed]
2. Mrohs, T.B.; Weichold, O. Multivalent Allylammonium-Based Cross-Linkers for the Synthesis of Homogeneous, Highly Swelling Diallyldimethylammonium Chloride Hydrogels. *Gels* **2022**, *8*, 100. [CrossRef] [PubMed]
3. Larrea-Wachtendorff, D.; Del Grosso, V.; Ferrari, G. Evaluation of the Physical Stability of Starch-Based Hydrogels Produced by High-Pressure Processing (HPP). *Gels* **2022**, *8*, 152. [CrossRef] [PubMed]
4. Suhail, M.; Vu, Q.L.; Wu, P.-C. Formulation, Characterization, and In Vitro Drug Release Study of β-Cyclodextrin-Based Smart Hydrogels. *Gels* **2022**, *8*, 207. [CrossRef] [PubMed]
5. Mahmood, A.; Erum, A.; Mumtaz, S.; Tulain, U.R.; Malik, N.S.; Alqahtani, M.S. Preliminary Investigation of *Linum usitatissimum* Mucilage-Based Hydrogel as Possible Substitute to Synthetic Polymer-Based Hydrogels for Sustained Release Oral Drug Delivery. *Gels* **2022**, *8*, 170. [CrossRef] [PubMed]
6. Harui, A.; Roth, M.D. Hyaluronidase Enhances Targeting of Hydrogel-Encapsulated Anti-CTLA-4 to Tumor Draining Lymph Nodes and Improves Anti-Tumor Efficacy. *Gels* **2022**, *8*, 284. [CrossRef] [PubMed]
7. Lee, S.; Liu, S.; Bristol, R.E.; Preul, M.C.; Blain Christen, J. Hydrogel Check-Valves for the Treatment of Hydrocephalic Fluid Retention with Wireless Fully-Passive Sensor for the Intracranial Pressure Measurement. *Gels* **2022**, *8*, 276. [CrossRef] [PubMed]
8. Ye, S.; Wei, B.; Zeng, L. Advances on Hydrogels for Oral Science Research. *Gels* **2022**, *8*, 302. [CrossRef] [PubMed]
9. Hamri, S.; Bouchaour, T.; Lerari, D.; Bouberka, Z.; Supiot, P.; Maschke, U. Cleaning of Wastewater Using Crosslinked Poly(Acrylamide-*co*-Acrylic Acid) Hydrogels: Analysis of Rotatable Bonds, Binding Energy and Hydrogen Bonding. *Gels* **2022**, *8*, 156. [CrossRef] [PubMed]
10. Stagno, V.; Ciccola, A.; Curini, R.; Postorino, P.; Favero, G.; Capuani, S. Non-Invasive Assessment of PVA-Borax Hydrogel Effectiveness in Removing Metal Corrosion Products on Stones by Portable NMR. *Gels* **2021**, *7*, 265. [CrossRef] [PubMed]
11. Xu, J.; Luo, Y.; Chen, Y.; Guo, Z.; Zhang, Y.; Xie, S.; Li, N.; Xu, L. Tough, Self-Recoverable, Spiropyran (SP3) Bearing Polymer Beads Incorporated PAM Hydrogels with Sole Mechanochromic Behavior. *Gels* **2022**, *8*, 208. [CrossRef]

Disclaimer/Publisher's Note: The statements, opinions and data contained in all publications are solely those of the individual author(s) and contributor(s) and not of MDPI and/or the editor(s). MDPI and/or the editor(s) disclaim responsibility for any injury to people or property resulting from any ideas, methods, instructions or products referred to in the content.

Review

Advances on Hydrogels for Oral Science Research

Shengjia Ye [1,2], Bin Wei [2,3,*] and Li Zeng [1,2,*]

[1] Department of Prosthodontics, Shanghai Ninth People's Hospital, Shanghai Jiao Tong University School of Medicine, Shanghai 200011, China; hello.ada@outlook.com

[2] College of Stomatology, Shanghai Jiao Tong University, National Center for Stomatology, National Clinical Research Center for Oral Diseases, Shanghai Key Laboratory of Stomatology, Shanghai Engineering Research Center of Advanced Dental Technology and Materials, Shanghai 200011, China

[3] Department of Stomatology Special Consultation Clinic, Shanghai Ninth People's Hospital, Shanghai Jiao Tong University School of Medicine, Shanghai 200011, China

* Correspondence: drweibin2003@126.com (B.W.); kqzengli@shsmu.edu.cn (L.Z.)

Abstract: Hydrogels are biocompatible polymer systems, which have become a hotspot in biomedical research. As hydrogels mimic the structure of natural extracellular matrices, they are considered as good scaffold materials in the tissue engineering area for repairing dental pulp and periodontal damages. Combined with different kinds of stem cells and growth factors, various hydrogel complexes have played an optimistic role in endodontic and periodontal tissue engineering studies. Further, hydrogels exhibit biological effects in response to external stimuli, which results in hydrogels having a promising application in local drug delivery. This review summarized the advances of hydrogels in oral science research, in the hopes of providing a reference for future applications.

Keywords: hydrogel; oral science; tissue regeneration; drug delivery system

1. Introduction

Hydrogel is a polymer network system formed by cross-linking the reaction of monomers and comprised of water-encapsulating networks [1]. It distinguishes itself from other biological materials by its unique characteristics in structure and performance. The polymer network formed by the hydrogel can bind water, which in turn shows good biocompatibility due to the high moisture content [2,3]. When the hydrogel is combined with biological tissue, its swelling property blurs the boundary between the hydrogel and the tissue, reduces the surface tension, and lessens the surface adhesion of cells and proteins, thus reducing the foreign body reactions [4,5]. Friction and mechanical damage to surrounding tissues can be relatively reduced after hydrogels absorbed water.

The three-dimensional network structure and viscoelasticity of the hydrogel are similar to the extracellular matrix (ECM), which can mimic the three-dimensional microenvironment of cells, support cells attachment, and induce cells proliferation and differentiation. Because of their favorable properties [3], hydrogels could meet the general requirements of scaffold and drug carriers. Previous studies have shown that hydrogels have been widely applied in biomedical studies of skin, vessels, cartilage, bone, and muscle tissue regeneration [6–10] (Summerized in Figure 1).

Oral health is considered an important part of general health and quality of life [11], and oral disease is still a major public health problem in developed countries and a growing burden for developing countries [12]. Common oral diseases include caries, periodontitis, pulp necrosis, oral mucositis, and so on. Oral science research has developed rapidly in recent years, and hydrogels have become a research hotspot in this field. To exert biological effects accurately and effectively, various hydrogels ranging from natural ones, and synthetic ones to composite hydrogels are being studied [13,14]. This paper reviews the application progress of hydrogels in oral tissue engineering and drug delivery, aiming to provide a reference for the subsequent research and application of biological materials.

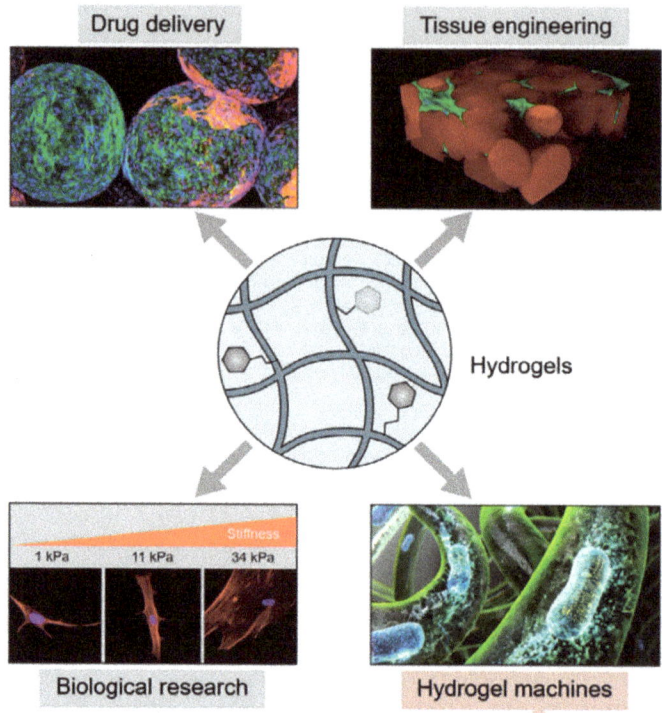

Figure 1. Schematic diagram of hydrogels in tissue engineering (Figure 1 is adapted from reference [10]).

2. Application of Hydrogels for Dental Pulp Regeneration

Dental pulp, also called endodontium, is located inside the pulp cavity of the tooth. Pulp tissue mainly contains nerves, blood vessels, lymphatic and connective tissues, as well as odontoblasts arranged in the outer periphery of the pulp, whose role is to produce dentin. Dental pulp plays an essential role in the maintenance of blood circulation and homeostasis, sensory transmission, and regeneration of dentin [15]. After conventional root canal treatment due to irreversible pulpitis and pulp necrosis, the pulpless teeth lose their natural biological defense, which may raise the risk of serious caries, apical periodontitis, and ultimately tooth loss [16,17]. Thus, the concept of dental pulp regeneration was put forward to recover the function of teeth and improve the prognosis of a pulpless tooth [14,18].

Pulp regeneration has raised great concerns in the treatment of pulp disease during the past few decades. The American Association of Endodontists (AAE) defines pulp regeneration as, "use biological means to replace damaged dental tissue, root, pulp–dentin complex and other structures to form functional pulp-like tissue". Studies on pulp regeneration mainly take histoengineering principles and means to induce differentiation of pulp–dentin complex through stem cells–scaffold–growth factor complexes [13], so as to repair damaged pulp tissue and restore physiological functions [14].

The scaffold materials play a variety of roles during this procedure [19], not only limited to providing a three-dimensional structural bracket for cell planting, adhesion, proliferation, and spatial distribution but also regulating cell behavior and intracellular signaling, simulating the recovery of the microenvironment of cell life, and the extra-cellular matrix. The action mechanism of hydrogels meets the mentioned requirements precisely [13,20], they act as carriers of stem/progenitor cells with odontogenic potential [21–27], carriers of local bioactive molecules [22,28–30], and release bioactive factors during degrading [31].

Pulp regeneration with hydrogels has become a reality and has been promoted and verified through molecular and developmental biology, as well as biomimetic principles and histological approaches [18,32]. The schematic illustration of the ideal pulp regeneration procedure is shown in Figure 2.

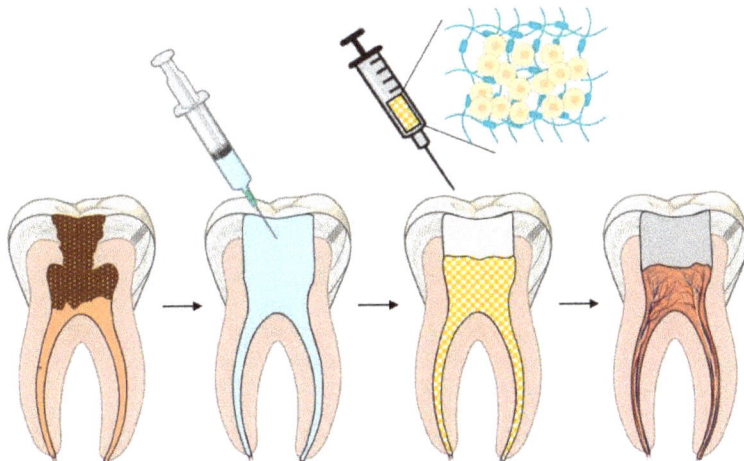

Figure 2. Schematic illustration of ideal pulp regeneration procedure.

Rosa et al. confirmed that stem cells from exfoliated deciduous teeth (SHED) can generate a functional dental pulp when injected into full-length root canals [26]. SHED survived and began to express specific molecules of odontoblastic differentiation when mixed with commercial peptide hydrogel and recombinant human collagen type I, respectively. Pulp-like tissues were observed with functional odontoblasts throughout the root canals in vivo, presenting similar cellularity and vascularization when compared with control human dental pulps. It appears that the physical properties of the scaffold [33,34], such as viscosity and mechanical capacity, play an important role in dental pulp tissue regeneration. A co-culture of dental pulp stem cells (DPSCs) and human umbilical vein endothelial cells (HUVECs) resulted in the formation of micro-vessels in the bio-printed collagen hydrogel structure within 2 weeks of in vitro culture. Excellent biocompatibility made collagen gel a good choice for the scaffold, while a potential drawback of this hydrogel is shrinkage and rapid degradation in vivo [35].

Chrepa et al. tested the hypothesis that a Food and Drug Administration-approved hyaluronic acid-based injectable gel may be a promising scaffold material for regenerative endodontics. Improvement of stem cells of the apical papilla (SCAP) survival, mineralization, and differentiation into an odontoblastic phenotype was observed in this research [36].

Chitosan, a natural biopolymer derived from chitin, was also found to be able to promote the differentiation and proliferation of dental pulp stromal/stem cells (DPSCs) [37]. Feng et al. [38] utilized small 3D porous chitosan scaffolds fabricated by freeze-drying to support neural differentiation of DPSCs in vitro. Chitosan hydrogel exhibits good conductivity and forms a suitable template. However, some other researchers observed that adding the additional chitosan scaffolds in regenerative procedures did not improve the formation of new mineralized tissues along the root canal walls and the pulp–dentin complex [39,40].

RGD-alginate hydrogels significantly enhance cell adhesion and proliferation [41]. An RGD-bearing alginate framework, that is simply shaped, was used to encapsulate DPSCs and HUVECs equally by Bhoj's team [28]. Adding dual growth factors to co-culture stem cells within RGD-alginate scaffolds led to the creation of micro-environments that significantly enhanced the proliferation of dental pulp stem cell/human umbilical vein endothelial cell combinations.

The above natural hydrogels are biocompatible, biodegradable, and optimistically bioactive with the ability to release bioactive molecules [13,42]. However, natural ones may carry the risk of disease transmission, immune response, batch variation, and poor mechanical properties [43]. Synthetic hydrogels were developed, characterized by easy standardization, large-scale production, adjustable mechanical properties, and microstructure without the risk of disease transmission. Synthetic hydrogels facilitate regeneration when cooperated with biologically active molecules and cell-binding sequences [44].

Currently, injectable composite hydrogels have become a promising application option in pulp tissue engineering [45,46]. UV light-crosslinked gelatin meth-acryloyl (GelMA) hydrogel has been used to create tissue-engineered pre-vascularized dental pulp-like constructs. Injectable GelMA for DPSCs/HUVECs can promote cell adhesion and proliferation, and meanwhile promote angiogenesis [47]. The survival rates of encapsulating dental pulp cells in GelMA were over 80% [48] and 90% [23] in different studies. Although the optical cross-linking procedure may reduce viability, the manufactured GelMA hydrogel combined with hDPSC/HUVECs posed well in the formation of the vasculature [47]. Studies have also shown that HyStem-C, an injectable composite hydrogel synthesized from polyethylene glycol diacrylate-hyaluronic acid-gelatin, also had good compatibility with DPSCs [49].

However, one of the main limitations of hydrogels is the spatial manipulation restriction, i.e., researchers are unable to fully control the organization and interactions of multiple cells, so the overall morphogenesis of tissues cannot be guaranteed totally. Luckily, combined with superior spatial control of 3D cell printing, this problem can be overcome in the near future [50,51]. The 3D cell printing technology will enable researchers to suspend and place various cells in a hydrogel. For instance, researchers can print odontoblasts along the dentin wall while having fibroblasts in the center of the pulp cavity. While the theoretical application of 3D cell printing in pulp tissue regeneration sounds feasible, there has been a lack of evidence so far. Several studies [50,52] have demonstrated the possibility of success in 3D printing capillaries, but in vivo angiogenesis has not been reported in this area. Although there are few in vivo studies currently, several studies have highlighted the potential application of 3D cell printing in pulp regeneration. For example, in a study by Athirasala and his colleagues, they showed that a novel hydrogel consisting of alginate and dentin (algn-dent) can support mouse odontoblast-like cell lines (OD21) [53].

Currently, in animal models, it appears to be possible to regenerate pulp and dentin, although challenges include the absence of dentin tubular formation, as well as difficulties in dealing with smaller tubes due to angiogenesis [54,55] (Summerized in Figure 3).

The use of growth factors and hydrogel scaffolds accelerates clinical translation and enhances dental tissue engineering, which is expected to be the best biological solution in endodontic medicine [32]. Hydrogel–cell complex-based regenerative endodontics is still in the experimental stage now. AAE (2018) and ESE (2016) have not yet recommended transplanting autologous or allogeneic stem cells in clinical regenerative treatment as the work relates to stem cell isolation, in vitro expansion, good manufacturing practice facilities, stem cell banks, government regulatory issues, clinician skills, training of chairside assistants, and relatively high costs [56].

Recent studies suggested that a hydrogel complex [13] may be a strategy to facilitate pulp tissue regeneration (Figure 4). However, in the field of pulp engineering, only a small amount of hydrogels with specific components have been studied in vivo, and there has been no clinical research report so far. In addition, there is still a lack of comparative studies of different hydrogels, further studies are required to enrich current knowledge in pulp tissue regeneration.

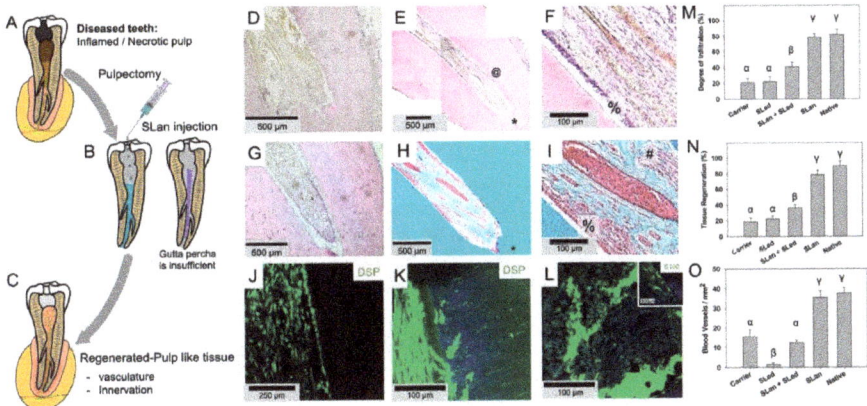

Figure 3. The application of hydrogel in dental root canals in animal model (Figure 3 is adapted from reference [55]). (**A**) Caries and trauma may lead to the inflammation and necrosis of the pulp. (**B**) After pulpectomy, implantation of injectable angiogenic SLan hydrogels help regenerate (**C**) vascularized pulp-like soft tissue in 28 days. In a canine pulpectomy model, disorganized blood clots form for over-instrumentation carrier filled (sucrose-HBSS) control (**D**). H&E staining of tooth roots of SLan filled teeth showed rapid infiltration of cells and tissue (**E**), and within crevices in the canal space (@), along with an odontoblast-like layer in apposition to the dentin wall (**F**-%). Control dentinogenic SLed hydrogels lead to disorganized tissue (**G**). Trichrome staining of SLan implants reveals blood vessels (**H,I**) with collagen deposition (blue); and an odontoblast-like layer (**I**-%) which stains with dental sialoprotein (DSP) (**J**) with cytoplasmic protrusions into dentinal tubules (**K**). S100+ Nerve bundles (Trichrome I-#) were regenerated along the length of the canal (**L** and inset). (**M**) Degree of infiltration, (**N**) degree of tissue regeneration, and (**O**) densities of blood vessels were similar for SLan and native teeth but significantly greater than controls.

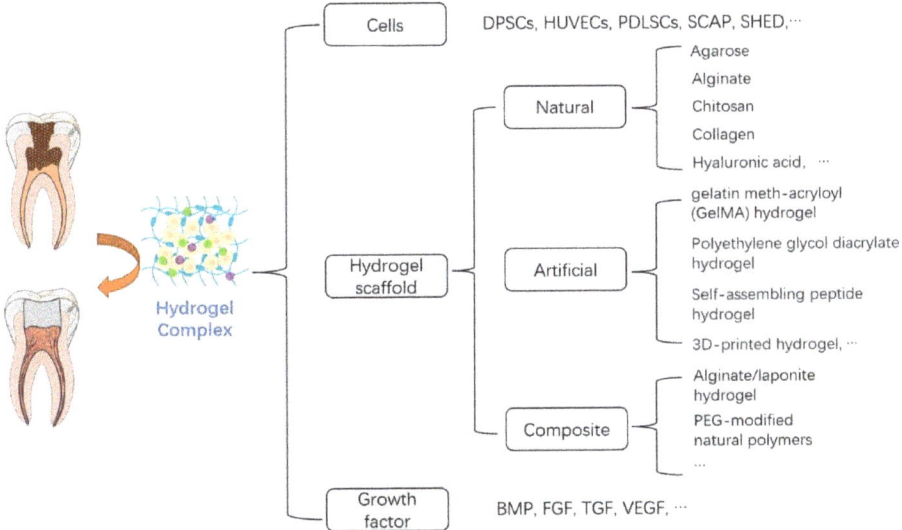

Figure 4. The classification of hydrogel complex in pulp tissue regeneration.

3. Application of Hydrogels for Periodontal Tissue Regeneration

Periodontal disease is a worldwide health problem that exerts a negative influence on patients. Periodontitis is a chronic inflammatory disease of the periodontal tissue caused by pathogenic microorganisms, with the characteristic of the destruction of teeth supporting structures [57]. Inflammation starts in gums, then penetrates deep, finally resulting in a periodontal pocket of bacteria that erodes the supporting ligaments of the teeth until they are lost [58]. Periodontal diseases lead to certain damage to the nearby tissues, such as loss of attachment, alveolar bone resorption, tooth loosening, etc., which eventually induces tooth loss and endangers oral health and even the whole body [59].

Traditional therapies, including mechanical plaque removal and scaling, are not effective enough in the long term [60]. This invasive method of scaling and root planning (SRP) may result in unpleasant side effects such as sensitivity and tooth topical damage [61]. Classical treatments for periodontitis are time-consuming, technically-sensitive but sub-optimal in the repair of tissue defects [62]. So, alternatives are being looked for in the scientific world. The ideal ultimate therapeutic purpose of periodontal disease is periodontal tissue regeneration to reconstruct both structures and functions.

Some strategies have been conducted to regenerate periodontal tissue, such as the guided tissue/bone regeneration membranes [63]. These applications are promising, while challenges still exist, including low cell transplantation, inaccurate cell localization, immune rejection, difficulty in effectively providing the required growth factors, and inability to control the tissue types that form. Defective areas may be deficient in cells and microvascular formation [1,64]. The main challenge, however, comes from the fact that the periodontal complex is a hybrid tissue unit [1] that consists of highly specialized neural and mechanical receptors, gingiva, alveolar bones, periodontal ligaments (PDL), and cementum. Current regeneration practices focus primarily on the regeneration of individual tissues, unable to simulate and regenerate such complex architectures yet.

Recently, hydrogels have been widely applied as a sustained-release system and scaffold materials in periodontal tissue regeneration research [65,66]. While different kinds of hydrogels can be used for dentoalveolar tissue regeneration, their modification or combination is often required for successful strategies [1]. The schematic illustration of the ideal periodontal regeneration procedure is shown in Figure 5.

In the field of alveolar bone regeneration, hydrogels based on hyaluronic acid (HA) have been used with different strategies to augment their mechanical properties. In Miranda's study, modified hyaluronic acid (HA) and chitosan (CS) were employed to create a hybrid CS-HA hydrogel scaffold [67], which combined the advantages of both ingredients. These porous structures proved suitable for periodontal tissue engineering because the cells migrated more when seeded. Polycaprolactone (mPCL) constructs combined with osteoblasts encapsulated in HA-hydrogel and bone morphogenetic protein-7 (BMP-7) have been proposed in Hamlet's study, and the constructs were proven to be suitable for mineral deposition in vivo implantation [68].

In addition to cell encapsulation, the combination of GelMA and polyethylene glycol (PEG) has been used for bioprinting regeneration of periodontal tissue [69]. Periodontal ligament stem cells (PDLSCs) encapsulated in this material exhibited higher viability and diffuseness in lower concentrations of PEG, while PEG enhanced the ability to control droplets. In a study on alveolar bone regeneration [70], the further performance of this material was analyzed and its stiffness was observed in the range of 4.5–23.5 kPa, and in vivo analysis results showed bone formation within 6 weeks after implantation. Still, due to the lack of further in-depth descriptions and degradation performance of the structure, structural integrity remains unknown in the long run and needs further studies.

Duarte Campus et al. [71] investigated the effects of the incorporation of collagen in a 3D bio-printed polysaccharide hydrogel on the regulation of cell morphology, osteogenesis potential, and mineralization. The mechanical properties and viscosity increased by combining thermo-responsive agarose hydrogel with collagen type I, which poses a better contour and construct than collagen individually. These composite hydrogels with a high-collagen ratio turned out to be more feasible for mesenchymal stem cells (MSCs) osteogenic differentiation. However, a hydrogel with a compression modulus lower than the natural bone may lead to complications of implant integration, particularly in the load-bearing region [72], which indicates the need to consider adding more mechanically strong materials [73].

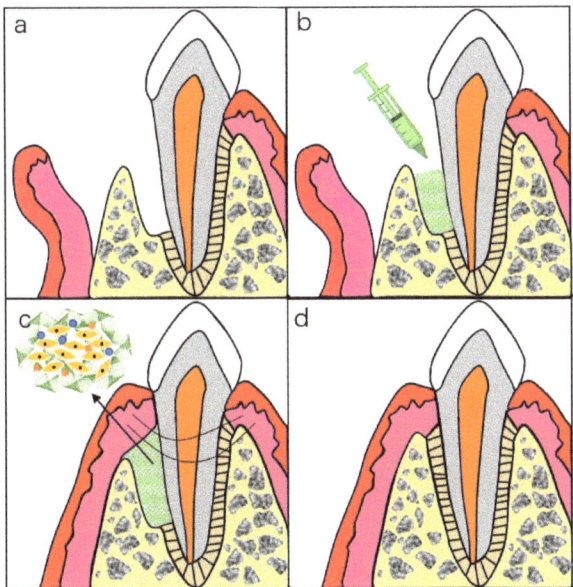

Figure 5. Schematic illustration of injectable hydrogels for periodontal repair. (**a**) Periodontal defect with loss of PDL and alveolar bone. (**b**) Inject hydrogel complexes into the defected site. (**c**) Sewing for closure of the wound. (**d**) Ideal repairment of periodontal tissue.

PDL, also known as the periodontal ligament, is a highly organized tissue between the cementum and alveolar bone. PDL is capable of taking extremely high forces, which poses a huge challenge for tissue engineering [74]. Constructs combined with hydrogel and stem cells are recommended because of the limited regenerative space. The 3D hydrogel complexes were proposed for a cell-laden array, and the GelMA/PEG composition could be used for periodontal regeneration based on PDLSCs [69]. Yan et al. demonstrated that enzymatically solidified chitosan hydrogels are highly biocompatible and biodegradable. Moreover, chitosan hydrogels without cell loading can improve periodontal regeneration in terms of functional ligament length, indicating the great potential of this hydrogel in clinical applications [75].

A major challenge in periodontal regeneration lies in the complexity [76] of tissue types and variation of repair speed. The introduction of a 3D-printed multiphase scaffold may make the constructs more similar to natural structures with tunable physicochemical and biological characteristics. Lee et al. [77] reported a multiphase matrix produced by bioprinting with different microchannel compartments, which can induce different tissue regeneration as assumed integration. Comprehensive strategies are required in need of regional tissue traits [1], while the network structure and crosslinking process are being dug into to enhance regeneration [78].

Scaffold materials support tissue regeneration to a certain extent, but they may not have the ability to induce tissue regeneration individually. Growth factors, a class of active signaling molecules, can regulate cell growth and other cellular functions by bonding to specific, high-affinity cell membrane receptors. Researchers combined collagen hydrogel scaffold with fibroblast growth factor-2 (FGF2) [79]. This growth factor is able to upregulate cell behaviors and accelerate wound healing to evaluate wound healing in furcation defects in vivo. This application promoted massive cellular and tissue in the growth containing blood vessel-like structure at day 10 and alveolar bone regeneration at 4 weeks. The periodontal attachment was also observed, showing that the FGF2-loaded scaffold was able to guide, reconstruct the function, and self-assemble periodontal organs without abnormal healing.

Chien's group applied an injectable and thermosensitive chitosan/gelatin/glycerol phosphate hydrogel to provide a 3D environment for transplanted induced pluripotent stem cells (iPSCs) and to enhance stem cell delivery and engraftment [66]. The iPSCs-BMP-6-hydrogel complex promoted osteogenesis, the differentiation of new connective tissue, and the periodontal ligament formation in vivo and reduced the levels of the inflammatory cytokine at the mean time. Hydrogel-encapsulated iPSCs combined with BMP-6 provided a new strategy to enhance periodontal regeneration versatilely.

Xu's group integrated chitosan, β-sodium glycerophosphate (β-GP), and gelatin to prepare an injectable and thermosensitive hydrogel, which intended to terminate the alveolar bone resorption with simultaneous anti-inflammation and promote periodontium regeneration [80]. The transition occurred at body temperature while seeding in vivo. After being drug-loaded, the hydrogel complex can continuously release aspirin and erythropoietin (EPO) to exert pharmacological effects of anti-inflammation and tissue regeneration, respectively. Both in vitro and in vivo study results demonstrated the potentialities of the hydrogel system in periodontal treatment applications (Smmerized in Figure 6).

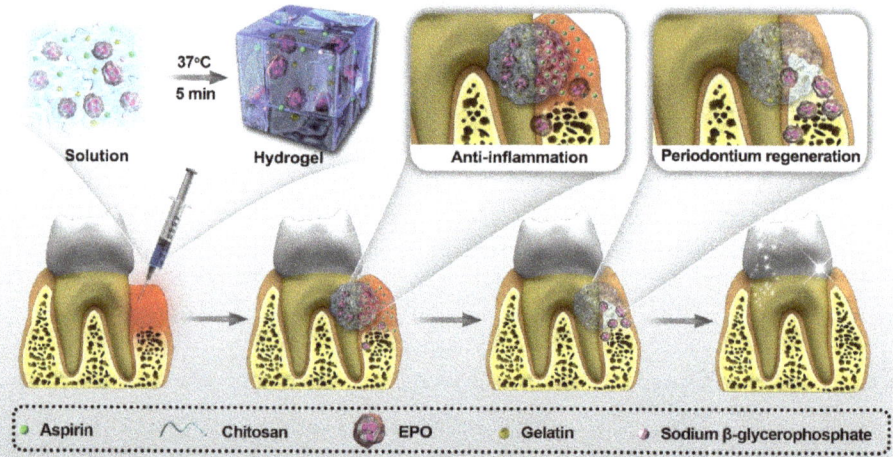

Figure 6. The application of hydrogel complex in periodontal treatment (Figure 6 is adapted from reference [80]).

A supramolecular hydrogel, SDF-1/BMP-2/NapFFY, was fabricated by combining NapFFY with SDF-1 and BMP-2 recently [81]. It was reported that the two bioactive factors released from the constructs ideally and continuously promote periodontal bone reconstruction both in vitro and in vivo. Specifically, a superior bone reconstruction rate of 56.7% was observed in the treatment of periodontal bone defect model rats after 8 weeks.

In short, periodontal tissue engineering with multiple kinds of hydrogels loaded with various mesenchymal stem cells or bioactive molecules is a promising therapy for an injured periodontal environment. Synthesized hydrogels have great potential for future clinical application, which urges more concerns and investigations in this field. No doubt these novel hydrogels could be able to alter transplantation in the clinic in the near future to repair periodontal defects [32].

4. Application of Hydrogels for Drug Delivery in Oral Science

As the common oral cavity diseases locate relatively superficially, the best therapy to control may be regional treatment. Conventional oral drug delivery systems (DDS), such as lozenges and oral spray, work to deliver active drugs topically, while disagreements aroused because of their short residence and instability in saliva [82]. Potential systemic toxicity and low accumulation at target sites are also significant drawbacks of the traditional ones [83,84]. In recent years, new DDSs have attracted the increasing attention of researchers [83,85] because of their ability to provide higher drug absorption and other routes of administration, efficient drug targeting, and lower systemic toxicity.

Different kinds of DDSs are being developed [86] including hydrogel, liposomes [87], electrospun nanofibers, mucoadhesive films, and micelles. A primary defect of the topical therapeutic administration is insufficient residence in the oral cavity. Take liposomal delivery systems, for instance, limitations of instability, drug leakage, and difficulties in large-scale manufacture cannot be ignored [88], although the liposomal antimicrobial agents targeting biofilms have proven effective.

As is depicted in the above text, hydrogels can absorb a large amount of liquid and swell due to their fantastic hydrophilicity, with good viscoelasticity and longer residence time. They are introduced as a novel DDSs to encapsulate various therapeutic agents/compounds and release them in a controlled manner [89]. A recent review discussed the environment-sensitive hydrogels as the "smart" ones, which are able to respond to various multiple stimuli, such as temperature, pH, light, enzymes, pressure, and so on, therefore, it is a promising approach to be used in clinic [90]. Despite releasing effective compounds with a controlled profile by hydrogel complex, some intelligent systems have been fabricated using physical and chemical stimuli as a sensor [91,92]. Temperature-sensitive hydrogels transform from the sol to gel phase at a body temperature of 37 °C [80,93–95] and facilitate drug release. Photosensitive hydrogels are supposed to be activated by a certain wavelength of light, generating ROS to kill microorganisms as well as phase transformation [96,97]. Several pH-sensitive drug delivery hydrogels with the ability to swell or shrink in response to pH changes have been reported, where the polymers could either accept or release protons in response to changes in pH in the microenvironment [98,99].

In pathological conditions, specific changes would occur in the local microenvironment of the tissue, such as local pH reduction under various conditions. As a drug delivery carrier, certain hydrogels complexes are fabricated to respond to local pathological stimuli and achieve delivery at a very point, affecting the biological distribution and toxicity of drugs.

Several researchers constructed an agarose hydrogel system for biomimetic mineralization of dentin [100] and enamel [101]. The designed systems displayed a good condition of mineralization in vitro, analyzed with scanning electron microscopy, X-ray diffraction, Fourier transform infrared spectroscopy, and the nanoindentation hardness test. Muşat's team first reported the simultaneous use of chitosan (CS) and agarose (A) in a biopolymer-based hydrogel for the biomimetic remineralization of an acid-etched native enamel surface [102]. They observed analogous Ca/P compound covered on natural tooth enamel, and found the microhardness recovery of the enamel-like layer under CS-A hydrogels by a 7-day remineralization process in artificial saliva. Ren's team designed a more clinically powerful anti-caries treatment by combining amelogenin-derived peptide QP5 with antibacterial chitosan in a hydrogel (CS-QP5 hydrogel), and reported an inhibition of cariogenic bacteria and the promotion of remineralization of initial caries lesions [103]. Therefore, these methods provide the experimental basis for remineralization and novel strategies to treat dentin hypersensitivity and dental caries.

Antimicrobial activity improves when hydrogels are loaded with antibiotics [104]. Aksel et al. found that the antibiotic-loaded chitosan-fibrin hydrogel enhanced the antibacterial property against E. faecalis biofilm [105]. Metronidazole and ciprofloxacin-loaded chitosan were found more suitable due to their perfect antibacterial property while maintaining cellular function. Yan et al. applied GelMA hydrogel as a carrier of metronidazole (MTR) and chlorhexidine (CHX) [106], and obvious antimicrobial effects against E. faecalis,

S. mutans, and P. intermedia were noticed. A similar application with GelMA and CHX was taken by Ribeiro et al. [107], they formulated injectable chlorhexidine (CHX)-loaded nanotube-modified GelMA hydrogel which provided the sustained release of CHX for dental infection ablation against E. faecalis. Ren et al. [103] designed CS-QP5 hydrogel which has a good antibacterial potency toward Streptococcus mutans by reducing adhesion and biofilm formation. Drug-loaded hydrogels might be a promising material for root canal disinfection and carious treatment to inhibit the dental interest of bacteria.

It is certain that periodontitis initiates from uncontrolled plaque which includes various microorganisms [108]. Bacterial infections are the main reason for the destruction of periodontal tissue. Local medication raised more attention instead of conventional systemic antibiotic therapy [86]. Periodontal sustained-release medications can prolong the duration of drug action and reduce the number of administrations [109–111]. An injectable and photo-cross-linkable gelatin methacryloyl (GelMA) hydrogel was engineered with ciprofloxacin (CIP)-eluting short nanofibers for oral infection ablation by Ribeiro et al. [110]. The hydrogels promoted localized, sustained, and effective cell-friendly antibiotic doses, meaning a good efficacy in inhibiting Enterococcus faecalis inflammation. Chang et al. designed a naringin-carrying CHC-β-GP-glycerol colloidal hydrogel [111], which can be used to inhibit experimental periodontitis with favorable handling and inflammation-responsive characteristics. A chitosan membrane containing polyphosphoester and minocycline hydrochloride (PPEM) was prepared in Li's research [112]. During the progression of the periodontitis, overexpressed ALP will promote the degradation of PPE and the release of antibiotics in the meantime. Liang's team came up with an optimal formulation of carbomer hydrogel, toluidine blue O (TBO) and NaOH, which improved the therapeutic effect of the original photodynamic therapy against Staphylococcus aureus and Escherichia coli [113]. Therefore, photodynamic therapy with the novel optimized TBO hydrogel formulations can be a promising strategy to treat periodontitis.

Hydrogel administration is conducted by injecting into the infected periodontal pocket [114–116], maintaining a controlled and constant concentration of the target drug, which cannot be removed by salivary flush. Side effects will be lessened with interesting potential for endogenous repair of alveolar bone [117].

Oral mucosal diseases such as lichen planus, aphthous stomatitis, oral mucositis, and wounds mostly require effective topical therapies. The primary problem in topical administration of therapeutic agents lies in the low residence time on the smooth and moist surface of oral soft tissue [86].

Hydrogels can be applied in mucosal injury as well for their elastic, adhesive, and degradable characteristics. Andreopoulos et al. [118] reported a method to prepare light-tunable PEG-NC gel scaffolds and the delivery of bFGF from the hydrogels could be controlled by altering the gel properties. They proposed that hydrogels can be applied as a wound healing membrane to treat chronic wounds. Carbomer hydrogels were also proven effective to promote greater residence time on the mucosa when the Carbopol® 980 was combined with lipid nanoparticles (NLC) for buccal administration [119]. Zhang et al. created a photo-triggered hydrogel adhesive [120], which operated on a fast S-nitrosylation coupling reaction and connected to host tissues. This novel hyaluronic acid gel was able to protect mucosal wounds for more than 24 h. The results from animal oral mucosa repair models demonstrated that this hydrogel adhesive created a favorable microenvironment for tissue repair and shortened tissue healing time, illustrating a promising therapy to advance the treatment of oral mucosal defects.

The proposal of a thermally sensitive mucoadhesive hydrogel aimed to facilitate the treatment of oral mucositis, which contained Trimethyl chitosan (TMC) and methylpyrrolidinone chitosan (MPC) [9,121]. Mixed with glycerophosphate (GP) according to different ratios, the best properties were shown. In addition, anti-inflammatory drugs such as benzydamine hydrochloride could be loaded on the complex, which showed good antimicrobial properties.

Antioxidants were also mixed with hydrogels to play roles in the oral cavity, and an isoguanosine–tannic acid (isoG-TA) supramolecular hydrogel was fabricated with leukoplakia (OLK) by Ding et al. [122]. Results showed that the proliferation of dysplastic oral keratinocytes (DOKs) was inhibited due to the antioxidant property of the complex. Azadikah and his colleagues developed a new antioxidant-photosensitizing hydrogel based on chitosan to control photodynamic therapy (PDT) activity in cancer treatment [123], which help to minimize the damage risk for normal cells. Hesperetin-loaded carbopol hydrogel can also be an effective therapy with a controlled release profile and could be used to treat topical oxidative conditions [124].

In conclusion, there are various formulations based on hydrogels in DDSs. Figure 7 illustrates the scope of the system in oral diseases. The advantages of such treatment are manifold, because they directly target the affected area, maintain relatively constant drug concentration levels, minimize systemic side effects as well as improve patient compliance.

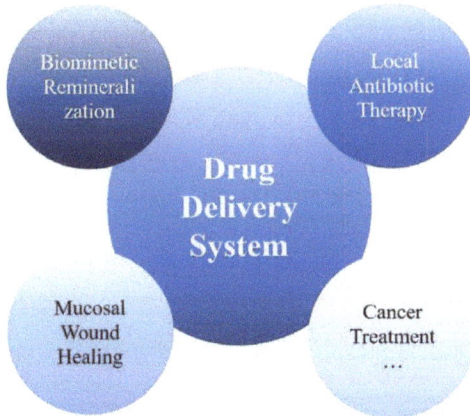

Figure 7. Drug delivery system based on hydrogel application for oral diseases.

5. Conclusions

This article reviews the potential of hydrogel to treat pathogenic oral cavity conditions. The applications of hydrogels for oral science research are wide, ranging from tissue reconstruction to oral disease therapy. The advantages of the application of hydrogel complexes include physical property [125], the straightforward chemistry procedure [126], ease of dental-derived MSC load, partially being condition-responsive, injectability, biodegradability, and the introduction of a three-dimensional delivery scaffold for tissue engineering. Although the results of most studies were promising, a larger number of clinical studies for determining the efficiency of prepared systems is required.

Author Contributions: Conceptualization, S.Y. and L.Z.; methodology, S.Y.; validation, B.W. and L.Z.; data curation, B.W.; writing—original draft preparation, S.Y.; writing—review B.W. and L.Z.; supervision, B.W.; project administration, L.Z.; funding acquisition, L.Z. All authors have read and agreed to the published version of the manuscript.

Funding: This research was funded by the Interdisciplinary Program of Shanghai Jiao Tong University, grant number YG2022QN052 and Fundamental research program funding of Ninth People's Hospital affiliated to Shanghai Jiao Tong University School of Medicine, grant number JYZZ136.

Institutional Review Board Statement: Not applicable.

Informed Consent Statement: Not applicable.

Conflicts of Interest: The authors declare no conflict of interest.

References

1. Amoli, M.S.; EzEldeen, M.; Jacobs, R.; Bloemen, V. Materials for Dentoalveolar Bioprinting: Current State of the Art. *Biomedicines* **2021**, *10*, 71. [CrossRef] [PubMed]
2. Ahmed, E.M. Hydrogel: Preparation, characterization, and applications: A review. *J. Adv. Res.* **2015**, *6*, 105–121. [CrossRef] [PubMed]
3. Gungor-Ozkerim, P.S.; Inci, I.; Zhang, Y.S.; Khademhosseini, A.; Dokmeci, M.R. Bioinks for 3D bioprinting: An overview. *Biomater. Sci.* **2018**, *6*, 915–946. [CrossRef] [PubMed]
4. Annabi, N.; Nichol, J.W.; Zhong, X.; Ji, C.; Koshy, S.; Khademhosseini, A.; Dehghani, F. Controlling the Porosity and Microarchitecture of Hydrogels for Tissue Engineering. *Tissue Eng. Part B Rev.* **2010**, *16*, 371–383. [CrossRef]
5. Bencherif, S.A.; Braschler, T.M.; Renaud, P. Advances in the design of macroporous polymer scaffolds for potential applications in dentistry. *J. Periodontal Implant Sci.* **2013**, *43*, 251–261. [CrossRef]
6. Gaspar, V.M.; Lavrador, P.; Borges, J.; Oliveira, M.B.; Mano, J.F. Advanced Bottom-Up Engineering of Living Architectures. *Adv. Mater.* **2019**, *32*, e1903975. [CrossRef]
7. Mabrouk, M.; Beherei, H.H.; Das, D.B. Recent progress in the fabrication techniques of 3D scaffolds for tissue engineering. *Mater. Sci. Eng. C* **2020**, *110*, 110716. [CrossRef]
8. Pathan, N.; Shende, P. Strategic conceptualization and potential of self-healing polymers in biomedical field. *Mater. Sci. Eng. C* **2021**, *125*, 112099. [CrossRef]
9. Ahsan, S.M.; Thomas, M.; Reddy, K.K.; Sooraparaju, S.G.; Asthana, A.; Bhatnagar, I. Chitosan as biomaterial in drug delivery and tissue engineering. *Int. J. Biol. Macromol.* **2018**, *110*, 97–109. [CrossRef]
10. Liu, X.; Liu, J.; Lin, S.; Zhao, X. Hydrogel machines. *Mater. Today* **2020**, *36*, 102–124. [CrossRef]
11. Petersen, P.E. World Health Organization global policy for improvement of oral health—World Health Assembly 2007. *Int. Dent. J.* **2008**, *58*, 115–121. [CrossRef] [PubMed]
12. Petersen, P.E. Global policy for improvement of oral health in the 21st century—Implications to oral health research of World Health Assembly 2007, World Health Organization. *Community Dent. Oral Epidemiol.* **2009**, *37*, 1–8. [CrossRef] [PubMed]
13. Abbass, M.M.S.; El-Rashidy, A.A.; Sadek, K.M.; El Moshy, S.; Radwan, I.A.; Rady, D.; Dörfer, C.E.; El-Sayed, K.M.F. Hydrogels and Dentin–Pulp Complex Regeneration: From the Benchtop to Clinical Translation. *Polymers* **2020**, *12*, 2935. [CrossRef] [PubMed]
14. Lin, L.M.; Huang, G.T.; Sigurdsson, A.; Kahler, B. Clinical cell-based versus cell-free regenerative endodontics: Clarification of concept and term. *Int. Endod. J.* **2021**, *54*, 887–901. [CrossRef]
15. Galler, K.; Weber, M.; Korkmaz, Y.; Widbiller, M.; Feuerer, M. Inflammatory Response Mechanisms of the Dentine–Pulp Complex and the Periapical Tissues. *Int. J. Mol. Sci.* **2021**, *22*, 1480. [CrossRef]
16. Jakovljevic, A.; Nikolic, N.; Jaćimović, J.; Pavlovic, O.; Milicic, B.; Beljic-Ivanovic, K.; Miletic, M.; Andric, M.; Milasin, J. Prevalence of Apical Periodontitis and Conventional Nonsurgical Root Canal Treatment in General Adult Population: An Updated Systematic Review and Meta-analysis of Cross-sectional Studies Published between 2012 and 2020. *J. Endod.* **2020**, *46*, 1371–1386.e8. [CrossRef]
17. Lempel, E.; Lovász, B.V.; Bihari, E.; Krajczár, K.; Jeges, S.; Tóth, Á.; Szalma, J. Long-term clinical evaluation of direct resin composite restorations in vital vs. endodontically treated posterior teeth—Retrospective study up to 13 years. *Dent. Mater.* **2019**, *35*, 1308–1318. [CrossRef]
18. Itoh, Y.; Sasaki, J.I.; Hashimoto, M.; Katata, C.; Hayashi, M.; Imazato, S. Pulp Regeneration by 3-dimensional Dental Pulp Stem Cell Constructs. *J. Dent. Res.* **2018**, *97*, 1137–1143. [CrossRef]
19. Shafiee, A.; Atala, A. Tissue engineering: Toward a new era of medicine. *Annu. Rev. Med.* **2017**, *68*, 29–40. [CrossRef]
20. Drury, J.L.; Mooney, D.J. Hydrogels for tissue engineering: Scaffold design variables and applications. *Biomaterials* **2003**, *24*, 4337–4351. [CrossRef]
21. Jang, J.-H.; Moon, J.-H.; Kim, S.G.; Kim, S.-Y. Pulp regeneration with hemostatic matrices as a scaffold in an immature tooth minipig model. *Sci. Rep.* **2020**, *10*, 12536. [CrossRef] [PubMed]
22. Park, J.H.; Gillispie, G.J.; Copus, J.S.; Zhang, W.; Atala, A.; Yoo, J.J.; Yelick, P.C.; Lee, S.J. The effect of BMP-mimetic peptide tethering bioinks on the differentiation of dental pulp stem cells (DPSCs) in 3D bioprinted dental constructs. *Biofabrication* **2020**, *12*, 035029. [CrossRef] [PubMed]
23. Athirasala, A.; Lins, F.; Tahayeri, A.; Hinds, M.; Smith, A.J.; Sedgley, C.; Ferracane, J.; Bertassoni, L.E. A Novel Strategy to Engineer Pre-Vascularized Full-Length Dental Pulp-like Tissue Constructs. *Sci. Rep.* **2017**, *7*, 3323. [CrossRef] [PubMed]
24. Zhang, S.; Thiebes, A.L.; Kreimendahl, F.; Rüetten, S.; Buhl, E.M.; Wolf, M.; Jockenhoevel, S.; Apel, C. Extracellular Vesicles-Loaded Fibrin Gel Supports Rapid Neovascularization for Dental Pulp Regeneration. *Int. J. Mol. Sci.* **2020**, *21*, 4226. [CrossRef] [PubMed]
25. Ha, M.; Athirasala, A.; Tahayeri, A.; Menezes, P.P.; Bertassoni, L.E. Micropatterned hydrogels and cell alignment enhance the odontogenic potential of stem cells from apical papilla in-vitro. *Dent. Mater.* **2019**, *36*, 88–96. [CrossRef]
26. Rosa, V.; Zhang, Z.; Grande, R.; Nör, J. Dental Pulp Tissue Engineering in Full-length Human Root Canals. *J. Dent. Res.* **2013**, *92*, 970–975. [CrossRef]
27. Ito, T.; Kaneko, T.; Sueyama, Y.; Kaneko, R.; Okiji, T. Dental pulp tissue engineering of pulpotomized rat molars with bone marrow mesenchymal stem cells. *Odontology* **2016**, *105*, 392–397. [CrossRef]

28. Bhoj, M.; Zhang, C.; Green, D.W. A First Step in De Novo Synthesis of a Living Pulp Tissue Replacement Using Dental Pulp MSCs and Tissue Growth Factors, Encapsulated within a Bioinspired Alginate Hydrogel. *J. Endod.* **2015**, *41*, 1100–1107. [CrossRef]
29. Pankajakshan, D.; Voytik-Harbin, S.L.; Nör, J.E.; Bottino, M.C. Injectable Highly Tunable Oligomeric Collagen Matrices for Dental Tissue Regeneration. *ACS Appl. Bio Mater.* **2020**, *3*, 859–868. [CrossRef]
30. Mu, X.; Shi, L.; Pan, S.; He, L.; Niu, Y.; Wang, X. A Customized Self-Assembling Peptide Hydrogel-Wrapped Stem Cell Factor Targeting Pulp Regeneration Rich in Vascular-Like Structures. *ACS Omega* **2020**, *5*, 16568–16574. [CrossRef]
31. Ishihara, M.; Obara, K.; Nakamura, S.; Fujita, M.; Masuoka, K.; Kanatani, Y.; Takase, B.; Hattori, H.; Morimoto, Y.; Ishihara, M.; et al. Chitosan hydrogel as a drug delivery carrier to control angiogenesis. *J. Artif. Organs* **2006**, *9*, 8–16. [CrossRef] [PubMed]
32. Haugen, H.J.; Basu, P.; Sukul, M.; Mano, J.F.; Reseland, J.E. Injectable Biomaterials for Dental Tissue Regeneration. *Int. J. Mol. Sci.* **2020**, *21*, 3442. [CrossRef] [PubMed]
33. Campos, D.F.D.; Zhang, S.; Kreimendahl, F.; Köpf, M.; Fischer, H.; Vogt, M.; Blaeser, A.; Apel, C.; Esteves-Oliveira, M. Hand-held bioprinting for de novo vascular formation applicable to dental pulp regeneration. *Connect. Tissue Res.* **2019**, *61*, 205–215. [CrossRef] [PubMed]
34. Moussa, D.G.; Aparicio, C. Present and future of tissue engineering scaffolds for dentin-pulp complex regeneration. *J. Tissue Eng. Regen. Med.* **2018**, *13*, 58–75. [CrossRef]
35. Suzuki, T.; Lee, C.H.; Chen, M.; Zhao, W.; Fu, S.; Qi, J.J.; Chotkowski, G.; Eisig, S.; Wong, A.; Mao, J. Induced Migration of Dental Pulp Stem Cells for in vivo Pulp Regeneration. *J. Dent. Res.* **2011**, *90*, 1013–1018. [CrossRef]
36. Chrepa, V.; Austah, O.; Diogenes, A. Evaluation of a Commercially Available Hyaluronic Acid Hydrogel (Restylane) as Injectable Scaffold for Dental Pulp Regeneration: An In Vitro Evaluation. *J. Endod.* **2016**, *43*, 257–262. [CrossRef]
37. Amir, L.R.; Suniarti, D.F.; Utami, S.; Abbas, B. Chitosan as a potential osteogenic factor compared with dexamethasone in cultured macaque dental pulp stromal cells. *Cell Tissue Res.* **2014**, *358*, 407–415. [CrossRef]
38. Feng, X.; Lu, X.; Huang, D.; Xing, J.; Feng, G.; Jin, G.; Yi, X.; Li, L.; Lu, Y.; Nie, D.; et al. 3D Porous Chitosan Scaffolds Suit Survival and Neural Differentiation of Dental Pulp Stem Cells. *Cell. Mol. Neurobiol.* **2014**, *34*, 859–870. [CrossRef]
39. Palma, P.J.; Ramos, J.C.; Martins, J.B.; Diogenes, A.; Figueiredo, M.H.; Ferreira, P.; Viegas, C.; Santos, J.M. Histologic Evaluation of Regenerative Endodontic Procedures with the Use of Chitosan Scaffolds in Immature Dog Teeth with Apical Periodontitis. *J. Endod.* **2017**, *43*, 1279–1287. [CrossRef]
40. Kim, N.R.; Lee, D.H.; Chung, P.-H.; Yang, H.-C. Distinct differentiation properties of human dental pulp cells on collagen, gelatin, and chitosan scaffolds. *Oral Surg. Oral Med. Oral Pathol. Oral Radiol. Endodontol.* **2009**, *108*, e94–e100. [CrossRef]
41. Rowley, J.A.; Madlambayan, G.; Mooney, D.J. Alginate hydrogels as synthetic extracellular matrix materials. *Biomaterials* **1999**, *20*, 45–53. [CrossRef]
42. Toh, W.S.; Loh, X.J. Advances in hydrogel delivery systems for tissue regeneration. *Mater. Sci. Eng. C* **2014**, *45*, 690–697. [CrossRef] [PubMed]
43. Chang, B.; Ahuja, N.; Ma, C.; Liu, X. Injectable scaffolds: Preparation and application in dental and craniofacial regeneration. *Mater. Sci. Eng. R Rep.* **2016**, *111*, 1–26. [CrossRef] [PubMed]
44. Guvendiren, M.; Burdick, J.A. Engineering synthetic hydrogel microenvirontments to instruct stem cells. *Curr. Opin. Biotechnol.* **2013**, *24*, 841–846. [CrossRef]
45. Jones, T.D.; Kefi, A.; Sun, S.; Cho, M.; Alapati, S.B. An Optimized Injectable Hydrogel Scaffold Supports Human Dental Pulp Stem Cell Viability and Spreading. *Adv. Med.* **2016**, *2016*, 7363579. [CrossRef]
46. Naghizadeh, Z.; Karkhaneh, A.; Khojasteh, A. Self-crosslinking effect of chitosan and gelatin on alginate based hydrogels: Injectable in situ forming scaffolds. *Mater. Sci. Eng. C* **2018**, *89*, 256–264. [CrossRef]
47. Khayat, A.; Monteiro, N.; Smith, E.; Pagni, S.; Zhang, W.; Khademhosseini, A.; Yelick, P. GelMA-Encapsulated hDPSCs and HUVECs for Dental Pulp Regeneration. *J. Dent. Res.* **2016**, *96*, 192–199. [CrossRef]
48. Monteiro, N.; Thrivikraman, G.; Athirasala, A.; Tahayeri, A.; Franca, C.; Ferracane, J.L.; Bertassoni, L.E. Photopolymerization of cell-laden gelatin methacryloyl hydrogels using a dental curing light for regenerative dentistry. *Dent. Mater.* **2018**, *34*, 389–399. [CrossRef]
49. Ravindran, S.; Zhang, Y.; Huang, C.-C.; George, A. Odontogenic Induction of Dental Stem Cells by Extracellular Matrix-Inspired Three-Dimensional Scaffold. *Tissue Eng. Part A* **2014**, *20*, 92–102. [CrossRef]
50. Murray, P.; Garcia-Godoy, F.; Hargreaves, K.M. Regenerative Endodontics: A Review of Current Status and a Call for Action. *J. Endod.* **2007**, *33*, 377–390. [CrossRef]
51. Ma, Y.; Xie, L.; Yang, B.; Tian, W. Three-dimensional printing biotechnology for the regeneration of the tooth and tooth-supporting tissues. *Biotechnol. Bioeng.* **2018**, *116*, 452–468. [CrossRef] [PubMed]
52. Obregon, F.; Vaquette, C.; Ivanovski, S.; Hutmacher, D.W.; Bertassoni, L. Three-Dimensional Bioprinting for Regenerative Dentistry and Craniofacial Tissue Engineering. *J. Dent. Res.* **2015**, *94*, 143S–152S. [CrossRef] [PubMed]
53. Athirasala, A.; Tahayeri, A.; Thrivikraman, G.; Franca, C.M.; Monteiro, N.; Tran, V.; Ferracane, J.; Bertassoni, L.E. A dentin-derived hydrogel bioink for 3D bioprinting of cell laden scaffolds for regenerative dentistry. *Biofabrication* **2017**, *10*, 024101. [CrossRef] [PubMed]
54. Huang, G.T.-J.; Liu, J.; Zhu, X.; Yu, Z.; Li, D.; Chen, C.-A.; Azim, A.A. Pulp/Dentin Regeneration: It Should Be Complicated. *J. Endod.* **2020**, *46*, S128–S134. [CrossRef]

55. Siddiqui, Z.; Sarkar, B.; Kim, K.K.; Kadincesme, N.; Paul, R.; Kumar, A.; Kobayashi, Y.; Roy, A.; Choudhury, M.; Yang, J.; et al. Angiogenic hydrogels for dental pulp revascularization. *Acta Biomater.* **2021**, *126*, 109–118. [CrossRef]
56. Huang, G.T.-J.; Al-Habib, M.; Gauthier, P. Challenges of stem cell-based pulp and dentin regeneration: A clinical perspective. *Endod. Top.* **2013**, *28*, 51–60. [CrossRef]
57. Könönen, E.; Gursoy, M.; Gursoy, U. Periodontitis: A Multifaceted Disease of Tooth-Supporting Tissues. *J. Clin. Med.* **2019**, *8*, 1135. [CrossRef]
58. Zięba, M.; Chaber, P.; Duale, K.; Maksymiak, M.M.; Basczok, M.; Kowalczuk, M.; Adamus, G. Polymeric Carriers for Delivery Systems in the Treatment of Chronic Periodontal Disease. *Polymers* **2020**, *12*, 1574. [CrossRef]
59. Khanuja, P.K.; Narula, S.; Rajput, R.; Sharma, R.; Tewari, S. Association of periodontal disease with glycemic control in patients with type 2 diabetes in Indian population. *Front. Med.* **2017**, *11*, 110–119. [CrossRef]
60. Baranov, N.; Popa, M.; Atanase, L.; Ichim, D. Polysaccharide-Based Drug Delivery Systems for the Treatment of Periodontitis. *Molecules* **2021**, *26*, 2735. [CrossRef]
61. Lamont, T.; Worthington, H.V.; Clarkson, J.E.; Beirne, P.V. Routine scale and polish for periodontal health in adults. *Cochrane Database Syst. Rev.* **2018**, *2020*, CD004625. [CrossRef] [PubMed]
62. Shaddox, L.M.; Walker, C.B. Treating chronic periodontitis: Current status, challenges, and future directions. *Clin. Cosmet. Investig. Dent.* **2010**, *2*, 79–91. [CrossRef]
63. Chen, F.-M.; Zhang, J.; Zhang, M.; An, Y.; Chen, F.; Wu, Z.-F. A review on endogenous regenerative technology in periodontal regenerative medicine. *Biomaterials* **2010**, *31*, 7892–7927. [CrossRef]
64. Neel, E.A.A.; Chrzanowski, W.; Salih, V.M.; Kim, H.-W.; Knowles, J.C. Tissue engineering in dentistry. *J. Dent.* **2014**, *42*, 915–928. [CrossRef] [PubMed]
65. Chen, L.; Shen, R.; Komasa, S.; Xue, Y.; Jin, B.; Hou, Y.; Okazaki, J.; Gao, J. Drug-Loadable Calcium Alginate Hydrogel System for Use in Oral Bone Tissue Repair. *Int. J. Mol. Sci.* **2017**, *18*, 989. [CrossRef] [PubMed]
66. Chien, K.-H.; Chang, Y.-L.; Wang, M.-L.; Chuang, J.-H.; Yang, Y.-C.; Tai, M.-C.; Wang, C.-Y.; Liu, Y.-Y.; Li, H.-Y.; Chen, J.-T.; et al. Promoting Induced Pluripotent Stem Cell-driven Biomineralization and Periodontal Regeneration in Rats with Maxillary-Molar Defects using Injectable BMP-6 Hydrogel. *Sci. Rep.* **2018**, *8*, 114. [CrossRef] [PubMed]
67. Miranda, D.G.; Malmonge, S.M.; Campos, D.M.; Attik, N.G.; Grosgogeat, B.; Gritsch, K. A chitosan-hyaluronic acid hydrogel scaffold for periodontal tissue engineering. *J. Biomed. Mater. Res. Part B Appl. Biomater.* **2016**, *104*, 1691–1702. [CrossRef]
68. Hamlet, S.M.; Vaquette, C.; Shah, A.; Hutmacher, D.W.; Ivanovski, S. 3-Dimensional functionalized polycaprolactone-hyaluronic acid hydrogel constructs for bone tissue engineering. *J. Clin. Periodontol.* **2017**, *44*, 428–437. [CrossRef]
69. Ma, Y.; Ji, Y.; Huang, G.; Ling, K.; Zhang, X.; Xu, F. Bioprinting 3D cell-laden hydrogel microarray for screening human periodontal ligament stem cell response to extracellular matrix. *Biofabrication* **2015**, *7*, 044105. [CrossRef]
70. Ma, Y.; Ji, Y.; Zhong, T.; Wan, W.; Yang, Q.; Li, A.; Zhang, X.; Lin, M. Bioprinting-Based PDLSC-ECM Screening for in Vivo Repair of Alveolar Bone Defect Using Cell-Laden, Injectable and Photocrosslinkable Hydrogels. *ACS Biomater. Sci. Eng.* **2017**, *3*, 3534–3545. [CrossRef]
71. Campos, D.F.D.; Blaeser, A.; Buellesbach, K.; Sen, K.S.; Xun, W.; Tillmann, W.; Fischer, H. Bioprinting Organotypic Hydrogels with Improved Mesenchymal Stem Cell Remodeling and Mineralization Properties for Bone Tissue Engineering. *Adv. Healthc. Mater.* **2016**, *5*, 1336–1345. [CrossRef] [PubMed]
72. Prasadh, S.; Wong, R.C.W. Unraveling the mechanical strength of biomaterials used as a bone scaffold in oral and maxillofacial defects. *Oral Sci. Int.* **2018**, *15*, 48–55. [CrossRef]
73. Yang, G.H.; Kim, M.; Kim, G. Additive-manufactured polycaprolactone scaffold consisting of innovatively designed microsized spiral struts for hard tissue regeneration. *Biofabrication* **2016**, *9*, 15005. [CrossRef] [PubMed]
74. De Jong, T.; Bakker, A.D.; Everts, V.; Smit, T.H. The intricate anatomy of the periodontal ligament and its development: Lessons for periodontal regeneration. *J. Periodontal Res.* **2017**, *52*, 965–974. [CrossRef] [PubMed]
75. Yan, X.-Z.; Beucken, J.J.V.D.; Cai, X.; Yu, N.; Jansen, J.A.; Yang, F. Periodontal Tissue Regeneration Using Enzymatically Solidified Chitosan Hydrogels With or Without Cell Loading. *Tissue Eng. Part A* **2015**, *21*, 1066–1076. [CrossRef] [PubMed]
76. Carter, S.-S.D.; Costa, P.; Vaquette, C.; Ivanovski, S.; Hutmacher, D.W.; Malda, J. Additive Biomanufacturing: An Advanced Approach for Periodontal Tissue Regeneration. *Ann. Biomed. Eng.* **2016**, *45*, 12–22. [CrossRef] [PubMed]
77. Lee, C.H.; Hajibandeh, J.; Suzuki, T.; Fan, A.; Shang, P.; Mao, J.J. Three-Dimensional Printed Multiphase Scaffolds for Regeneration of Periodontium Complex. *Tissue Eng. Part A* **2014**, *20*, 1342–1351. [CrossRef]
78. Chai, Q.; Jiao, Y.; Yu, X. Hydrogels for Biomedical Applications: Their Characteristics and the Mechanisms behind Them. *Gels* **2017**, *3*, 6. [CrossRef]
79. Momose, T.; Miyaji, H.; Kato, A.; Ogawa, K.; Yoshida, T.; Nishida, E.; Murakami, S.; Kosen, Y.; Sugaya, T.; Kawanami, M. Collagen Hydrogel Scaffold and Fibroblast Growth Factor-2 Accelerate Periodontal Healing of Class II Furcation Defects in Dog. *Open Dent. J.* **2016**, *10*, 347–359. [CrossRef]
80. Xu, X.; Gu, Z.; Chen, X.; Shi, C.; Liu, C.; Liu, M.; Wang, L.; Sun, M.; Zhang, K.; Liu, Q.; et al. An injectable and thermosensitive hydrogel: Promoting periodontal regeneration by controlled-release of aspirin and erythropoietin. *Acta Biomater.* **2019**, *86*, 235–246. [CrossRef]

81. Tan, J.; Zhang, M.; Hai, Z.; Wu, C.; Lin, J.; Kuang, W.; Tang, H.; Huang, Y.; Chen, X.; Liang, G. Sustained Release of Two Bioactive Factors from Supramolecular Hydrogel Promotes Periodontal Bone Regeneration. *ACS Nano* **2019**, *13*, 5616–5622. [CrossRef] [PubMed]
82. Hearnden, V.; Sankar, V.; Hull, K.; Juras, D.V.; Greenberg, M.; Kerr, A.R.; Lockhart, P.B.; Patton, L.L.; Porter, S.; Thornhill, M.H. New developments and opportunities in oral mucosal drug delivery for local and systemic disease. *Adv. Drug Deliv. Rev.* **2012**, *64*, 16–28. [CrossRef] [PubMed]
83. Chivere, V.T.; Kondiah, P.P.D.; Choonara, Y.E.; Pillay, V. Nanotechnology-Based Biopolymeric Oral Delivery Platforms for Advanced Cancer Treatment. *Cancers* **2020**, *12*, 522. [CrossRef] [PubMed]
84. Calixto, G.; Fonseca-Santos, B.; Chorilli, M.; Bernegossi, J. Nanotechnology-based drug delivery systems for treatment of oral cancer: A review. *Int. J. Nanomed.* **2014**, *9*, 3719–3735. [CrossRef]
85. Babadi, D.; Dadashzadeh, S.; Osouli, M.; Daryabari, M.S.; Haeri, A. Nanoformulation strategies for improving intestinal permeability of drugs: A more precise look at permeability assessment methods and pharmacokinetic properties changes. *J. Control. Release* **2020**, *321*, 669–709. [CrossRef]
86. Hosseinpour-Moghadam, R.; Mehryab, F.; Torshabi, M.; Haeri, A. Applications of Novel and Nanostructured Drug Delivery Systems for the Treatment of Oral Cavity Diseases. *Clin. Ther.* **2021**, *43*, e377–e402. [CrossRef]
87. De Leo, V.; Mattioli-Belmonte, M.; Cimmarusti, M.T.; Panniello, A.; Dicarlo, M.; Milano, F.; Agostiano, A.; De Giglio, E.; Catucci, L. Liposome-modified titanium surface: A strategy to locally deliver bioactive molecules. *Colloids Surf. B Biointerfaces* **2017**, *158*, 387–396. [CrossRef]
88. Wang, Y. Liposome as a delivery system for the treatment of biofilm-mediated infections. *J. Appl. Microbiol.* **2021**, *131*, 2626–2639. [CrossRef]
89. De Freitas, L.M.; Calixto, G.M.F.; Chorilli, M.; Giusti, J.S.M.; Bagnato, V.S.; Soukos, N.S.; Amiji, M.M.; Fontana, C.R.; De Freitas, L.M.; Calixto, G.M.F.; et al. Polymeric Nanoparticle-Based Photodynamic Therapy for Chronic Periodontitis in Vivo. *Int. J. Mol. Sci.* **2016**, *17*, 769. [CrossRef]
90. Huang, J.; Wang, Z.; Krishna, S.; Hu, Q.; Xuan, M.; Xie, H. Environment-sensitive hydrogels as potential drug delivery systems for the treatment of periodontitis. *Mater. Express* **2020**, *10*, 975–985. [CrossRef]
91. Guo, J.; Sun, H.; Lei, W.; Tang, Y.; Hong, S.; Yang, H.; Tay, F.; Huang, C. MMP-8-Responsive Polyethylene Glycol Hydrogel for Intraoral Drug Delivery. *J. Dent. Res.* **2019**, *98*, 564–571. [CrossRef] [PubMed]
92. Aminu, N.; Chan, S.-Y.; Yam, M.-F.; Toh, S.-M. A dual-action chitosan-based nanogel system of triclosan and flurbiprofen for localised treatment of periodontitis. *Int. J. Pharm.* **2019**, *570*, 118659. [CrossRef] [PubMed]
93. Boonlai, W.; Tantishaiyakul, V.; Hirun, N.; Sangfai, T.; Suknuntha, K. Thermosensitive Poloxamer 407/Poly(Acrylic Acid) Hydrogels with Potential Application as Injectable Drug Delivery System. *AAPS PharmSciTech* **2018**, *19*, 2103–2117. [CrossRef] [PubMed]
94. Morelli, L.; Cappelluti, M.A.; Ricotti, L.; Lenardi, C.; Gerges, I. An Injectable System for Local and Sustained Release of Antimicrobial Agents in the Periodontal Pocket. *Macromol. Biosci.* **2017**, *17*. [CrossRef]
95. Fitzpatrick, S.D.; Fitzpatrick, L.; Thakur, A.; Mazumder, M.A.J.; Sheardown, H. Temperature-sensitive polymers for drug delivery. *Expert Rev. Med. Devices* **2012**, *9*, 339–351. [CrossRef]
96. González-Delgado, J.A.; Kennedy, P.J.; Ferreira, M.; Tomé, J.P.C.; Sarmento, B. Use of Photosensitizers in Semisolid Formulations for Microbial Photodynamic Inactivation. *J. Med. Chem.* **2015**, *59*, 4428–4442. [CrossRef]
97. Abduljabbar, T.; Vohra, F.; Javed, F.; Akram, Z. Antimicrobial photodynamic therapy adjuvant to non-surgical periodontal therapy in patients with diabetes mellitus: A meta-analysis. *Photodiagn. Photodyn. Ther.* **2017**, *17*, 138–146. [CrossRef]
98. Xie, J.; Li, A.; Li, J. Advances in pH-Sensitive Polymers for Smart Insulin Delivery. *Macromol. Rapid Commun.* **2017**, *38*. [CrossRef]
99. Qiu, Y.; Park, K. Environment-sensitive hydrogels for drug delivery. *Adv. Drug Deliv. Rev.* **2012**, *64*, 49–60. [CrossRef]
100. Ning, T.-Y.; Xu, X.-H.; Zhu, L.-F.; Zhu, X.-P.; Chu, C.H.; Liu, L.-K.; Li, Q.-L. Biomimetic mineralization of dentin induced by agarose gel loaded with calcium phosphate. *J. Biomed. Mater. Res. Part B Appl. Biomater.* **2011**, *100B*, 138–144. [CrossRef]
101. Cao, Y.; Mei, M.L.; Li, Q.-L.; Lo, E.C.M.; Chu, C.H. Agarose Hydrogel Biomimetic Mineralization Model for the Regeneration of Enamel Prismlike Tissue. *ACS Appl. Mater. Interfaces* **2013**, *6*, 410–420. [CrossRef] [PubMed]
102. Muşat, V.; Anghel, E.M.; Zaharia, A.; Atkinson, I.; Mocioiu, O.C.; Buşilă, M.; Alexandru, P. A Chitosan–Agarose Polysaccharide-Based Hydrogel for Biomimetic Remineralization of Dental Enamel. *Biomolecules* **2021**, *11*, 1137. [CrossRef] [PubMed]
103. Ren, Q.; Ding, L.; Li, Z.; Wang, X.; Wang, K.; Han, S.; Li, W.; Zhou, X.; Zhang, L. Chitosan hydrogel containing amelogenin-derived peptide: Inhibition of cariogenic bacteria and promotion of remineralization of initial caries lesions. *Arch. Oral Biol.* **2019**, *100*, 42–48. [CrossRef] [PubMed]
104. Bekhouche, M.; Bolon, M.; Charriaud, F.; Lamrayah, M.; Da Costa, D.; Primard, C.; Costantini, A.; Pasdeloup, M.; Gobert, S.; Mallein-Gerin, F.; et al. Development of an antibacterial nanocomposite hydrogel for human dental pulp engineering. *J. Mater. Chem. B* **2020**, *8*, 8422–8432. [CrossRef]
105. Aksel, H.; Mahjour, F.; Bosaid, F.; Calamak, S.; Azim, A.A. Antimicrobial Activity and Biocompatibility of Antibiotic-Loaded Chitosan Hydrogels as a Potential Scaffold in Regenerative Endodontic Treatment. *J. Endod.* **2020**, *46*, 1867–1875. [CrossRef]
106. Yan, Y.; Zhou, P.; Lu, H.; Guan, Y.; Ma, M.; Wang, J.; Shang, G.; Jiang, B. Potential apply of hydrogel-carried chlorhexidine and metronidazole in root canal disinfection. *Dent. Mater. J.* **2021**, *40*, 986–993. [CrossRef]

107. Ribeiro, J.S.; Bordini, E.A.F.; Ferreira, J.A.; Mei, L.; Dubey, N.; Fenno, J.C.; Piva, E.; Lund, R.G.; Schwendeman, A.; Bottino, M.C. Injectable MMP-Responsive Nanotube-Modified Gelatin Hydrogel for Dental Infection Ablation. *ACS Appl. Mater. Interfaces* **2020**, *12*, 16006–16017. [CrossRef]
108. Socransky, S.S.; Haffajee, A.D.; Cugini, M.A.; Smith, C.; Kent, R.L., Jr. Microbial complexes in subgingival plaque. *J. Clin. Periodontol.* **1998**, *25*, 134–144. [CrossRef]
109. Pakzad, Y.; Ganji, F. Thermosensitive hydrogel for periodontal application: In vitro drug release, antibacterial activity and toxicity evaluation. *J. Biomater. Appl.* **2016**, *30*, 919–929. [CrossRef]
110. Ribeiro, J.S.; Daghrery, A.; Dubey, N.; Li, C.; Mei, L.; Fenno, J.C.; Schwendeman, A.; Aytac, Z.; Bottino, M.C. Hybrid Antimicrobial Hydrogel as Injectable Therapeutics for Oral Infection Ablation. *Biomacromolecules* **2020**, *21*, 3945–3956. [CrossRef]
111. Chang, P.-C.; Chao, Y.-C.; Hsiao, M.-H.; Chou, H.-S.; Jheng, Y.-H.; Yu, X.-H.; Lee, N.; Yang, C.; Liu, D.-M. Inhibition of Periodontitis Induction Using a Stimuli-Responsive Hydrogel Carrying Naringin. *J. Periodontol.* **2017**, *88*, 190–196. [CrossRef] [PubMed]
112. Li, N.; Jiang, L.; Jin, H.; Wu, Y.; Liu, Y.; Huang, W.; Wei, L.; Zhou, Q.; Chen, F.; Gao, Y.; et al. An enzyme-responsive membrane for antibiotic drug release and local periodontal treatment. *Colloids Surfaces B Biointerfaces* **2019**, *183*, 110454. [CrossRef] [PubMed]
113. Liang, H.; Xu, J.; Liu, Y.; Zhang, J.; Peng, W.; Yan, J.; Li, Z.; Li, Q. Optimization of hydrogel containing toluidine blue O for photodynamic therapy by response surface methodology. *J. Photochem. Photobiol. B Biol.* **2017**, *173*, 389–396. [CrossRef]
114. Rajeshwari, H.R.; Dhamecha, D.; Jagwani, S.; Rao, M.; Jadhav, K.; Shaikh, S.; Puzhankara, L.; Jalalpure, S. Local drug delivery systems in the management of periodontitis: A scientific review. *J. Control. Release* **2019**, *307*, 393–409. [CrossRef]
115. Petit, C.; Batool, F.; Stutz, C.; Anton, N.; Klymchenko, A.; Vandamme, T.; Benkirane-Jessel, N.; Huck, O. Development of a thermosensitive statin loaded chitosan-based hydrogel promoting bone healing. *Int. J. Pharm.* **2020**, *586*, 119534. [CrossRef] [PubMed]
116. Shen, Z.; Kuang, S.; Zhang, Y.; Yang, M.; Qin, W.; Shi, X.; Lin, Z. Chitosan hydrogel incorporated with dental pulp stem cell-derived exosomes alleviates periodontitis in mice via a macrophage-dependent mechanism. *Bioact. Mater.* **2020**, *5*, 1113–1126. [CrossRef]
117. Li, H.; Ji, Q.; Chen, X.; Sun, Y.; Xu, Q.; Deng, P.; Hu, F.; Yang, J. Accelerated bony defect healing based on chitosan thermosensitive hydrogel scaffolds embedded with chitosan nanoparticles for the delivery of BMP2 plasmid DNA. *J. Biomed. Mater. Res. Part A* **2016**, *105*, 265–273. [CrossRef]
118. Andreopoulos, F.M.; Persaud, I. Delivery of basic fibroblast growth factor (bFGF) from photoresponsive hydrogel scaffolds. *Biomaterials* **2006**, *27*, 2468–2476. [CrossRef]
119. Marques, A.C.; Rocha, A.I.; Leal, P.; Estanqueiro, M.; Lobo, J.M.S. Development and characterization of mucoadhesive buccal gels containing lipid nanoparticles of ibuprofen. *Int. J. Pharm.* **2017**, *533*, 455–462. [CrossRef]
120. Zhang, W.; Bao, B.; Jiang, F.; Zhang, Y.; Zhou, R.; Lu, Y.; Lin, S.; Lin, Q.; Jiang, X.; Zhu, L. Promoting Oral Mucosal Wound Healing with a Hydrogel Adhesive Based on a Phototriggered S-Nitrosylation Coupling Reaction. *Adv. Mater.* **2021**, *33*, e2105667. [CrossRef]
121. Rossi, S.; Marciello, M.; Bonferoni, M.C.; Ferrari, F.; Sandri, G.; Dacarro, C.; Grisoli, P.; Caramella, C. Thermally sensitive gels based on chitosan derivatives for the treatment of oral mucositis. *Eur. J. Pharm. Biopharm.* **2010**, *74*, 248–254. [CrossRef] [PubMed]
122. Ding, T.; Zou, J.; Qi, J.; Dan, H.; Tang, F.; Zhao, H.; Chen, Q. Mucoadhesive Nucleoside-Based Hydrogel Delays Oral Leukoplakia Canceration. *J. Dent. Res.* **2022**, 220345221085192. [CrossRef] [PubMed]
123. Azadikhah, F.; Karimi, A.R.; Yousefi, G.H.; Hadizadeh, M. Dual antioxidant-photosensitizing hydrogel system: Cross-linking of chitosan with tannic acid for enhanced photodynamic efficacy. *Int. J. Biol. Macromol.* **2021**, *188*, 114–125. [CrossRef] [PubMed]
124. Vaz, V.M.; Jitta, S.R.; Verma, R.; Kumar, L. Hesperetin loaded proposomal gel for topical antioxidant activity. *J. Drug Deliv. Sci. Technol.* **2021**, *66*, 102873. [CrossRef]
125. Rossi, S.; Sandri, G.; Caramella, C.M. Buccal drug delivery: A challenge already won? *Drug Discov. Today Technol.* **2005**, *2*, 59–65. [CrossRef]
126. Ansari, S.; Seagroves, J.T.; Chen, C.; Shah, K.; Aghaloo, T.; Wu, B.M.; Bencharit, S.; Moshaverinia, A. Dental and orofacial mesenchymal stem cells in craniofacial regeneration: The prosthodontist's point of view. *J. Prosthet. Dent.* **2017**, *118*, 455–461. [CrossRef]

Brief Report

Hyaluronidase Enhances Targeting of Hydrogel-Encapsulated Anti-CTLA-4 to Tumor Draining Lymph Nodes and Improves Anti-Tumor Efficacy

Airi Harui * and Michael D. Roth

Division of Pulmonary & Critical Care Medicine, Geffen School of Medicine at UCLA, Los Angeles, CA 90095-1690, USA; mroth@mednet.ucla.edu
* Correspondence: aharui@mednet.ucla.edu

Abstract: Immunotherapy targeting checkpoint inhibitors, such as CTLA-4 and/or PD-1, has emerged as a leading cancer therapy. While their combination produces superior efficacy compared to monotherapy, it also magnifies inflammatory and autoimmune toxicity that limits clinical utility. We previously reported that a peri-tumor injection of low-dose hydrogel-encapsulated anti-CTLA-4 produced anti-tumor responses that were equal to, or better than, systemic dosing despite a >80% reduction in total dose. Injection of hydrogel-encapsulated anti-CTLA-4 was associated with low serum exposure and limited autoimmune toxicity, but still synergized with anti-PD-1. In this report, we employ live and ex vivo imaging to examine whether peri-tumor administration specifically targets anti-CTLA-4 to tumor-draining lymph nodes (TDLN) and whether the incorporation of hyaluronidase enhances this effect. Tumor-free survival analysis was also used to measure the impact of hyaluronidase on tumor response. Compared to systemic dosing, peri-tumor injection of hydrogel-encapsulated anti-CTLA-4/DyLight 800 resulted in preferential labeling of TDLN. Incorporating hyaluronidase within the hydrogel improved the rapidity, intensity, and duration of TDLN labeling and significantly improved tumor-free survival. We conclude that hydrogel-encapsulated anti-CTLA acts as a localized antibody reservoir and that inclusion of hyaluronidase optimizes the blockade of CTLA-4 in TDLN and thereby imparts superior anti-tumor immunity.

Keywords: checkpoint inhibitor blockade; immunotherapy; hydrogel; anti-CTLA-4; peri-tumor injection; hyaluronidase; antibody delivery; tumor draining lymph nodes

Citation: Harui, A.; Roth, M.D. Hyaluronidase Enhances Targeting of Hydrogel-Encapsulated Anti-CTLA-4 to Tumor Draining Lymph Nodes and Improves Anti-Tumor Efficacy. *Gels* **2022**, *8*, 284. https://doi.org/10.3390/gels8050284

Academic Editors: Wei Ji and Kunpeng Cui

Received: 14 March 2022
Accepted: 25 April 2022
Published: 3 May 2022

Publisher's Note: MDPI stays neutral with regard to jurisdictional claims in published maps and institutional affiliations.

Copyright: © 2022 by the authors. Licensee MDPI, Basel, Switzerland. This article is an open access article distributed under the terms and conditions of the Creative Commons Attribution (CC BY) license (https://creativecommons.org/licenses/by/4.0/).

1. Introduction

Monoclonal antibodies (mAbs) that block immune checkpoint inhibitors are rapidly expanding the treatment options for patients with soft tissue cancers [1,2]. However, individual mAbs are effective in only a minority of patients while combination therapy that blocks both CTLA-4 and PD-1 yields a significantly greater response and progression-free survival [3–7]. Unfortunately, systemic administration of combination therapy also increases acute inflammatory and autoimmune toxicity that can limit clinical utility [8,9]. Hydrogel-encapsulated checkpoint inhibitor therapy has recently been identified as an approach for improving this narrow benefit-to-risk ratio [10–13]. Using a mouse model, we previously reported that peri-tumor injection of a controlled-release hydrogel containing low-dose anti-CTLA-4 produces equal or greater tumor control than does high-dose systemic therapy [13]. This targeted low-dose approach still synergizes with systemic anti-PD-1 therapy and promotes lasting systemic protection against tumor re-challenge [13]. Local injection of anti-CTLA-4 alone (without a controlled-release hydrogel) produces only a limited response. Furthermore, using a NOD.H-2h4 model to assess anti-CTLA-4-associated autoimmune thyroiditis, systemic administration of anti-CTLA-4 induced high titers of anti-thyroglobulin antibodies while hydrogel-based administration did not. At the conclusion of that work, we hypothesized that controlled regional delivery of low-dose anti-CTLA-4 results in a sustained perfusion of tumor draining lymph nodes (TDLN) that selectively expands and activates tumor-specific T cells while sparing autoimmune T cell activation at distant nodal sites. In related work, we demonstrated that the addition of hyaluronidase (HAse) to the hydrogel allowed us to fine tune the rate of mAb release and, by promoting autolysis of the hydrogel reservoir, facilitates repeated administration at the same location [14].

In the current brief report, we evaluate whether the incorporation of HAse into the hydrogel matrix has other beneficial effects on therapy. Injection of recombinant human HAse is approved by the FDA to permeabilize soft-tissue matrix in order to promote the uptake of fluids, drugs, and mAb therapies through existing lymphatic pathways [15]. Given our focus on regional perfusion of TDLN with anti-CTLA-4, we hypothesized that HAse might enhance lymphatic uptake and nodal targeting of administered anti-CTLA-4 and therefore promote even greater anti-tumor efficacy. In this study, live animal imaging is used to track anti-CTLA-4 distribution following either systemic injection, targeted peri-tumor injection of a standard hydrogel, or administration of a hydrogel that contains incorporated HAse. Lymph nodes were also excised and imaged ex vivo. Our established pre-clinical tumor immunotherapy model was then employed to evaluate the effects of HAse on anti-tumor activity and tumor-free survival [13].

2. Results and Discussion

2.1. Delivery of Hydrogel-Encapsulated Anti-CTLA-4 by Peri-Tumor Injection Preferentially Targets TDLN

A whole-animal optical imaging technique was adapted to investigate whether a peri-tumor injection of hydrogel-encapsulated anti-CTLA-4 creates a subcutaneous (SQ) mAb reservoir that preferentially targets regional TDLN. A near-infrared anti-CTLA-4/DyLight 800 conjugate was developed in order to allow both in vivo live imaging and ex vivo quantitation of mAb trafficking. When tumor-bearing mice were treated with 100 µg of anti-CTLA-4/DyLight 800 by intraperitoneal (IP) injection, the labeled mAb rapidly dispersed throughout the peritoneal cavity as expected (Figure 1a). In contrast, a peri-tumor SQ injection of low-dose hydrogel-encapsulated anti-CTLA-4/DyLight 800 (25 µg) resulted in a localized reservoir of mAb at the site of injection. These primary distribution patterns remained relatively constant when serially imaged over the course of 48 h. Expression of the label within lymph nodes was not observed in live animals, likely due to the overwhelming intensity of fluorescence from the primary injection and the deeper location of lymph nodes. In order to improve the sensitivity and specificity for detecting the micro-distribution of anti-CTLA-4 DyLight 800, sets of animals that had been treated in an identical manner were sacrificed at 24 and 48 h after injection. Bilateral axillary and inguinal lymph nodes (LN) were surgically excised and simultaneously examined by ex vivo optical imaging (Figure 1b). No fluorescent signal was detected at 24 h in any LN regardless of whether the animals received IP or SQ (hydrogel-based) dosing with anti-CTLA-4/DyLight 800. However, at 48 h, a fluorescent signal was always observed in axillary TDLN from animals that had received low-dose hydrogel-encapsulated anti-CTLA-4/DyLight 800. No signal was detected in animals that had received IP dosing even though they had received four-times the mAb dose. In addition, no fluorescent signals were detected from inguinal or contralateral LN in either group. These findings directly support our primary hypothesis that encapsulating anti-CTLA-4 within the hydrogel matrix and delivering it by peri-tumor injection produces a localized reservoir of mAb that preferentially targets TDLN. Minimizing systemic exposure to anti-CTLA-4 is particularly important due to the role of CTLA-4 in autoimmune-related toxicity [16]. As demonstrated previously, high-dose systemic administration of anti-CTLA-4 results in high serum concentrations that trigger/enhance the production of autoimmune antibodies [13]. These results also explain why targeted low-dose anti-CTLA-4, as delivered by the hydrogel, can produce equal/greater anti-tumor efficacy than does systemic dosing [13].

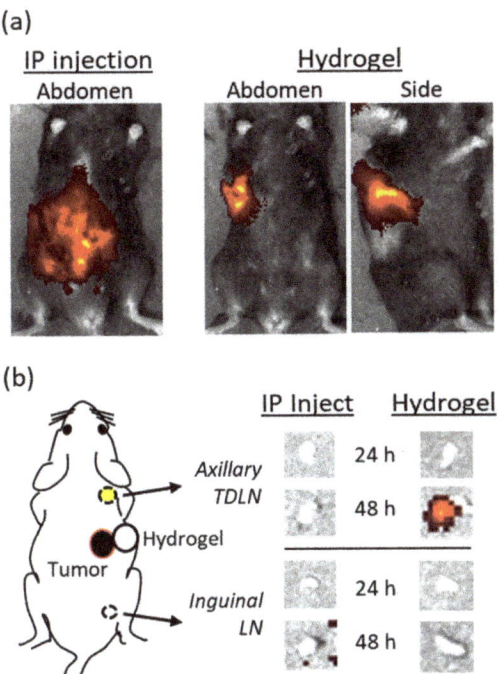

Figure 1. Biodistribution of fluorescent-labeled anti-CTLA-4 to tumor draining lymph nodes (TDLN) following either systemic (intraperitoneal; IP) or hydrogel (subcutaneous; SQ) injection. C57BL/6 mice bearing palpable MC-38 tumors implanted in the right posterior flank were treated with either 100 μg of anti-CTLA-4/DyLight 800 by IP injection or 25 μg of hydrogel-encapsulated anti-CTLA-4/DyLight 800 by peri-tumor SQ injection. Biodistribution of the injected anti-CTLA-4 was determined (**a**) immediately following administration by whole-animal in vivo optical fluorescence imaging and (**b**) at 24 and 48 h by ex vivo optical fluorescence imaging performed on surgically resected axillary TDLN and inguinal LN.

2.2. Incorporation of HAse into the Hydrogel Matrix Enhances Antibody Delivery to TDLN

We had previously reported that incorporating HAse into the hydrogel mixture results in a number of beneficial effects. The reactive moieties on thiolated carboxymethyl hyaluronic acid (CMHA-S) and poly-(ethylene glycol)-diacrylate (PEG-DA), designed to promote spontaneous cross-linking and formation of the hydrogel matrix, can also interact with incorporated proteins with denaturing effects. The presence of HAse reduces this effect, likely by acting as an alternative protein target or impairing the interaction of these reactive moieties with incorporated proteins [11]. This protective effect from HAse was also observed for incorporated anti-CTLA-4. In an in vitro antibody release assay, total recovery of anti-CTLA-4 from the standard hydrogel preparation was 86.8% and this increased to 93.6% with addition of HAse 50 U and 97.8% with HAse 250 U (Figure S1). In addition, incorporated HAse breaks down the hyaluronic acid backbone of the matrix in a dose- and time-dependent manner. This can be used to fine-tune the rate of mAb release from the hydrogel matrix and assures complete release of encapsulated mAb payloads [14,17] (Figure S1). HAse also promotes more rapid resorption of the hydrogel matrix, promoting repeated injections at the same site for future cycles of immunotherapy.

HAse is clinically approved for patient use based on its ability to break down SQ tissue barriers and facilitate the access of fluids, medications, and mAb into tissue lymphatic pathways [15,18]. Existing data suggest that this promotes faster and more efficient trafficking to draining LN [19]. We therefore hypothesized that addition of HAse to the hydrogel formulation could enhance targeting and exposure of TDLN to anti-CTLA-4. To test this hypothesis, tumor-bearing mice were injected with hydrogel-encapsulated anti-CTLA-4/DyLight 800 (50 µg) in the presence or absence of 250 U HAse and then imaged for trafficking of the fluorescent label (Figure 2). In vivo live imaging shows similar localization and concentration of the fluorescent label in both groups at the time of injection. However, from 24 through 72 h, there is a significant and time-dependent reduction in the anti-CTLA-4/DyLight 800 signal at the injection site when HAse was incorporated into the injected hydrogels. Our conclusion is that anti-CTLA-4/DyLight 800 is released faster and/or more effectively in vivo when HAse is present. Consistent with this, when implanted hydrogels were recovered by surgical excision at 2 weeks after injection (Figure 2b), the hydrogels containing HAse were dramatically smaller and appeared to be infiltrated by host cells, consistent with accelerated degradation.

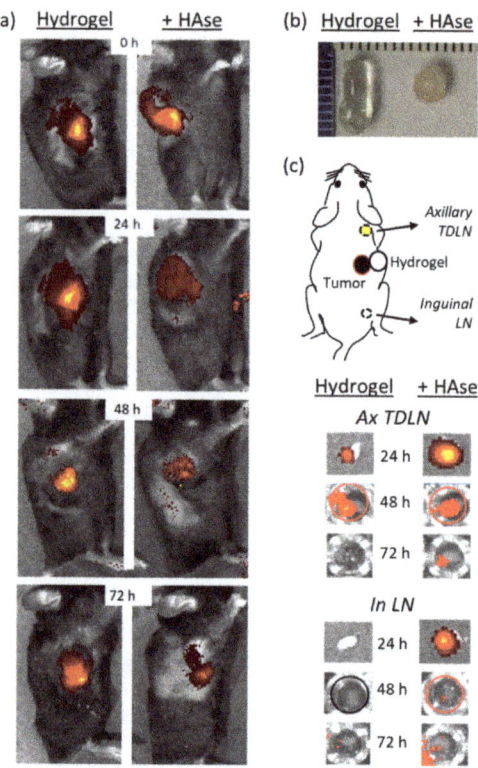

Figure 2. Impact of hyaluronidase (HAse) on the delivery of hydrogel-encapsulated anti-CTLA-4 to TDLN. C57BL/6 mice bearing palpable MC-38 tumors were treated with 50 µg of hydrogel-encapsulated anti-CTLA-4/DyLight 800 by peri-tumor SQ injection. Anti-CTLA-4 biodistribution was compared in animals receiving a standard hydrogel formulation to one that incorporated HAse (250 U). (**a**) Whole-animal in vivo optical fluorescence imaging of the injection site immediately following injection (0 h) and at 24, 48, and 72 h. (**b**) Hydrogels from the same animals were surgically recovered after 2 weeks for visual inspection. (**c**) Axillary TDLN and ipsilateral inguinal LN were surgically resected at 24, 48, and 72 h from a cohort of animals treated in the same manner and subjected to ex vivo optical fluorescence imaging.

The fundamental question addressed in the next study was whether this results in more effective perfusion of TDLN by anti-CTLA-4. As previously described, sets of tumor-bearing animals that had been treated in an identical manner were sacrificed at 24, 48, and 72 h after injection of anti-CTLA-4/DyLight 800. Recovered axillary and inguinal LN were simultaneously examined for the presence of labeled anti-CTLA-4 by ex vivo optical imaging (Figure 2c). As hypothesized, a higher level of fluorescence was detected in axillary TDLN at 24 h when the hydrogel contained HAse. At 48 h, the fluorescence level was similar in both groups while the fluorescent signal was only detected at 72 h from animals that received hydrogel containing HAse. This visual trend was confirmed by measuring total fluorescent emission from LN recovered at these time points (Table S1). Ex vivo imaging of recovered inguinal LN showed a similar—but less intense—pattern in animals that received hydrogels containing HAse but no uptake by inguinal LN at any time point in the absence of HAse (Figure 2c). This interesting finding suggests that the increased tissue permeability resulting from HAse can spread anti-CTLA-4 to a wider tissue distribution. Given the size difference between mice and humans, it is not yet clear whether this would have a meaningful effect in a clinical setting. We did not detect anti-CTLA-4 uptake by other organs by whole-body live imaging, nor was a signal detected in isolated spleens. This likely reflects the imaging threshold of our approach and it is very possible that more sensitive approaches—such as immunohistochemistry or flow cytometry—might yield findings that are beyond the focus of our targeted studies.

Taken together, these results suggest that incorporating HAse into the hydrogel formulation enhances the release of functional anti-CTLA-4 from the hydrogel matrix; opens lymphatic barriers in surrounding SQ tissues; and produces faster, higher, and more prolonged binding to target sites in TDLN while still sparing mAb accumulation in distant (e.g., contralateral) LN.

2.3. Incorporation of HAse into the Hydrogel Matrix Enhances the Anti-Tumor Efficacy of Low Dose Anti-CTLA-4

Targeting of TDLN occurs when a peri-tumor injection of hydrogel-incorporated anti-CTLA-4 is delivered, and this promotes effective tumor killing while limiting systemic exposure [13]. The addition of HAse improves anti-CTLA-4 release from the hydrogel, promotes mAb access to the lymphatic pathway, and results in higher and more prolonged targeting of CTLA-4 binding sites in TDLN. Anti-CTLA-4 results in an expansion of tumor-reactive T cells, reduction in CTLA-4-expressing cells, and a relative expansion of cytotoxic CD8 cells [13]. The remaining question is whether greater TDLN perfusion by anti-CTLA-4 translates into better therapeutic outcomes. To test the hypothesis that HAse can improve immune activation resulting from low-dose anti-CTLA-4, tumor-bearing mice (N = 9 or 10) were treated with hydrogel-encapsulated anti-CTLA-4 (50 µg) that contained either 0, 50, or 250 U of HAse. Peri-tumor injections were delivered on days 6 and 11 after tumor implantation [13]. Tumor growth was monitored for 28 days (Figure 3). All tumors, regardless of assigned treatment group, grew in the first few days after starting therapy but thereafter the response was significantly different in the treatment group receiving 250 U HAse. At the end of 28 days, 5 mice (56%) in the control group (no HAse) were tumor free; 5 mice (50%) in the group receiving gels containing 50 U HAse were tumor free; but tumor-free survival was present in 8 mice (89%) that had received hydrogels containing 250 U HAse ($p = 0.04$; log-rank test for tumor-free survival; Figure S2). Additional controls demonstrated that these responses required the presence of anti-CTLA-4 as the difference in treatment response was highly significant when comparing animals treated with hydrogels containing only 250 U HAse and those receiving anti-CTLA-4 plus 250 U HAse ($p < 0.001$ by ANOVA; data not shown). The synergism that occurs between the delivery of anti-CTLA-4 and HAse, when delivered together in a peri-tumor injection of self-polymerizing hydrogel, therefore spans from enhanced effects on anti-CTLA-4 release, lymphatic uptake, and targeting of TDLN to a positive impact on anti-tumor efficacy.

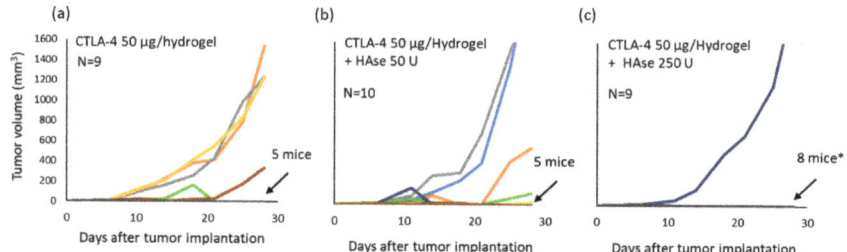

Figure 3. Incorporation of HAse enhances the anti-tumor activity of hydrogel-encapsulated anti-CTLA-4. C57BL/6 mice injected with MC-38 tumor cells were treated with 50 µg of hydrogel-encapsulated anti-CTLA-4 by peri-tumor SQ injection at days 6 and 11 after tumor implantation. Groups of animals received either a standard hydrogel formulation (**a**), or a hydrogel formulated with either 50 U (**b**) or 250 U (**c**) of HAse. Tumor volume was measured every 3 or 4 days over a course of 28 days. Each line represents tumor growth in a single mouse. The number of tumor-free animals at day 28 are indicated (* $p < 0.05$ compared to no HAse control by log-rank test).

3. Conclusions

In this study, a combination of whole-animal in vivo imaging and ex vivo imaging of recovered LNs provided direct evidence that a peri-tumor injection of hydrogel-encapsulated anti-CTLA-4 creates a local drug reservoir that targets mAb delivery to TDLN. While systemic anti-CTLA-4 therapy has the potential to disrupt tumor-associated immunosuppression through multiple pathways and at multiple sites, targeted delivery to TDLN has been shown in animal models to promote proportionally greater T-cell activation and systemic tumor control [11]. These outcomes occur even in the absence of direct perfusion of the tumor or depletion of intratumoral regulatory T cells [11]. This mechanism of action likely explains the capacity for hydrogel-delivered anti-CTLA-4 to induce effective anti-tumor immunity while sparing systemic exposure and drug-associated autoimmune toxicity [13]. Furthermore, we demonstrated that incorporating HAse into the hydrogel facilitates anti-CTLA-4 release, as well as the speed, intensity, and duration of its binding within TDLN. The end result is a significant improvement in systemic anti-tumor efficacy and tumor-free survival without an increase in delivered dose. According to convention, the dose of systemic anti-CTLA-4 therapy is directly linked to both its efficacy and toxicity [20,21]. The capacity to target TDLNs via a peri-tumor injection disrupts this relationship. As demonstrated here, incorporation of HAse into the hydrogel matrix further enhances the effectiveness of low-dose anti-CTLA-4 therapy and should further improve the benefit-to-risk ratio associated with its use. As such, these pre-clinical investigations define a clear pathway for employing hydrogel-encapsulated anti-CTLA-4 as a strategy for improving the tolerability and potency of cancer immunotherapy. As previously noted, the greatest promise from such an approach would be the ability to safely administer combination therapy that included hydrogel-encapsulated anti-CTLA-4, targeting TDLN, and a systemic PD-1 checkpoint inhibitor.

4. Materials and Methods

4.1. Animals

C57BL/6 mice were purchased from the Charles River Laboratory (Wilmington, MA, USA) and housed at the UCLA Division of Laboratory Animal Medicine facility. All protocols and procedures were approved by the UCLA Animal Research Committee.

4.2. Reagents

Mouse colorectal cancer cell line, MC-38, was obtained from the Division of Cancer Treatment and Diagnosis Tumor Repository, National Cancer Institute. Hydrogel matrix components, CMHA-S and PEG-DA, were provided by Lineage Cell Therapeutics (Alameda, CA, USA). The anti-mouse CTLA-4 (clone # 9H10) was purchased from BioXcell (West Lebanon, NH, USA) and a matching fluorescent-conjugated anti-mouse CTLA-4/Dylight800 was prepared specifically for this study by Leinco Technology (St. Louis, MO). Purified bovine hyaluronidase (HAse) was from MP Biomedicals (Santa Ana, CA, USA).

4.3. Hydrogel Formulation

Hydrogels were formulated by first mixing the anti-CTLA-4 9H10 clone (50 µg/gel) or the anti-CTLA-4/Dylight 800 fluorescent construct (25 or 50 µg/gel) with the solubilized PEG-DA and then combining with CMHA-S to initiate spontaneous cross-linking as described previously [13]. Briefly, CMHA-S and PEGDA were individually dissolved in degassed deionized water (pH 7.4) to prepare solutions of 1.25% (w/v) and 6% (w/v), respectively. Final component concentrations within 150 µL hydrogels were 0.8% w/v for CMHA-S and 1.2% w/v for PEG-DA as detailed in Table 1. Hydrogels that incorporated HAse (50 U–250 U) were formulated in the same manner except that the HAse component was pre-mixed with the PEG-DA solution prior to adding the other components.

Table 1. Hydrogel composition.

Hydrogel Component	Final Concentration in a 150 µL Injection
CMHA-S	0.8% w/v
PEG-DA	1.2% w/v
Hyaluronidase (HAse)	0, 50, or 250 Units

4.4. Tumor Model and Treatment with Anti-CTLA-4

The immunotherapeutic activity of anti-CTLA-4 in tumor-bearing mice was assessed as prescribed previously [13]. In brief, C57BL/6 mice were implanted with MC-38 cells (3×10^5/mouse) by SQ injection into the right upper flank. For optical fluorescent imaging experiments, mice (2 mice/group) were treated with a single dose of anti-CTLA-4/Dylight 800 delivered by either IP injection (100 µg in PBS) or by peri-tumor SQ injection (25 or 50 µg) using a hydrogel formulation between 11 and 14 days after tumor implantation. For assessing the impact of hydrogel-encapsulated anti-CTLA-4 (50 µg/dose) on tumor growth, all animals were implanted with tumor on the same day and those with palpable tumors at Day 6 were randomly divided into treatment groups (N = 9–10/group). All three groups received a 150 µL injection of hydrogel-encapsulate anti-CTLA-4 delivered by peri-tumor SQ dosing on Days 6 and 11. One group received a standard formulation without HAse, one containing 50 U HAse, and one with 2500 U HAse. Tumor volumes were measured by calipers every 3 to 4 days up to day 28.

4.5. Optical Fluorescence Imaging

In vivo and ex vivo optical imaging were performed using an IVIS Lumina II system (Caliper Life Sciences, Inc.; Hopkinton, MA, USA) at the Crump Preclinical Imaging Technology Center at UCLA. In vivo whole-body imaging of mice was carried out under isoflurane anesthesia with serial assessments in the same animals. For ex vivo imaging of LN, LN were recovered by surgical resection in replicate cohorts of animals that had been euthanized at various time points (0–72 h) after administration of anti-CTLA-4/DyLight 800.

4.6. Statistical Evaluations

Biodistribution studies were performed in replicates, with all animals imaged under identical settings at the same session to facilitate comparison of fluorescent intensity, as indicated by a continuous red to yellow intensity spectrum. Where indicated, visual findings were quantitated by assessing the fluorescent emission (photons/second; 800 nm wavelength) within the region of interest. Mean values are represented. Tumor immunotherapy responses are presented by individual spaghetti plots showing measured tumor volume over time for each animal in each treatment group. Comparison between groups was carried out by a tumor-free survival analysis employing a log-rank test. Impact on tumor growth over time was compared between groups using an ANOVA (two-factors with replication). $p < 0.05$ was considered as significant.

Supplementary Materials: The following supporting information can be downloaded at: https://www.mdpi.com/article/10.3390/gels8050284/s1, Figure S1: Effect of hyaluronidase (HAse) on the release of fluorescent-labeled anti-CTLA-4 from hydrogels; Figure S2: Inclusion of HAse within the hydrogel formulation enhances tumor-free survival; Table S1: Impact of hyaluronidase (HAse) on the delivery of hydrogel-encapsulated anti-CTLA-4 to TDLN.

Author Contributions: Conceptualization, methodology, formal analysis, resources, data curation, writing, reviewing, editing, project administration, and funding acquisition were all carried out by M.D.R. and A.H. Investigations were carried out by A.H. All authors have read and agreed to the published version of the manuscript.

Funding: This research was funded in part by the National Center for Advancing Translational Sciences, National Institutes of Health, which supported optical imaging experiments through UCLA CTSI grant no. UL1TR001881.

Institutional Review Board Statement: The study was conducted under approval from the UCLA Institutional Animal Care and Use Committee (protocol 2016-031; date of approval: 21 April 2016).

Informed Consent Statement: Not applicable.

Data Availability Statement: The data presented in this study are available on request from the corresponding author.

Acknowledgments: The authors thank Pat Leinert, Jr. and the production team at Leinco Technologies for the generation of anti-CTLA-4/DyLight 800; Thomas I. Zarembinski and BioTime Inc., for providing hydrogel reagents; Shili Xu and the Crump Preclinical Imaging Technology Center, Crump Institute for Molecular Imaging at UCLA (NIH/NCI 2P30CA016042-44).

Conflicts of Interest: The authors declare no financial conflict of interest. The funders had no role in the design of the study; in the collection, analyses, or interpretation of data; in the writing of the manuscript, or in the decision to publish the results.

References

1. Twomey, J.D.; Zhang, B. Cancer Immunotherapy Update: FDA-Approved Checkpoint Inhibitors and Companion Diagnostics. *AAPS J.* **2021**, *23*, 39. [CrossRef]
2. Liang, F.; Zhang, S.; Wang, Q.; Li, W. Clinical benefit of immune checkpoint inhibitors approved by US Food and Drug Administration. *BMC Cancer* **2020**, *20*, 823. [CrossRef]
3. Larkin, J.; Chiarion-Sileni, V.; Gonzalez, R.; Grob, J.J.; Cowey, C.L.; Lao, C.D.; Schadendorf, D.; Dummer, R.; Smylie, M.; Rutkowski, P.; et al. Combined Nivolumab and Ipilimumab or Monotherapy in Untreated Melanoma. *N. Engl. J. Med.* **2015**, *373*, 23–34. [CrossRef] [PubMed]
4. Wolchok, J.D.; Chiarion-Sileni, V.; Gonzalez, R.; Grob, J.J.; Rutkowski, P.; Lao, C.D.; Cowey, C.L.; Schadendorf, D.; Wagstaff, J.; Dummer, R.; et al. Long-Term Outcomes With Nivolumab Plus Ipilimumab or Nivolumab Alone Versus Ipilimumab in Patients With Advanced Melanoma. *J. Clin. Oncol.* **2022**, *40*, 127–137. [CrossRef] [PubMed]
5. Paz-Ares, L.; Ciuleanu, T.E.; Cobo, M.; Schenker, M.; Zurawski, B.; Menezes, J.; Richardet, E.; Bennouna, J.; Felip, E.; Juan-Vidal, O.; et al. First-line nivolumab plus ipilimumab combined with two cycles of chemotherapy in patients with non-small-cell lung cancer (CheckMate 9LA): An international, randomised, open-label, phase 3 trial. *Lancet Oncol.* **2021**, *22*, 198–211. [CrossRef]
6. Motzer, R.J.; Tannir, N.M.; McDermott, D.F.; Arén Frontera, O.; Melichar, B.; Choueiri, T.K.; Plimack, E.R.; Barthélémy, P.; Porta, C.; George, S.; et al. CheckMate 214 Investigators. Nivolumab plus Ipilimumab versus Sunitinib in Advanced Renal-Cell Carcinoma. *N. Engl. J. Med.* **2018**, *378*, 1277–1290. [CrossRef] [PubMed]
7. Doki, Y.; Ajani, J.A.; Kato, K.; Xu, J.; Wyrwicz, L.; Motoyama, S.; Ogata, T.; Kawakami, H.; Hsu, C.H.; Adenis, A.; et al. CheckMate 648 Trial Investigators. Nivolumab Combination Therapy in Advanced Esophageal Squamous-Cell Carcinoma. *N. Engl. J. Med.* **2022**, *386*, 449–462. [CrossRef] [PubMed]
8. Wang, D.Y.; Salem, J.E.; Cohen, J.V.; Chandra, S.; Menzer, C.; Ye, F.; Zhao, S.; Das, S.; Beckermann, K.E.; Ha, L.; et al. Fatal Toxic Effects Associated With Immune Checkpoint Inhibitors: A Systematic Review and Meta-analysis. *JAMA Oncol.* **2018**, *4*, 1721–1728. [CrossRef]

9. Lebbé, C.; Meyer, N.; Mortier, L.; Marquez-Rodas, I.; Robert, C.; Rutkowski, P.; Menzies, A.M.; Eigentler, T.; Ascierto, P.A.; Smylie, M.; et al. Evaluation of Two Dosing Regimens for Nivolumab in Combination With Ipilimumab in Patients With Advanced Melanoma: Results From the Phase IIIb/IV CheckMate 511 Trial. *J. Clin. Oncol.* **2019**, *37*, 867–875. [CrossRef]
10. Chung, C.K.; Fransen, M.F.; van der Maaden, K.; Campos, Y.; García-Couce, J.; Kralisch, D.; Chan, A.; Ossendorp, F.; Cruz, L.J. Thermosensitive hydrogels as sustained drug delivery system for CTLA-4 checkpoint blocking antibodies. *J. Control. Release* **2020**, *323*, 1–11. [CrossRef]
11. Francis, D.M.; Manspeaker, M.P.; Schudel, A.; Sestito, L.F.; O'Melia, M.J.; Kissick, H.T.; Pollack, B.P.; Waller, E.K.; Thomas, S.N. Blockade of immune checkpoints in lymph nodes through locoregional delivery augments cancer immunotherapy. *Sci. Transl. Med.* **2020**, *12*, eaay3575. [CrossRef] [PubMed]
12. Kim, J.; Francis, D.M.; Thomas, S.N. In Situ Crosslinked Hydrogel Depot for Sustained Antibody Release Improves Immune Checkpoint Blockade Cancer Immunotherapy. *Nanomaterials* **2021**, *11*, 471. [CrossRef] [PubMed]
13. Harui, A.; McLachlan, S.M.; Rapoport, B.; Zarembinski, T.I.; Roth, M.D. Peri-tumor administration of controlled release anti-CTLA-4 synergizes with systemic anti-PD-1 to induce systemic antitumor immunity while sparing autoimmune toxicity. *Cancer Immunol. Immunother.* **2020**, *69*, 1737–1749. [CrossRef] [PubMed]
14. Harui, A.; Roth, M.D. Employing a glutathione-s-transferase-tag and hyaluronidase to control cytokine retention and release from a hyaluronic acid hydrogel matrix. *J. Biomater. Appl.* **2019**, *34*, 631–639. [CrossRef]
15. Locke, K.W.; Maneval, D.C.; LaBarre, M.J. ENHANZE®drug delivery technology: A novel approach to subcutaneous administration using recombinant human hyaluronidase PH20. *Drug Deliv.* **2019**, *26*, 98–106. [CrossRef]
16. Martins, F.; Sofiya, L.; Sykiotis, G.P.; Lamine, F.; Maillard, M.; Fraga, M.; Shabafrouz, K.; Ribi, C.; Cairoli, A.; Guex-Crosier, Y.; et al. Adverse effects of immune-checkpoint inhibitors: Epidemiology, management and surveillance. *Nat. Rev. Clin. Oncol.* **2019**, *16*, 563–580. [CrossRef]
17. Xu, K.; Lee, F.; Gao, S.; Tan, M.H.; Kurisawa, M. Hyaluronidase-incorporated hyaluronic acid-tyramine hydrogels for the sustained release of trastuzumab. *J. Control. Release* **2015**, *216*, 47–55. [CrossRef]
18. Bookbinder, L.H.; Hofer, A.; Haller, M.F.; Zepeda, M.L.; Keller, G.A.; Lim, J.E.; Edgington, T.S.; Shepard, H.M.; Patton, J.S.; Frost, G.I. A recombinant human enzyme for enhanced interstitial transport of therapeutics. *J. Control. Release* **2006**, *114*, 230–241. [CrossRef]
19. Gomi, M.; Sakurai, Y.; Okada, T.; Miura, N.; Tanaka, H.; Akita, H. Development of Sentinel LN Imaging with a Combination of HAase Based on a Comprehensive Analysis of the Intra-lymphatic Kinetics of LPs. *Mol. Ther.* **2021**, *29*, 225–235. [CrossRef]
20. Wolchok, J.D.; Neyns, B.; Linette, G.; Negrier, S.; Lutzky, J.; Thomas, L.; Waterfield, W.; Schadendorf, D.; Smylie, M.; Guthrie, T., Jr.; et al. Ipilimumab monotherapy in patients with pretreated advanced melanoma: A randomised, double-blind, multicentre, phase 2, dose-ranging study. *Lancet Oncol.* **2010**, *11*, 155–164. [CrossRef]
21. Bertrand, A.; Kostine, M.; Barnetche, T.; Truchetet, M.E.; Schaeverbeke, T. Immune related adverse events associated with anti-CTLA-4 antibodies: Systematic review and meta-analysis. *BMC Med.* **2015**, *13*, 211. [CrossRef] [PubMed]

Article

Hydrogel Check-Valves for the Treatment of Hydrocephalic Fluid Retention with Wireless Fully-Passive Sensor for the Intracranial Pressure Measurement

Seunghyun Lee [1,2], Shiyi Liu [1], Ruth E. Bristol [3], Mark C. Preul [4] and Jennifer Blain Christen [1,*]

1. School of Electrical Computer and Energy Engineering, Arizona State University, Tempe, AZ 85281, USA; slee346@asu.edu (S.L.); shiyi.liu.1@asu.edu (S.L.)
2. Children's Hospital of Orange County, Orange, CA 92868, USA
3. Phoenix Children's Hospital, Phoenix, AZ 85016, USA; rbristol@phoenixchildrens.com
4. Barrow Neurological Institute, Phoenix, AZ 85013, USA; mark.preul@dignityhealth.org
* Correspondence: jennifer.blainchristen@asu.edu

Abstract: Hydrocephalus (HCP) is a neurological disease resulting from the disruption of the cerebrospinal fluid (CSF) drainage mechanism in the brain. Reliable draining of CSF is necessary to treat hydrocephalus. The current standard of care is an implantable shunt system. However, shunts have a high failure rate caused by mechanical malfunctions, obstructions, infection, blockage, breakage, and over or under drainage. Such shunt failures can be difficult to diagnose due to nonspecific systems and the lack of long-term implantable pressure sensors. Herein, we present the evaluation of a fully realized and passive implantable valve made of hydrogel to restore CSF draining operations within the cranium. The valves are designed to achieve a non-zero cracking pressure and no reverse flow leakage by using hydrogel swelling. The valves were evaluated in a realistic fluidic environment with ex vivo CSF and brain tissue. They display a successful operation across a range of conditions, with negligible reverse flow leakage. Additionally, a novel wireless pressure sensor was incorporated alongside the valve for in situ intracranial pressure measurement. The wireless pressure sensor successfully replicated standard measurements. Those evaluations show the reproducibility of the valve and sensor functions and support the system's potential as a chronic implant to replace standard shunt systems.

Keywords: hydrocephalus; hydrogel; MEMS; 3D printing; brain implant

1. Introduction

Hydrocephalus (HCP) is a chronic neurological disorder characterized by the inability to automatically adjust the drainage physiology of cerebrospinal fluid (CSF). Nearly 1 in 500 infants born in the United States suffer from hydrocephalus. CSF is a vital supporting liquid that flows through and around the cerebral cortex, functioning as (1) a "cushion" for protecting the brain and spinal cord from external shocks; (2) a "vehicle" for nutrients necessary for the brain and removing waste; and (3) a "regulator" to adjust intracranial pressures (ICP) by flowing between the cranium and spine [1]. With regard to ICP adjustment, CSF is partially drained through the arachnoid granulations (AG), which act as one-way biological valves from the subarachnoid space (SAS) to the superior sagittal sinus (SSS). In the case of hydrocephalic patients, this draining pathway of AG is disrupted, which causes an imbalance between the amount of produced and drained CSF, leading to excess accumulation of CSF in the ventricles, the subarachnoid space, and other cisternal regions of the brain [2–4].

The current standard treatment for HCP is the implantation of a drainage tube, termed a shunt, between the ventricles of the brain and the abdominal cavity or atrium of the heart for providing drainage of excessively accumulated CSF. These are implanted such that an

Citation: Lee, S.; Liu, S.; Bristol, R.E.; Preul, M.C.; Blain Christen, J. Hydrogel Check-Valves for the Treatment of Hydrocephalic Fluid Retention with Wireless Fully-Passive Sensor for the Intracranial Pressure Measurement. *Gels* 2022, *8*, 276. https://doi.org/10.3390/gels8050276

Academic Editor: Wei Ji

Received: 3 March 2022
Accepted: 24 April 2022
Published: 29 April 2022

Publisher's Note: MDPI stays neutral with regard to jurisdictional claims in published maps and institutional affiliations.

Copyright: © 2022 by the authors. Licensee MDPI, Basel, Switzerland. This article is an open access article distributed under the terms and conditions of the Creative Commons Attribution (CC BY) license (https://creativecommons.org/licenses/by/4.0/).

outflow catheter directs CSF away from the ventricle through the skull and into a valve that controls CSF flow into another outflow catheter from the valve, which then directs CSF into distal drainage spaces [5,6].

Unfortunately, shunts in current usage have a notoriously high failure rate. Fifty percent of shunts fail within the first two years of implantation, even though the shunts are the most popular and primary treatment method for hydrocephalus. The majority of shunt complications are catheter-based occlusion failures. Catheters of an implanted shunt are exposed to blood and cellular debris, both containing proteins that bind to the shunt, which further cause macrophages and monocytes to produce growth factors. Such accumulations of debris can cause clogging of the tubing and valve. Accumulations on the outside of the tubing and valve, in turn, can attract astrocytes and microglia, potentiating an inflammatory response. Generally, occlusion-based shunt failures are caused by foreign body response, infection, and cellular growth [5–9].

Therefore, a catheter-less approach to reduce the overall implant surface area would be appealing to reduce these complications. Alternative shunt methods have been studied to improve current shunt systems, including MicroElectroMechanical System (MEMS)-based devices. MEMS valves have been extensively researched and developed, with fluidic valves proposed to control CSF as an implantable valve in various forms, such as cantilever, bridge, perforated membrane, or spherical ball-type valves [10–18]. However, critical challenges still exist, such as reverse flow leakage [10], valve deformation in long-term operations [14], valve stiction, imperfect sealing [10,13,14], etc., leading to low reproducibility and durability of the valves. To overcome these challenges, our earlier work has proposed the hydrogel check valve, meeting the two required features: (1) a non-zero cracking pressure (from 20 to 110 mmH$_2$O) and (2) negligible reverse flow leakage [17–19]. However, the previous work suffered from a deficiency in the accurate measurement of fluid pressure across the valve as measured inside the setup and a lack of verification of the valve in realistic environments.

Herein, we report a physiologically and biologically realistic evaluation of the hydrogel check-valve to therapeutically manage hydrocephalus and propose a wireless measurement of fluid pressure across the valve, all designed to reduce current problems with shunt management for hydrocephalus (Figure 1). We tested the hydrogel check-valve in increasingly realistic fluidic conditions while maintaining ~37 °C during evaluation. These evaluations showed that the hydrogel check-valve maintains the functionality in each condition within the specifications of traditional shunt systems. The long-term behavior of the valve was also evaluated through automated loop functional tests, and it demonstrates improved repeatability and durability of the valve compared to our earlier work [17–19].

Additionally, we developed a fully-passive wireless pressure sensor to measure the pressure applied to the valve. Currently, clinical diagnosis of HCP is based on symptoms such as headache, sleepiness, vomiting, etc., which could be unrelated to HCP if there are other disease processes or comorbidities. Brain imaging usually shows the increased ventricular size, but not always. The intracranial pressure may be measured, and for those that have a valve in place, the valve may be checked for its operational status. If their HCP treatment device is malfunctioning, the ICP range will likely be abnormal, leading to HCP or collapsed CSF spaces (i.e., ventricles). In the worst case, the HCP patients will undergo surgery to interrogate their implanted device, remove it, and, if necessary, replace the valve and tubing. In order to avoid this situation, wireless pressure measurement has distinct potential as a non-invasive and non-surgical method. The fully-passive wireless pressure sensor was designed using a resistive pressure sensor and RF backscattering to transmit the ICP measurement to a receiver outside of the brain. It shows comparable output to wired pressure measurements.

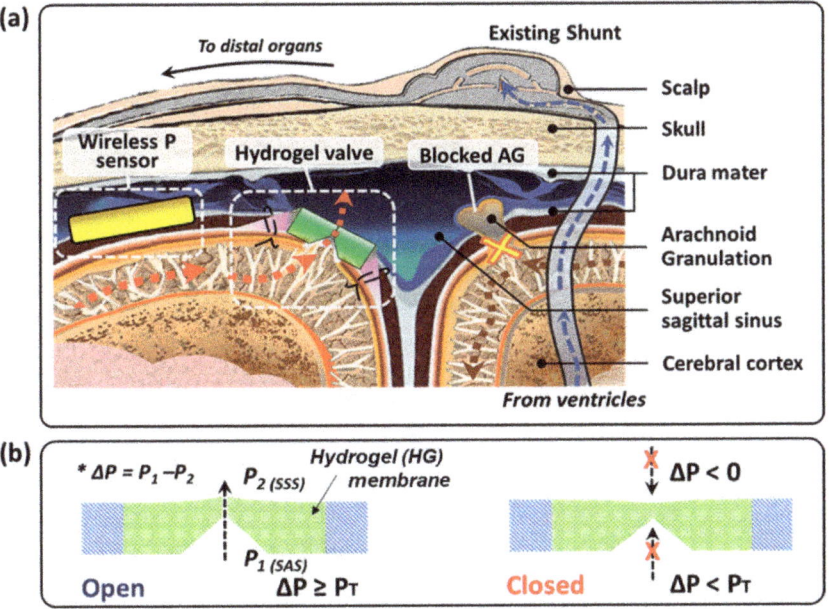

Figure 1. (a) Illustration of alternative cerebrospinal fluid (CSF) draining methods (existing shunt system and hydrogel valve of this work) for hydrocephalus treatment. Existing shunts include catheters connected from the brain ventricles to distal body spaces in order to drain CSF outwards. The proposed hydrogel valve is implanted directly in the intradural space of the superior sagittal sinus and directs CSF drainage from the subarachnoid space (SAS) into the superior sagittal sinus (SSS). This method of directing CSF allows the CSF draining process to be confined within the cranium without the use of catheters that cause many complications. (b) Basic operation of the valve at a cross-sectional view. When the hydrogel becomes hydrated, the swollen hydrogel structure closes the hole, forming the closed valve. When the pressure in the SAS reaches higher than the SSS over the threshold, namely, cracking pressure (P_T), $\Delta P > P_T$, the swollen hydrogel valve becomes open and CSF can flow unidirectionally from SAS to SSS. When the SAS has lower pressure than the SSS by less than a differential threshold, P_T, $\Delta P < P_T$, the valve is closed and blocks the CSF flow as the pressure difference cannot open the valve.

2. Results and Discussion

2.1. Experimental Setup

2.1.1. Bench-Top Setup

The hydrogel check-valve was evaluated for basic functionality in the bench-top fluidic circulatory setup. The functional tests were performed to ensure that the valve would open under forward flow and that the reverse flow leakage was properly sealed by the hydrogel swelling phenomena. The setup consists of the hydrogel valve, fluid source, and pressure & flow rate sensors (Figure 2). For the fluid circulatory system, we used a few types of source—syringe pump (Model 33 syringe pump, Harvard Apparatus, Holliston, MA, USA) and peristaltic pump (P-70, Harvard Apparatus, Holliston, MA, USA), for short-term and long-term functionality tests, respectively. Resistive pressure sensors (PX26-001DV, Omega, Biel/Bienne, Switzerland), which have 1 mmH$_2$O resolution, were used for measurement of differential pressure and flow rate and calibrated for the actual experiments by a customized setting. All data were recorded by a data acquisition board (NI USB 6216, National Instruments, Austin, TX, USA) and SignalExpress 2015 software [17–19].

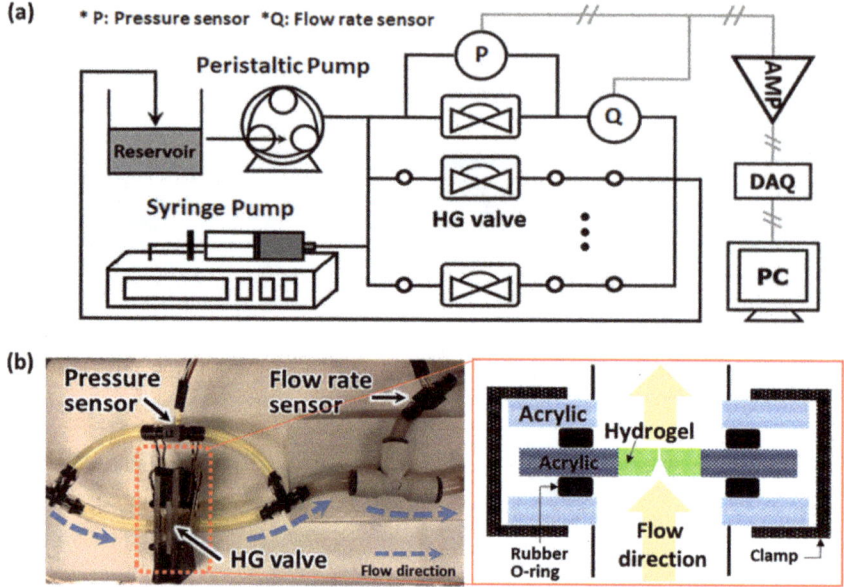

Figure 2. (a) Experimental setup for evaluation of valve functionality. For the fluid sources, a syringe pump (Model 33 syringe pump, Harvard Apparatus, Holliston, MA, USA) and peristaltic pump (P-70, Harvard Apparatus, Holliston, MA, USA) were used depending on the protocol for the experiments. Commercial resistive pressure sensors (PX26-001DV, Omega, Biel/Bienne, Switzerland) were used for measuring the pressure and flow rate with customized calibrations and the electric signals, voltage, from the sensors were transmitted through a voltage amplifier and data acquisition board (NI USB 6216, National Instruments, Austin, TX, USA) and recorded by SignalExpress software on a PC. (b) Photograph of the hydrogel valve setup (**left**) and diagram of the sandwich-type connecting module (**right**). The module was constructed using two acrylic plates with a hole at the center and two rubber O-rings, each placed between the acrylic plates and the respective side of the valve. The module sealing was performed by aligning the holes of the acrylic plates with rubber O-rings and valve and compressing the module on both sides of the acrylic plate using binder clips.

Long-term valve testing was performed to check the valve's reliability and durability by automated loop testing setup using the programmable peristaltic pump to control fluidic flow with the number of flow cycles, flow rate, and flow direction. The valve was evaluated in a qualitative way to measure the number of valves operating instead of a quantitative way in a real-time duration due to the limits of a lab-based environment.

2.1.2. In Vitro Animal Model Setup

The in vitro evaluation setting for the valve was built on a fixed sheep brain (Bio corporation, Alexandria, MN, USA) (Figure 3). The sheep brain was preserved with the dura mater and enclosed by Polydimethylsiloxane (PDMS) molding to prevent fluid leakage. The hydrogel valve and customized wireless pressure sensor were placed on the locations we punched through the dura mater manually. CSF was injected into the cavity created by the valve and SAS, and the differential pressure across the valve was measured by the resistive pressure sensor.

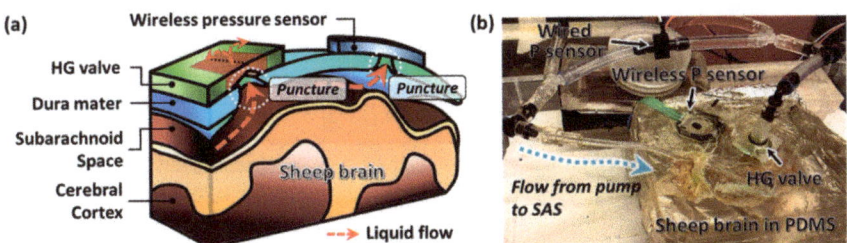

Figure 3. The hydrogel valve was evaluated in an in vitro setting using a fixed sheep brain. (**a**) Schematic of the experiment setup. The valve and customized wireless pressure sensor were set in the designated location on the dura mater where we manually punctured through. (**b**) Actual configuration of the experimental setup. A syringe pump injected CSF into the cavity formed by SAS and the valve. The excess portion of the manual cut was sealed by PDMS to prevent any leakage. The differential pressure was measured across the valve using conventional wired and novel wireless sensors simultaneously.

2.2. Hydrodynamic Valve Characteristics

The basic hydrodynamic response of the hydrogel valve was measured and displayed the valve operation within the target range in terms of normal CSF drainage in bench-top testing (Figure 4a) and in an in vitro fixed sheep brain (Figure 4b). To evaluate the valve in more realistic patient conditions, the valve was tested in the emulated biological fluids, all based on CSF with various additives known to generate occlusion-based failures in traditional shunts: Water, CSF, CSF + Calcium (1.1 mM), CSF + Blood (5% v/v), CSF + Fibronectin (7.5 µg/mL), and CSF + All additives (Calcium + Fibronectin + Blood) at ~37 °C [20–22]. These additives were used to simulate worst-case scenarios such as fibrous encapsulation, excessive foreign body response, and device calcification: all situations where current shunts are known to fail. In general, the implanted shunt is stained with blood and cellular debris because the brain tissue and blood barrier are damaged during shunt implantation. The proteins from the blood and cells can bind to shunt-induced macrophages and monocytes to produce growth factors and consequently attract astrocytes and enhance the inflammatory response. Mineralization or calcification cases are also often mentioned in studies of shunt failure. This results in mechanical stress and barium sulfate added to the catheter causing nucleation of mineral deposits, ultimately leading to disintegration and the cracking of the catheter. The respective concentrations for each additive were based on concentrations from studies analyzing patient CSF [23–25].

Table 1. Summary of specifications for hydrogel valves tested in worst-case environments.

Fluid Type	Bench-Top		Sheep Brain	
	P_T [mmH$_2$O]	Q_R [µL/min]	P_T [mmH$_2$O]	Q_R [µL/min]
CSF	46.0 ± 7.3	1.1 ± 0.9	113.0 ± 9.8	3.7 ± 1.0
Water	51.5 ± 5.5	1.3 ± 0.7	81.2 ± 4.7	2.6 ± 1.1
CSF + Calcium	38.5 ± 5.4	1.7 ± 0.7	135.1 ± 5.1	3.3 ± 0.9
CSF + Fibronectin	68.3 ± 4.7	1.7 ± 0.7	117.5 ± 5.9	1.9 ± 0.9
CSF + Blood	62.9 ± 3.2	2.1 ± 1.1	152.7 ± 4.9	2.7 ± 1.3
CSF + All	72.4 ± 6.1	1.4 ± 0.5	92.2 ± 7.0	3.0 ± 0.6

(P_T: Cracking pressure, Q_R: Flow rate).

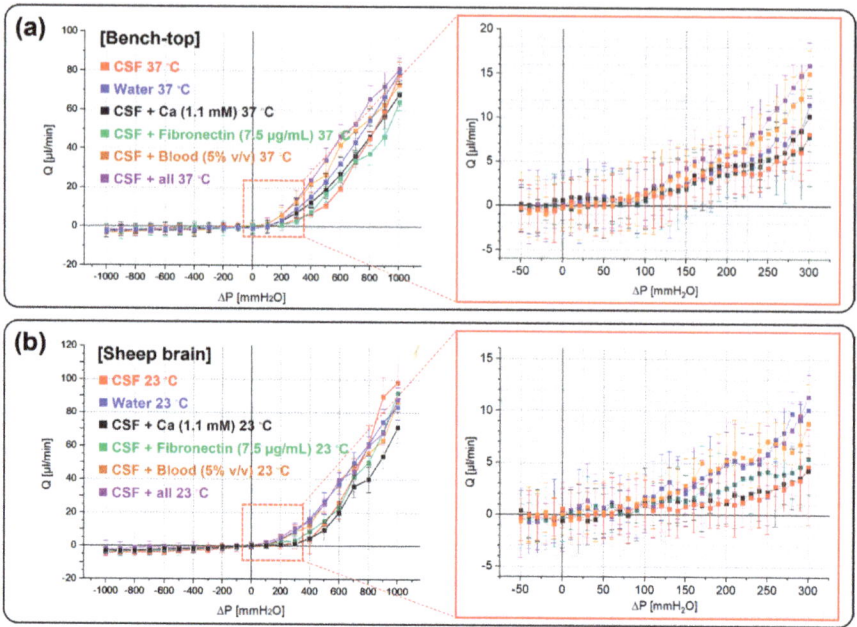

Figure 4. A series of increasingly worst-case fluidic conditions were used to emulate short-term conditions related to various failure modes seen in traditional shunt systems, namely occlusion due to foreign body response, fibrous encapsulation, and mineralization due to calcification. The functional tests were performed on (**a**) bench-top and (**b**) in vitro fixed sheep brain setups. The hydrogel check-valve remained functional, with a reasonable cracking pressure range and negligible reverse flow leakage. The detailed results of the valve's functionalities are described in Table 1.

Note that many of these tissue-based occlusion mechanisms are not tested, just the direct response of the hydrogel valve to the fluid and additives. The hydrogel valve remained functional, showing high diodicity without a noticeable reverse flow leakage while being tested under the six fluid conditions. The measured average P_T and leakage flow were 56.6 ± 5.4 mmH$_2$O and 1.55 µL/min in the bench-top experiments and 115.3 ± 6.2 mmH$_2$O and 2.0 µL/min in the sheep brain experiments. The measured P_T from the sheep brain experiments is higher than that of the bench-top evaluation. This may be because the sheep brain is deformable (higher compliance), and the valve was placed on the thin flexible dura mater, possibly leading to reduced pressure across the valve [19]. Detailed measurements of the valve's flow response are shown in Table 1.

2.3. Fully-Passive Wireless Pressure Sensor

Wireless monitoring of intracranial pressure (ICP) was successfully accomplished in a completely passive manner. The overall operating principle of the wireless fully-passive pressure sensor is demonstrated in Figure 5a. Herein, two different kinds of energy, namely infrared light (IR) and radiofrequency (RF) electromagnetic wave, are adopted to enable the wireless acquisition of pressure. An external LED IR emitter radiates a beam of IR light whose intensity is modulated by an external pulse signal to enable chopping. The photodiode D_1 on the sensor senses the variation of IR energy and converts it to an electrical voltage signal, which has the same waveform as the external pulse. The generated voltage signal is divided by a voltage divider circuit consisting of resistors R_1–R_3, capacitor C_1, and the pressure-sensitive resistor. Figure 5b shows the structure of the pressure-sensitive resistor. A pair of interdigitated electrodes is printed on a flexible polymer film. The film is placed upside down on top of a resistive sheet, with a spacer sandwiched in the middle.

Under external pressure, the flexible polymer will deform, causing the interdigitated electrodes to contact the resistive sheet, thereby reducing the overall resistance between the two interdigitated electrodes. This forms a pressure sensing resistor whose resistance decreases as external pressure increases. As a proof-of-concept prototype, the pressure-sensitive resistor is assembled using a commercially available force-sensitive resistor (FSR). The polymer film with interdigitated electrodes is removed from the commercial FSR and is adhered to the resistive sheet (Velostat) using double-sided tape, which also functions as a spacer. Assembled pressure-sensitive resistor is sealed with sealing tape. Two electrical leads connect the pressure-sensitive resistor to the fully-passive wireless sensor.

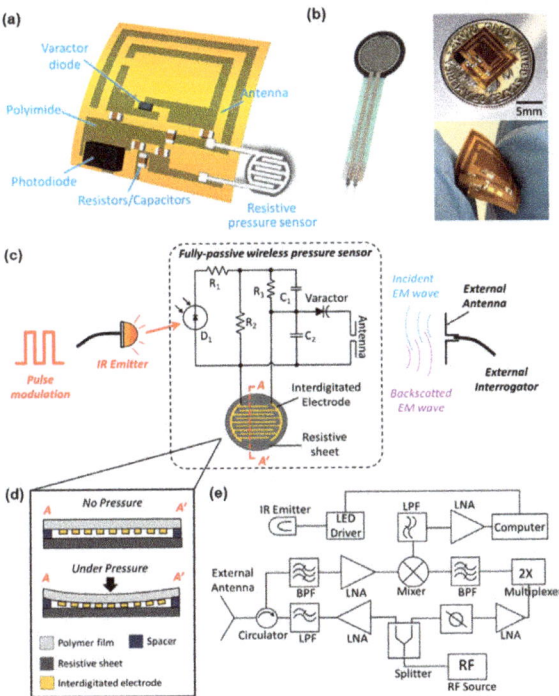

Figure 5. Wireless intracranial pressure (ICP) monitoring through a fully-passive method. (**a**) Illustration of the fabricated wireless fully-passive pressure sensor. (**b**) Photograph of the thin film force sensing register (FSR 402, Interlink Electronics, Camarillo, CA, USA) used for the resistive pressure sensor (Left) and the wireless fully-passive pressure sensor fabricated on a flexible polyimide substrate (Right). (**c**) Overall view of the system operating principle. Wireless fully-passive pressure sensing is accomplished using both infrared (IR) and radiofrequency (RF) electromagnetic energy. The photodiode (D_1) on the sensor receives the modulated IR light from the external LED IR emitter and generates an electrical voltage signal, which has the same waveform as the external modulation pulse. The generated pulse is voltage divided by the resistors R_1–R_3 and the pressure-sensitive resistor (**d**), and outputs to the varactor diode. The voltage across the varactor diode is mixed with incident RF energy and wirelessly transmitted to the external interrogator using the RF backscattering method. The external interrogator then extracts the pressure information through a series of demodulation and calculation processes. (**d**) Operation principle of the pressure-sensitive resistor. The resistor comprises a pair of interdigitated electrodes which are printed on a flexible polymer film. The polymer film is placed on top of a resistive sheet with a spacer sandwiched between. Under external pressure, the polymer film deforms and results in contact between the interdigitated electrodes and the resistive sheet, lowering the resistance of the resistor. (**e**) Detailed structure of the external interrogator for extracting the pressure value.

The voltage divider circuit measures the resistance of the pressure-sensitive resistor and outputs an electrical signal to the varactor diode. Thus, the external pressure value can be obtained by measuring the amplitude of voltage signal, V_m, across the varactor diode, which is accomplished utilizing RF backscattering [26]. In short, the external interrogator generates and radiates an RF electromagnetic wave (f_0). The sensor receives the incident EM wave through its integrated antenna. The varactor diode then mixes the RF signal (f_0) with the target signal (f_m) and produces the third-order mixing products ($2f_0 \pm f_m$), which are scattered back and picked up by the external interrogator. The external interrogator extracts the target signal (f_m) through a series of filtering and demodulating procedures. The extracted signal is then processed by a computer to calculate the pressure value. The fully-passive wireless sensor is fabricated using a copper-clad polyimide pad, and discrete surface mount electronic components, including photodiode, varactor diode, resistors, and capacitors, are soldered onto the exposed pad.

2.4. Wireless Pressure Measurements

The pressure measurements were also performed by using a wireless pressure sensor. The wireless pressure sensor required unit calibration first with arbitrary units since the pressure to resistance relationship is highly fabrication dependent. The wired pressure sensor was used for the calibration as a reference measurement, while the wireless pressure sensor measured the pressure variance. The relationship between the wireless and wired sensor showed linear behavior in the positive pressure range and revealed a slope of 0.0012 a.u./mmH$_2$O with a coefficient of determination value of $R^2 = 0.9687$, obtained from a fitted linear regression model. In the negative pressure range, the wireless pressure sensor could not measure the pressure variance due to its geometrical limit of resistance. The sensor can measure pressure only in one direction currently; bi-directional wireless pressure measurement will serve as the subject of future studies.

The calibrated wireless pressure sensor was used for valve functional tests in sheep brain experiments, and the results shown in Figure 6 display reasonable hydrodynamic behavior of the valve within the target specification. The valve has a P_T of 108.1 ± 3.4 mmH$_2$O, which is clearly comparable to the result of wired measurement, a P_T of 113.0 ± 9.8 mmH$_2$O.

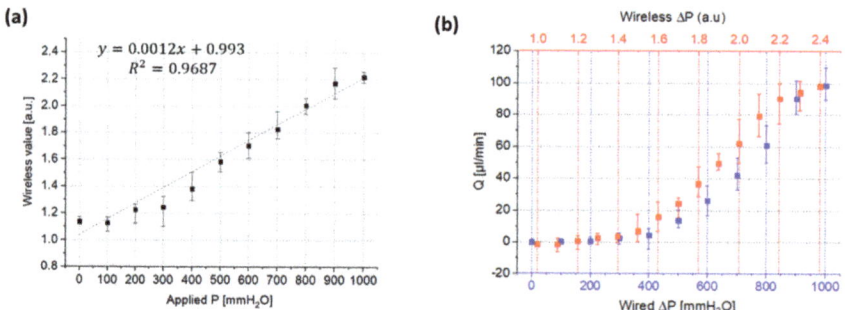

Figure 6. The calibrated wireless pressure sensor was used for valve functional tests in the positive pressure range and served to provide reasonable hydrodynamic behavior of the valve. (**a**) Calibration of the wireless sensor in terms of the applied pressure. (**b**) Flow response of the valve in the wired/wireless measurement. The valve demonstrated reasonable flow response in both modes with high flow diodicity and no flow leakage. By the wired measurement, the valve shows flow rates of 0~98.5 µL/min, P_T of 113.0 ± 9.8 mmH$_2$O, and negligible reverse flow leakage of 3.7 µL/min. The wireless measurement also displays that the valve has a comparable flow response to the wired measurement with a P_T of 108.1 ± 3.4 mmH$_2$O.

2.5. Long-Term Functional Tests

We performed repetitive functional tests to evaluate the valve's long-term feasibility in terms of the run cycles. The test was controlled by a programmable peristaltic pump for repetitive sequences of forward/reverse cycles in the range of −50 < ΔP < 300 mmH$_2$O [27]. In order to advance the reliability and durability of the valve, glass was used as the supporting material instead of acrylic. Because the hydrophilicity of glass is higher than acrylic, it provides more robust adhesion strength with hydrogel than acrylic-hydrogel. A total of 5 devices were used for the evaluation and maintained comparable cracking pressure and negligible reverse flow leakage for >2000 running times within the range of 52.3 < P$_T$ < 67.1 mmH$_2$O and < 0.5 μL/min, respectively (Figure 7a,b). Compared to the results of previous devices using acrylic, glass-based hydrogel valves display much enhanced long-term functionality, increased by ~50% as a function of run cycles. Throughout the repetitive testing, the device under test #4 (DUT 4[G]) out of the 5 devices remained within design specifications over the course of 2256 cycles of running with relatively constant cracking pressure and little reverse flow leakage of 48.3 < P$_T$ < 75.1 mmH$_2$O and <0.4 μL/min, respectively (Figure 7c).

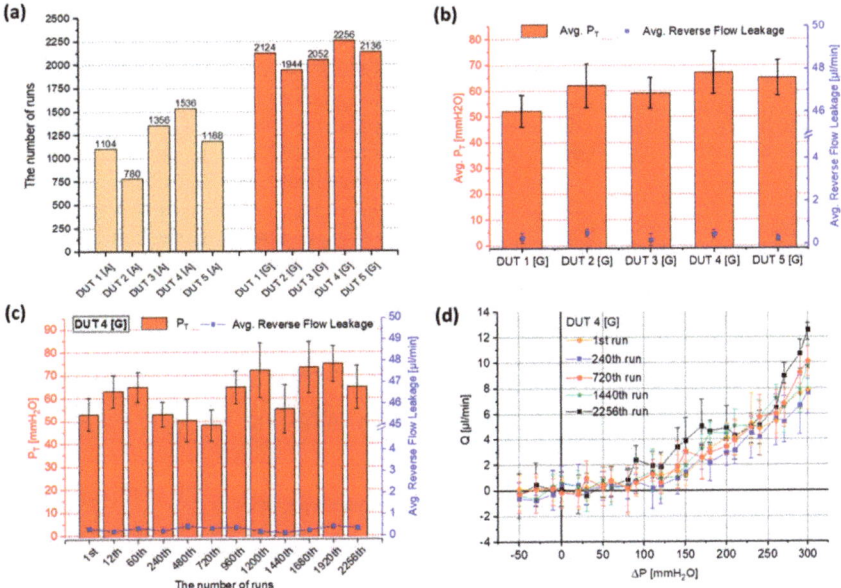

Figure 7. Automated repetitive tests were performed by a programmable peristaltic pump to evaluate the valve's long-term functionality. In order to enhance the reliability and durability of the valve, glass was used for the surrounding substrate instead of the acrylic plate because glass has higher adhesion strength than acrylic due to its superior hydrophilicity. One cycle consists of flow response measurements in forward and reverse flow and is analogous to a one-day operation of the valve [20]. (**a**) Comparison of the automated repetitive test results between acrylic-based (DUT [A]) and glass-based devices (DUT [G]). The results of the acrylic-based devices (DUT [A]) are referenced from our previous study [19] for comparison purposes. The 4 out of the 5 devices with glass surrounding substrate (DUT 1[G], 3[G], 4[G], and 5[G]) marked much higher run cycles, >2000. Generally, glass-based devices display ~50% higher running times than acrylic-based devices. (**b**) The average value of P$_T$ and reverse flow leakage of the devices during the repetitive tests. The devices show P$_T$ and reverse flow leakage in the range of 52.3~67.1 mmH$_2$O and < 0.5 μL/min, respectively. Error bar: standard deviation. (**c**) P$_T$ and reverse flow leakage of DUT 4[G] over 2256 cycles. The measurements show the valve functions consistently at 48.3 < P$_T$ < 75.1 mmH$_2$O and reverse flow leakage of <0.3 μL/min. Error bar: standard deviation. (**d**) The flow response of the valve (DUT 4[G]) shows high diodicity behavior with negligible reverse flow leakage throughout 2256 runs. Error bar: min to max.

3. Conclusions

The hydrogel check valve presented here provides near-identical flow characteristics with an above zero cracking pressure and little to no reverse flow leakage as an alternative HCP treatment method. Ultimately, we have developed an implantable valve which is capable of accurately replicating the function of arachnoid granulations in a variety of in vitro models of hydrocephalus with biologically and physiologically realistic conditions.

The hydrogel swelling effect is a critical component in determining non-zero cracking pressure and countercurrent leakage, which allowed us to manufacture a valve that can meet the required parameters. Therefore, characterizing and optimizing the hydrogel in terms of stiffness, roughness, and swelling ratio and correlating it with valve performance would help improve the robustness and performance of our systems, which will remain as our future work. The wireless pressure measurement was also performed using a customized fully-passive pressure sensor. By comparison with the wired measurement, the results validate the valve's potential for non-invasive and non-surgical ICP measurement.

4. Materials and Methods

4.1. Valve Design and Fabrication

The valve was designed by considering the actual SSS of the human skull for ideal future implantation; the valve needs to be comparable to or smaller than the SSS. The SSS roughly has a mean diameter of 7.3–8.8 mm [28]; thus, the valve was designed with a 7 mm diameter. The target specification of the valve was designed by referencing traditional shunt systems, which are evaluated by industry standards, ISO 7197 and ASTM F647, in terms of cracking pressure, fluidic resistance, and maximum reverse flow leakage [29,30]. Moreover, we considered the actual condition of HCP patients; the normal ICP in humans ranges from −100 to 350 mmH$_2$O, and it increases to >500 mmH$_2$O for hydrocephalus patients. Overall, the valve aims to operate at the range of −200 < ΔP < 600 mmH$_2$O [3,8,9]. The target cracking pressure, P_T, is 10 to 230 mmH$_2$O [3,6]. The valve is composed of a hydrogel and surrounding substrate to support the valve structure. 3D-printed molds are assembled in the punctured hole at the center of the surrounding plate then the liquid-state hydrogel is cured inside the hole by UV light to be solidified (Figure 8).

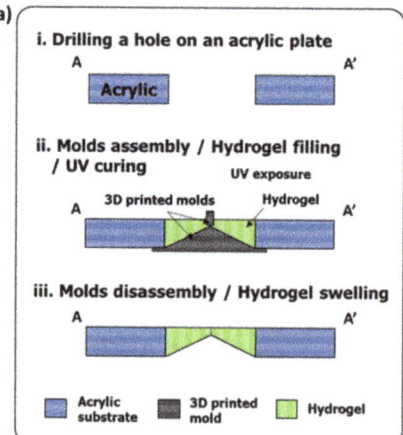
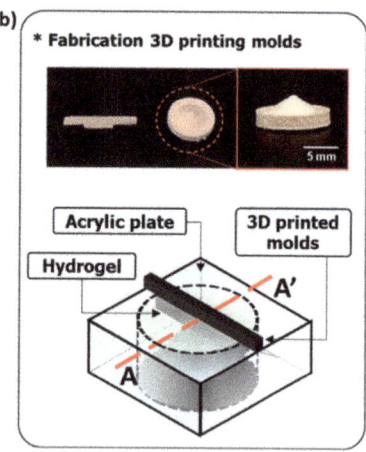

Figure 8. (**a**) Fabrication process: i. Drilling a hole in the center of an acrylic plate. ii. 3D printed molds are assembled in the hole, then the prepared hydrogel solution is poured and cured by UV light (365 nm, 400 mJ/cm^2). iii. The 3D printed molds are disassembled and the device is immersed in water to swell the hydrogel. (**b**) The actual 3D printed molds and illustration of the parts assembled for the device fabrication.

4.1.1. Hydrogel

Hydrogels are hydrophilic and absorbent polymeric networks which can contain over 90% water. Therefore, the mechanical properties of a hydrogel can be very similar to natural tissue; this, along with their excellent biocompatibility, has led to their frequent use in medical devices. In this study, the preparation of the hydrogel solution was performed by mixing the base (2-hydroxyethyl methacrylate, Sigma Aldrich, St. Louis, MO, USA), crosslinker (ethylene glycol dimethacrylate, Sigma Aldrich, St. Louis, MO, USA), and photoinitiator (2,2-dimethoxy-2-phenylacetophenone, Sigma Aldrich, St. Louis, MO, USA) at a ratio of 1: 0.04: 0.1, respectively [17–19]. This solution was poured onto the prepared substrate and cross-linked by exposure to UV of ~400 mJ/cm^2 and a 365 nm wavelength for curing before the hydration step and immersed in water for 24 h at room temperature, 23 °C.

4.1.2. 3D-Printed Mold

The tapered hydrogel structure was fabricated using 3D printed molds (Figure 8b). 3D printing implements a technology known as additive manufacturing involving iteratively depositing horizontal cross-sections of the desired 3D object [31]. This technology allows for the rapid development of custom 3D models through reducing fabrication costs and the time required to prototype custom models [32,33]. To create the models, 3D printers deposit layers of molten thermoplastic, a process called Fused Filament Fabrication (FFF) [31]. The design of these 3D models requires 3D modelling software. The models used in this research were designed in Autodesk's Fusion 360 isometric modeling software and were printed on the Ultimaker 2 Go 3D printer (Utrecht, Netherlands) with polylactic acid (PLA) filament. The Ultimaker 2 Go has a positioning precision of 12.5 µm in the horizontal direction and a precision of 5 µm in the vertical direction.

4.1.3. Cerebrospinal Fluid Preparation

CSF was collected from hydrocephalic patients at the Phoenix Children's Hospital with an approved materials transfer agreement (MTA) and used under an Institutional Biosafety Committee (IBC) approved protocol. The CSF solutions were from an aggregate of multiple CSF samples and maintained in a 4 °C environment when not in use. Five CSF sample solutions were generated from an aggregate of patient CSF, four of which were spiked with various additives known to generate occlusions in traditional shunts: calcium (1.1 mM), fibronectin (7.5 µg/mL), blood (5%), and all three combined [20–22].

4.2. Wireless Pressure Sensor

One of the main symptoms of shunt failure is headache [4,5]. As this is a non-specific symptom with many causes, it can be difficult to determine whether this, or any symptom, is due to uncontrolled hydrocephalus. In the worst case, shunt patients will need exploratory surgery to determine whether their device is working appropriately. The most straightforward way to check the valve functionality is to observe the CSF pressure in the brain, the intracranial pressure (ICP), because ICP is the most related factor for showing HCP symptoms. If the HCP treatment device is not working properly, the ICP will be out of the normal range—usually elevated. Through ICP measurement, we can confirm that the symptoms must be due to issues with the device.

To this end, we need to observe the status of the device when the HCP patients show any related symptoms of device failure in a more patient-friendly way. Therefore, we developed a fully-passive wireless pressure sensor to measure ICP non-invasively. The sensor was designed using a resistive pressure sensor and RF backscattering to transmit the ICP measurement transcranially.

The RF components of the device use the varactor and antenna to encode information about the sensor voltage onto the second harmonic of a backscattered RF signal [26]. Additionally, the present design uses a secondary input from an external LED powering an internal photodiode. The LED and photodiode operate at infrared (IR) wavelength in order to pass through tissue. The LED is pulsed at two different frequencies in order to encode information from the pressure sensor in a manner which is resilient to natural variations in the RF and IR attenuation.

4.2.1. Pressure Value Calculation

Wireless intracranial pressure (ICP) monitoring through a fully-passive method is shown in Figure 5. External pressure affects the resistance of the pressure-sensitive resistor. Increasing the pressure decreases the resistance. To read out the resistance change, R_3, C_1, and the resistive sensor form a voltage divider circuit, which divides the output voltage of the photodiode based on the impedance ratio between $R_3 \| C_1$ and the resistance of the pressure-sensitive resistor. For simplicity, suppose the photodiode generates a sine wave to the voltage divider circuits. The sine wave has a frequency of f1 and an amplitude of A_{i1}. The resistance of the pressure sensor is R_x. Then the amplitude of the output signal, A_{o1}, can be written as:

$$A_{o1} = \frac{R_x}{Z_{t1}} \times A_{i1} \quad (1)$$

where Z_{t1} represents the impedance of $R_3 \| C_1$ (the impedance of R_3 in parallel with C_1). For simplicity, suppose R_3 is 100 kΩ and C_1 is 1 nF, then Z_{t1} can be expressed as:

$$Z_{t1} = \sqrt{10^{10}\left[Im(\frac{1}{1+i\frac{\pi f_1}{5000}})\right]^2 + \left[R_x + 10^5 Re(\frac{1}{1+i\frac{\pi f_1}{5000}})\right]^2} \quad (2)$$

where $Re(f)$ and $Im(f)$ denote the real and imaginary parts of f. The amplitude of the voltage divider output signal A_{o1} is a function of the pressure-sensitive resistor (R_x), photodiode output voltage A_{i1}, and the modulation frequency f_1. The diode output voltage A_{i1} is greatly affected by the external environment, making the output A_{o1} unstable. To overcome such an effect, a second modulation frequency, f_2, is introduced. Under f_2, the output signal amplitude can be written as:

$$A_{o2} = \frac{R_x}{Z_{t2}} \times A_{i2} \quad (3)$$

where Z_{t2} is the impedance of $R_3 \| C_1$ at f_2, which can be expressed as:

$$Z_{t2} = \sqrt{10^{10}\left[Im(\frac{1}{1+i\frac{\pi f_2}{5000}})\right]^2 + \left[R_x + 10^5 Re(\frac{1}{1+i\frac{\pi f_2}{5000}})\right]^2} \quad (4)$$

The ratio between A_{o1} and A_{o2} is:

$$\text{Ratio} = \frac{A_{o1}}{A_{o2}} = \frac{A_{i1} Z_{t2}}{A_{i2} Z_{t1}} \quad (5)$$

Since the voltage output of the diode detector is not affected by the frequency, $A_{i1} = A_{i2}$. Therefore, the ratio is:

$$\text{Ratio} = \frac{Z_{t2}}{Z_{t1}} \quad (6)$$

The above equation shows that the ratio is only a function of R_x (resistance of the pressure-sensitive resistor), whose value is only related to the external pressure. Therefore, the pressure value can be obtained by calculating the ratio between A_{o1} and A_{o2}.

The two frequencies (f_1 and f_2) are chosen to be 500 Hz and 2000 Hz, respectively. It should be noted that the actual signal outputted by the photodiode is a pulse wave instead of a sinewave; therefore, additional digital filters need to be applied during the post-processing steps. During testing, the DAQ output alternates between two frequencies

(f_1 and f_2) of the square wave to modulate the emission of IR light. The ratio of the two signal amplitudes is measured to calculate the real-time pressure value.

4.2.2. The External Interrogator

Figure 5e shows the structure of the external interrogator. The RF source (RF function generator E4432B, Agilent, Santa Clara, CA, USA) produces a 2.33 GHz RF carrier (f_0) signal, which is equally divided into two paths through a power splitter [26]. The first path doubles the frequency to be 4.66 GHz ($2f_0$) via a frequency multiplexer for the local oscillator (LO) of the down-converter. The second path amplifies and filters the RF carrier and radiates the signal through a dual-band (2.4 GHz/5 GHz) ceramic chip antenna (A10,194, Antenova, Cambridgeshire, UK). Concurrently, the antenna picks up 4.66 GHz ($2f_0 \pm f_m$) backscattered third-order mixing products which carry target pressure information. The circulator isolates the backscattered signal from the RF carrier. After amplifying and filtering, the third-order mixing products ($2f_0 \pm f_m$) mix with the LO ($2f_0$) to down-convert the output to be f_m. The demodulated signal (f_m) goes through filtering and amplifying (SR560, Stanford Research System, Sunnyvale, CA, USA) and is sampled at 40,000 bit/s using a Data Acquisition Card (DAQ, NI-6361, National Instrument, Austin, TX, USA). The Labview (National Instrument, Austin, TX, USA) program is developed to post-process the signal and calculate the pressure value.

Author Contributions: S.L. (Seunghyun Lee) performed the overall research investigation, including experiments and data analysis, and prepared the main manuscript text and all figures. S.L. (Shiyi Liu), R.E.B. and M.C.P. assisted with preparing the experiments and edited the manuscript. J.B.C. oversaw the project and assisted with the writing of the manuscript. All authors have read and agreed to the published version of the manuscript.

Funding: This research was funded by Army Medical Research Acquisition Activity (USAMRAA), grant number W81XWH2010805.

Data Availability Statement: Not applicable.

Conflicts of Interest: The authors declare no competing interest.

References

1. Verrees, M.; Selman, W.R. Management of normal pressure hydrocephalus. *Am. Fam. Physician* **2004**, *70*, 1071–1078.
2. Brodbelt, A.; Stoodley, M. CSF pathways: A review. *Br. J. Neurosurg.* **2007**, *21*, 510–520. [CrossRef]
3. Chabrerie, A.; Black, P.M. Ventricular shunts. *J. Intens. Care Med.* **2002**, *17*, 218–229. [CrossRef]
4. Kirkpatrick, M.; Engleman, H.; Minns, A.R. Symptoms and signs of progressive hydrocephalus. *Arch. Dis. Child.* **1989**, *64*, 124–128. [CrossRef]
5. Drake, J.M.; Kestle, J.R.W.; Tuli, S. CSF shunts 50 years on—Past, present and future. *Child's Nerv. Syst.* **2000**, *16*, 800–804. [CrossRef]
6. Drake, J.M.; Kestle, J.R.; Milner, R.; Cinalli, G.; Boop, F.; Piatt, J., Jr.; Haines, S.; Schiff, S.J.; Cochrane, D.D.; Steinbok, P.; et al. Randomized Trial of Cerebrospinal Fluid Shunt Valve Design in Pediatric Hydrocephalus. *Neurosurgery* **1998**, *43*, 294–303. [CrossRef]
7. Stone, J.J.; Walker, C.T.; Jacobson, M.; Phillips, V.; Silberstein, H.J. Revision rate of pediatric ventriculoperitoneal shunts after 15 years. *J. Neurosurg. Pediatr.* **2013**, *11*, 15–19. [CrossRef]
8. Czosnyka, Z.H.; Cieslicki, K.; Czosnyka, M.; Pickard, J.D. Hydrocephalus shunts and waves of intracranial pressure. *Med. Biol. Eng. Comput.* **2005**, *43*, 71–77. [CrossRef]
9. Czosnyka, M.; Whitehouse, H.; Pickard, J.D. Hydrodynamic properties of hydrocephalus shunts: United Kingdom Shunt Evaluation Laboratory. *J. Neurol. Neurosurg. Psychiatry* **1997**, *62*, 43–50. [CrossRef]
10. Oh, J.; Kim, G.; Kralick, F.; Noh, H. Design and Fabrication of a PDMS/Parylene Microvalve for the Treatment of Hydrocephalus. *J. Microelectromech. Syst.* **2011**, *20*, 811–818. [CrossRef]
11. Oh, K.W.; Ahn, C.H. A review of microvalves. *J. Micromech. Microeng.* **2006**, *16*, R13–R39. [CrossRef]
12. Chen, P.-J.; Rodger, D.; Meng, E.M.; Humayun, M.S.; Tai, Y.-C. Surface-Micromachined Parylene Dual Valves for On-Chip Unpowered Microflow Regulation. *J. Microelectromech. Syst.* **2007**, *16*, 223–231. [CrossRef]
13. Tir'en, J.; Tenerz, L.; Hok, B. A batch-fabricated non-reverse valve with cantilever beam manufactured by micromachining of silicon. *Sens. Actuators* **1989**, *18*, 389–396. [CrossRef]
14. Kim, D.; Beebe, D.J. A bi-polymer micro one-way valve. *Sens. Actuators A Phys.* **2007**, *136*, 426–433. [CrossRef]

15. Lo, R.; Li, P.-Y.; Saati, S.; Agrawal, R.N.; Humayun, M.S.; Meng, E. A passive MEMS drug delivery pump for treatment of ocular diseases. *Biomed. Microdevices* **2009**, *11*, 959–970. [CrossRef] [PubMed]
16. Moon, S.; Im, S.; An, J.; Park, C.J.; Kim, H.G.; Park, S.W.; Kim, H.I.; Lee, J.-H. Selectively bonded polymeric glaucoma drainage device for reliable regulation of intraocular pressure. *Biomed. Microdevices* **2011**, *14*, 325–335. [CrossRef]
17. Schwerdt, H.N.; Bristol, R.E.; Chae, J. Miniaturized Passive Hydrogel Check Valve for Hydrocephalus Treatment. *IEEE Trans. Biomed. Eng.* **2013**, *61*, 814–820. [CrossRef]
18. Schwerdt, H.N.; Amjad, U.; Appel, J.; Elhadi, A.M.; Lei, T.; Preul, M.C.; Bristol, R.E.; Chae, J. In Vitro Hydrodynamic, Transient, and Overtime Performance of a Miniaturized Valve for Hydrocephalus. *Ann. Biomed. Eng.* **2015**, *43*, 603–615. [CrossRef]
19. Lee, S.; Bristol, R.E.; Preul, M.C.; Chae, J. Three-Dimensionally Printed Microelectromechanical-System Hydrogel Valve for Communicating Hydrocephalus. *ACS Sens.* **2020**, *5*, 1398–1404. [CrossRef]
20. Del, B.; Marc, R.; Domenico, L.; Di, C. Nonsurgical therapy for hydrocephalus: A comprehensive and critical review. *Fluids Barriers CNS* **2015**, *13*, 3.
21. Beems, T.; Simons, K.S.; Van Geel, W.J.A.; De Reus, H.P.M.; Vos, P.E.; Verbeek, M.M. Serum- and CSF-concentrations of brain specific proteins in hydrocephalus. *Acta Neurochir.* **2003**, *145*, 37–43. [CrossRef] [PubMed]
22. Brydon, H.L.; Bayston, R.; Hayward, R.; Harkness, W. Reduced bacterial adhesion to hydrocephalus shunt catheters mediated by cerebrospinal fluid proteins. *J. Neurol. Neurosurg. Psychiatry* **1996**, *60*, 671–675. [CrossRef] [PubMed]
23. Jones, H.C.; Keep, R. Brain fluid calcium concentration and response to acute hypercalcaemia during development in the rat. *J. Physiol.* **1988**, *402*, 579–593. [CrossRef] [PubMed]
24. Rutter, N.; Smales, O.R. Calcium, magnesium, and glucose levels in blood and CSF of children with febrile convulsions. *Arch. Dis. Child.* **1976**, *51*, 141–143. [CrossRef]
25. Schutzer, S.E.; Liu, T.; Natelson, B.H.; Angel, T.E.; Schepmoes, A.A.; Purvine, S.; Hixson, K.K.; Lipton, M.S.; Camp, D.G.; Coyle, P.K.; et al. Establishing the Proteome of Normal Human Cerebrospinal Fluid. *PLoS ONE* **2010**, *5*, e10980. [CrossRef]
26. Schwerdt, H.N.; Xu, W.; Shekhar, S.; Abbaspour-Tamijani, A.; Towe, B.C.; Miranda, F.A.; Chae, J. A Fully-Passive Wireless Microsystem for Recording of Neuropotentials using RF Backscattering Methods. *J. Microelectromech. Syst.* **2011**, *20*, 1119–1130. [CrossRef]
27. Klarica, M.; Radoš, M.; Erceg, G.; Petošić, A.; Jurjević, I.; Orešković, D. The Influence of Body Position on Cerebrospinal Fluid Pressure Gradient and Movement in Cats with Normal and Impaired Craniospinal Communication. *PLoS ONE* **2014**, *9*, e95229.
28. Boddu, S.R.; Gobin, P.; Oliveira, C.; Dinkin, M.; Patsalides, A. Anatomic measurements of cerebral venous sinuses in idiopathic intracranial hypertension patients. *PLoS ONE* **2018**, *13*, e0196275. [CrossRef]
29. Viker, T.; Stice, J. Modular Redundancy for Cerebrospinal Fluid Shunts: Reducing Incidence of Failure due to Catheter Obstruction. In Proceedings of the 2019 Design of Medical Devices Conference, Minneapolis, MA, USA, 15–18 April 2019; Volume 41037.
30. Chung, S.; Kim, J.K.; Wang, K.C.; Han, D.-C.; Chang, J.-K. Development of MEMS-based Cerebrospinal Fluid Shunt System. *Biomed. Microdev.* **2003**, *5*, 311–321. [CrossRef]
31. Rengier, F.; Mehndiratta, A.; von Tengg-Kobligk, H.; Zechmann, C.M.; Unterhinninghofen, R.; Kauczor, H.-U.; Giesel, F.L. 3D printing based on imaging data: Review of medical applications. *Int. J. Comput. Assist. Radiol. Surg.* **2010**, *5*, 335–341. [CrossRef]
32. Bose, S.; Vahabzadeh, S.; Bandyopadhyay, A. Bone Tissue Engineering Using 3D Printing. *Mater. Today* **2013**, *16*, 496–504. [CrossRef]
33. Morgan, A.J.L.; Jose, L.H.S.; Jamieson, W.; Wymant, J.M.; Song, B.; Stephens, P.; Barrow, D.A.; Castell, O. Simple and Versatile 3D Printed Microfluidics Using Fused Filament Fabrication. *PLoS ONE* **2016**, *11*, e0152023. [CrossRef] [PubMed]

Article

Tough, Self-Recoverable, Spiropyran (SP3) Bearing Polymer Beads Incorporated PAM Hydrogels with Sole Mechanochromic Behavior

Jianxiong Xu, Yuecong Luo, Yin Chen, Ziyu Guo, Yutong Zhang, Shaowen Xie, Na Li and Lijian Xu *

Hunan Key Laboratory of Biomedical Nanomaterials and Devices, College of Life Sciences and Chemistry, Hunan University of Technology, Zhuzhou 412007, China; 13658@hut.edu.cn (J.X.); luohuahua2021@163.com (Y.L.); chenyin96110@163.com (Y.C.); xiaoguo19970314@163.com (Z.G.); zyt17836227933@163.com (Y.Z.); 13502431554@163.com (S.X.); lina6980955@163.com (N.L.)
* Correspondence: xlj235@hut.edu.cn; Tel.: +86-731-22182107

Abstract: Spiropyran-containing hydrogels that can respond to external stimuli such as temperature, light, and stress have attracted extensive attention in recent years. However, most of them are generally dual or multiple stimuli-responsive to external stimuli, and the interplay of different stimulus responses is harmful to their sensitivity. Herein, spiropyran bearing polymer beads incorporated PAM (poly(AM–co–MA/DMSP3)) hydrogels with sole mechanochromic properties were synthesized by emulsion polymerization of acrylamide (AM) and methyl acrylate (MA) in the presence of spiropyran dimethacrylate mechanophore (DMSP3) crosslinker. Due to the hydrophobic nature of MA and DMSP3, the resultant hydrogel afforded a rosary structure with DMSP3 bearing polymer beads incorporated in the PAM network. It is found that the chemical component (e.g., AM, MA, and DMSP3 concentrations) significantly affect the mechanical and mechanoresponsive properties of the as-obtained poly(AM–co–MA/DMSP3) hydrogel. Under optimal conditions, poly(AM–co–MA/DMSP3) hydrogel displayed high mechanical properties (tensile stress of 1.91 MPa, a tensile strain of 815%, an elastic modulus of 0.67 MPa, and tearing energy of 3920 J/m^2), and a good self-recovery feature. Owing to the mechanoresponsive of SP3, the hydrogels exhibited reversible color changes under force-induced deformation and relaxed recovery states. More impressive, the poly(AM–co–MA/DMSP3) hydrogel showed a linear correlation between tensile strain and chromaticity (x, y) as well as a stain and resting time-dependent color recovery rate. This kind of hydrogel is believed to have great potential in the application of outdoor strain sensors.

Keywords: mechanophore; mechanoresponsive hydrogels; emulsion polymerization; spiropyran

Citation: Xu, J.; Luo, Y.; Chen, Y.; Guo, Z.; Zhang, Y.; Xie, S.; Li, N.; Xu, L. Tough, Self-Recoverable, Spiropyran (SP3) Bearing Polymer Beads Incorporated PAM Hydrogels with Sole Mechanochromic Behavior. *Gels* **2022**, *8*, 208. https://doi.org/10.3390/gels8040208

Academic Editor: Wei Ji

Received: 27 February 2022
Accepted: 23 March 2022
Published: 27 March 2022

Publisher's Note: MDPI stays neutral with regard to jurisdictional claims in published maps and institutional affiliations.

Copyright: © 2022 by the authors. Licensee MDPI, Basel, Switzerland. This article is an open access article distributed under the terms and conditions of the Creative Commons Attribution (CC BY) license (https://creativecommons.org/licenses/by/4.0/).

1. Introduction

In recent years, we have witnessed the prosperity of stimuli-responsive materials that can change their physical and/or chemical properties in response to external stimulation, e.g., temperature [1–3], pH [4], light [5], ionic strength [6,7], and magnetic/electric [8,9] fields. Such intelligent stimuli-responsive materials have shown significant potential in drug delivery [10–12], environmental remediation [13–15], artificial intelligence [16,17], wearable electronic devices [18–20], and so on. Inspired by the mechanical-induced deformation behavior of Mimosa in nature, the mechanoresponsive materials are particularly attractive due to their promising application in materials damage determination, human motion monitoring, and smart robots [21]. The key component of mechanoresponsive materials is the mechanophores that can change their output performance, such as electronic signals, luminescence, and appearance color under external force [22]. Usually, the detection of electronic signals and luminescence needs additional sophisticated instruments. In comparison, mechanical discoloration is intuitive visualization and is considered to be more convenient. Therefore, mechanochromic molecules have been regarded as the extremely attractive primitive in the fabrication of mechanoresponsive materials [23–27].

Up to now, several mechanochromic molecules, including diarylethenes, stilbenes, azobenzenes, fulgides, and spiropyrans (SP), have been reported [28]. Among them, spiropyran has received enormous attention because of its unique mechanochemistry, fast dynamic responsiveness, high fatigue resistance, and ease of functionalization [29–33]. Under external force-stimuli, SP can be able to undergo a reversible 6-π ring-opening reaction (spiropyran (SP) ↔ merocyanine (MC)) due to the cleavage and reformation of the C–O bond on the spiro ring [34]. During the past decades, SP mechanophores have been widely used in solid-state switches. However, the main threat in these materials is the tight molecular packing in solid states, which has significantly impeded the transformation occurring between the SP and MC forms due to the limited free volumes in the solid state.

Consequently, the incorporation of SP into polymer networks is a valuable method to improve their stimulus responsiveness. To achieve covalent incorporation, it is necessary that the SP mechanophores contain polymerizable groups as substituents on the aromatic ring of SP. Generally, the SP mechanophores can be divided into three types (SP1, SP2, and SP3) according to the anchoring sites located at both sides of the C–O bond on the spiro ring, as shown in Scheme 1. The chemical component of SP1 and SP2 is similar. Both of them contain a nitro substituent on the aromatic ring of SP moiety. The anchoring sites of SP1 are located at the aromatic rings of each side, while the anchoring sites of SP2 are located at the aromatic ring and the nitrogen of dimethylaniline, respectively. For the SP3 mechanophore, the anchoring sites are similar to that of SP2 but lack of nitro substituent on the aromatic ring. These structural differences make them different stimulus responsiveness performances. Usually, the SP1 and SP2 mechanophores showed discoloration at the stimulus of external force, UV light, and heat, while the SP3 mechanophore only exhibited mechanochromic property [34–38]. Vidavsky and co-workers reported the introduction of the mechanoactive spiropyran into the polycarbonate backbone. The synthesized spiropyran–bisphenol A polycarbonate (SP–BPA–PC) was a hard glassy polymer with an elastic modulus of 1.9 GPa and a maximum elongation of only about 100% [39]. In order to further improve the ductility, Wang et al. synthesized a waterborne polyurethane polymer membrane (SP–MSPU) by grafting SP on the polyurethane chain with a maximum tensile strain of 400% and maximum tensile stress of 4 MPa [40].

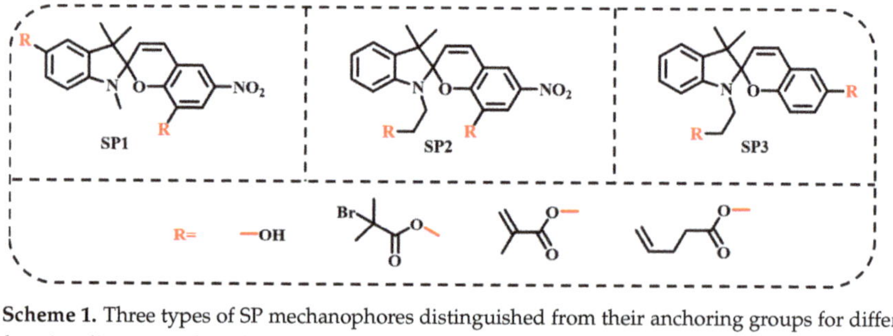

Scheme 1. Three types of SP mechanophores distinguished from their anchoring groups for different functionalization and applications.

Hydrogels as three-dimensional networks swollen polymers have outstanding performance (e.g., 3D porous networks, high stretchability, and excellent elastic deformation) [41–45]. Therefore, the fabrication of SP mechanophores encapsulated hydrogels is very important for the development of advanced strain sensors. One challenge is how to incorporate the highly hydrophobic SP mechanophores into a hydrophilic hydrogel framework. In our previous work, we have demonstrated a micelle polymerization method to encapsulate the SP1 type crosslinker into the PAM hydrogels [34]. The as-prepared SP1 containing hydrogels exhibited stimulus responsiveness under external force, UV light, and heat. However, the interplay of different stimulus responses seriously affects the response sensitivity of the hydrogel-based sensors.

Herein, a novel SP3 type crosslinker of dimethacrylate spiropyran (DMSP3) was synthesized and used for the preparation of poly(AM–co–MA/DMSP3) mechanically responsive hydrogels by the emulsion copolymerization of acrylamide (AM) and methyl acrylate (MA). The effect of MA:AM weight ratios and the concentration of DMSP3 crosslinker on the mechanical and mechanochromic performances were investigated to obtain the poly(AM–co–MA/DMSP3) hydrogels with excellent mechanical properties. Moreover, the stimulus responsiveness and self-recovery ability were also tested. Ascribed to the presence of SP3, the hydrogel exhibited reversibility of discoloration in the stretched and original state. The discoloration behavior was only responded to a single mechanical force stimulation and not disturbed by ultraviolet light and thermal stimulation, which ensured the accuracy of the sensing signal. Furthermore, mechanochromic properties were further researched by building a quantitative relationship between the external force stimulation and color change and evaluating the reversible color recovery times.

2. Results and Discussion

The tough, self-recoverable, spiropyran bearing polymer beads incorporated poly(AM–co–MA/DMSP3) hydrogels were synthesized via photo-initiated emulsion polymerization. The preparing process was schematically shown in Figure 1. Initially, hydrophobic MA monomer, DMSP3 crosslinker, and PBPO photo-initiator were homo-dispersed in the aqueous solution of AM monomer with the help of TWEEN 80 to form a stable emulsion. In this state, the hydrophobic species mainly existed in surfactant stabilized oil droplets and surfactant micelles. Once the polymerization was triggered by photo-irradiation, well-dispersed spiropyran bearing polymer beads (P(MA/DMSP3)) was formed. With the progress of the polymerization, hydrophilic PAM chains were produced and covalently attached on the surface of the P(MA/DMSP3) bead due to the copolymerization of AM and MA. After the completion of the polymerization, rosary-like three-dimensional poly(AM–co–MA/DMSP3) polymer hydrogels with soft PAM as "threads" and hard P(MA/DMSP3) polymer microspheres as "beads" were obtained. Due to the unique structure and the presence of DMSP3 mechanochromic probes, the poly(AM–co–MA/DMSP3) hydrogels were expected to have excellent mechanical properties and mechanochromic characteristics.

Herein, we choose the AM_{25}–$DMSP3_{0.4}$–MA_{25} hydrogels (the preparation condition can be found in Table 1) as a typical example to check their structure and properties. AM_{25}–$DMSP3_{0.4}$–MA_{25} hydrogels displayed extraordinary mechanical and flexible properties. As shown in Figure 2A, the hydrogels can withstand knotted stretching, original stretching, and crossover stretching up to six times their original length without any observable damage. In parallel, the hydrogels can bear up to 1 kg of weight, which is about 1000 times their own weight (Figure 2B). Figure 2C displays the crack propagation process of the AM_{25}–$DMSP3_{0.4}$–MA_{25} hydrogels with a cut notch (~5 mm). The result showed that with the increase in tensile strain (λ), the notch was obviously passivated and gradually developed into a semicircular crack, indicating the excellent toughness of the hydrogels. It was interesting to note that in all stretching states, a pronounced color change of the hydrogels from light yellow to blue-gray was observed in areas of stress concentration and deformation, confirming the mechanochromic property of the hydrogels. The mechanochromic mechanism of the hydrogels was due to the mechanical activation of SP-to-MC transition in the polymer network, as schematically shown in Figure 1. The inner structure of the hydrogels was observed by SEM. As shown in the SEM image (Figure 2D), the AM_{25}–$DMSP3_{0.4}$–MA_{25} hydrogels presented as a porous honeycomb structure. At the SEM image of high magnification (Figure 2E), it can be seen that polymer beads with an average size of 200~500 nm were uniformly embedded in the framework of the hydrogels. We called this structure a three-dimensional rosary interpenetrating polymer network.

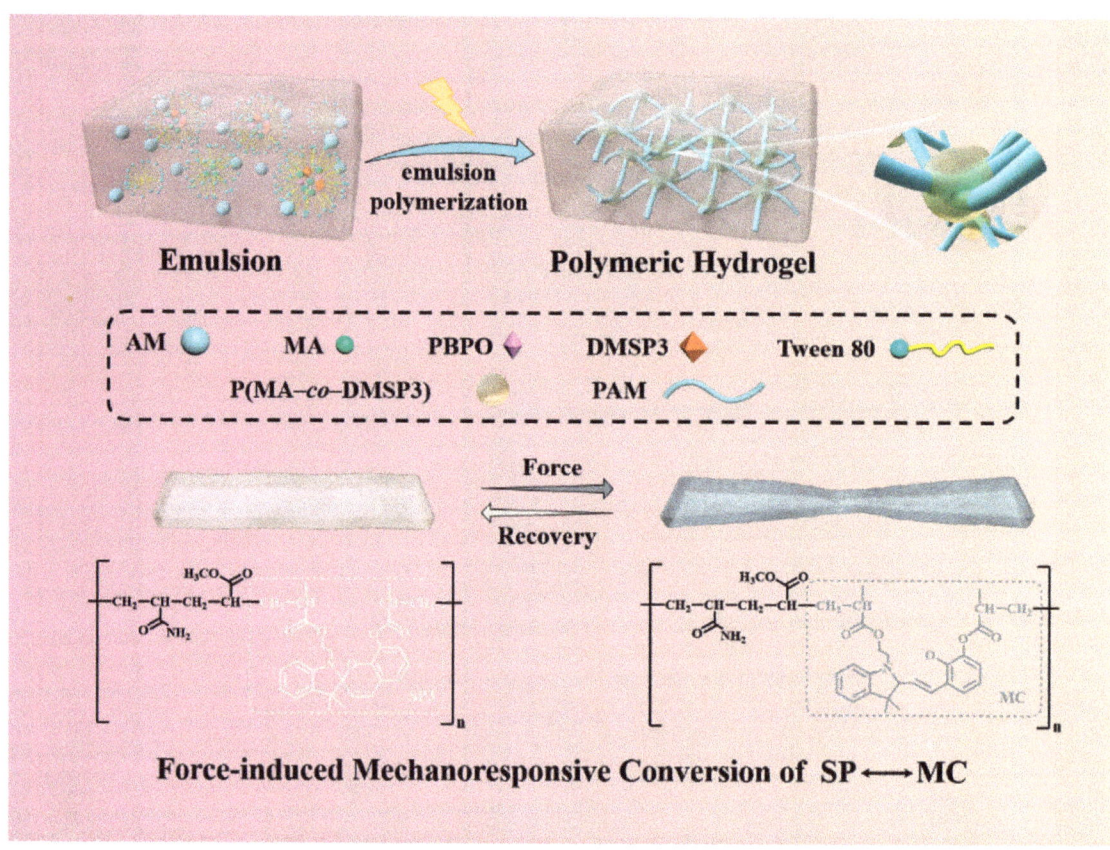

Figure 1. Scheme for the preparation of poly(AM–co–MA/DMSP3) hydrogels and the mechanism of their mechanochromic characteristic.

Table 1. Synthetic formula and mechanical properties of the poly(AM–co–MA/DMSP3) hydrogels prepared under different conditions.

poly(AM–co–MA/SP3) Hydrogels	Water Contents (wt%)	MA:AM Ratio	MA–SP3–AM Concentrations (mol%)	σ (MPa)	λ (mm/mm)	E (KPa)
MA_{10}-$SP3_{0.4}$-AM_{40}	50	1:4	0.4	0.43	1.64	1.45
MA_{20}-$SP3_{0.4}$-AM_{30}	50	2:3	0.4	0.52	4.29	0.57
MA_{25}-$SP3_{0.4}$-AM_{25}	50	1:1	0.4	1.91	8.15	0.67
MA_{30}-$SP3_{0.4}$-AM_{20}	50	3:2	0.4	1.39	8.30	0.58
MA_{40}-$SP3_{0.4}$-AM_{10}	50	4:2	0.4	0.32	8.90	0.20
MA_{25}-$SP3_{0.1}$-AM_{25}	50	1:1	0.1	0.49	9.13	0.27
MA_{25}-$SP3_{0.2}$-AM_{25}	50	1:1	0.2	0.85	9.04	0.28
MA_{25}-$SP3_{0.3}$-AM_{25}	50	1:1	0.3	1.47	9.90	0.39
MA_{25}-$SP3_{0.5}$-AM_{25}	50	1:1	0.5	1.45	5.64	0.71

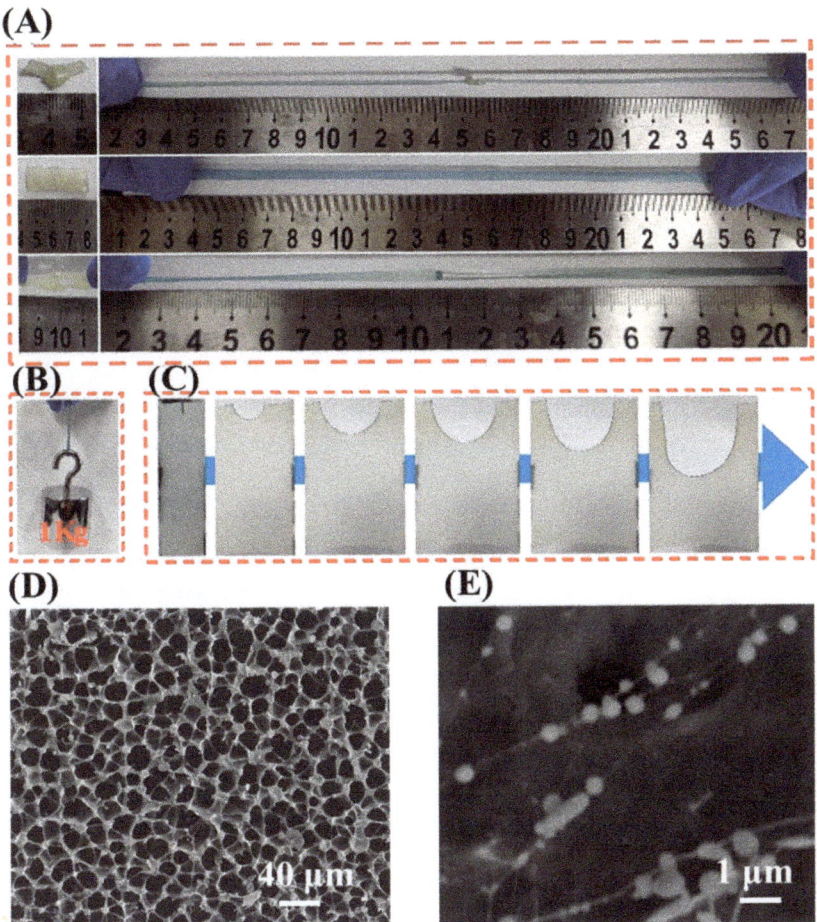

Figure 2. Visual photograph of the mechanical and mechanoresponsive properties of as-prepared poly(AM–co–MA/DMSP3) hydrogels by (**A**) knotted, original, and crossover stretching; (**B**) holding 1 kg of weight; (**C**) resisting crack propagation, (**D**,**E**) SEM images of poly(AM–co–MA/DMSP3) hydrogels with micellar structures at different magnifications.

The rheological property of the hydrogels was evaluated. Figure 3A showed the strain amplitude sweep test of AM_{25}–$DMSP3_{0.4}$–MA_{25} hydrogels at a fixed angular frequency (10 rad/s) at 25 °C. As shown, the storage modulus G′ and loss modulus G″ are independent of the applied strain at lower strain ($\lambda = 0\sim10\%$). Moreover, the G′ is always larger than the G″. The results suggested that the AM_{25}–$DMSP3_{0.4}$–MA_{25} hydrogels exhibited a typical elastic response with a linear viscoelastic region at $\lambda = 0\sim10\%$. The rheological property of A AM_{25}–$DMSP3_{0.4}$–MA_{25} and AM_{25}–MA_{25} hydrogels were compared by checking the G′ and G″ variation as a function of frequency at a fixed strain of $\lambda = 1\%$ (Figure 3B). It can be observed that the G′ and G″ of AM_{25}–$DMSP3_{0.4}$–MA_{25} hydrogels were larger than those of AM_{25}–MA_{25} hydrogels in all frequency ranges, implying that the addition of DMSP3 significantly improved the viscosity and elasticity of the hydrogel. The results also demonstrated the crosslinked structure of the AM_{25}–$DMSP3_{0.4}$–MA_{25} hydrogels.

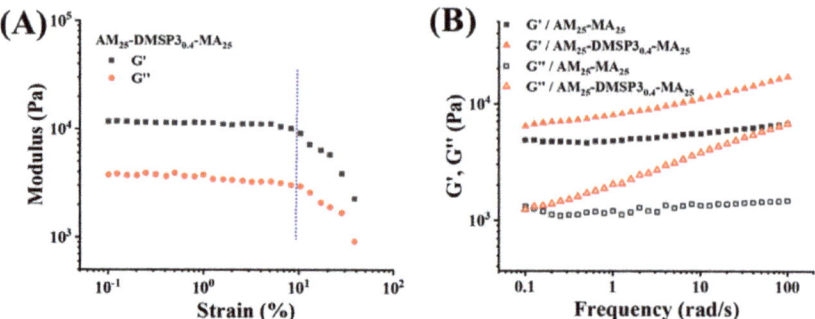

Figure 3. (**A**) The strain amplitude sweep test of AM_{25}–$DMSP3_{0.4}$–MA_{25} hydrogel at a fixed angular frequency (10 rad/s) at 25 °C; (**B**) Variation of storage modulus G' and loss modulus G" as a function of frequency for AM_{25}–MA_{25} hydrogels and AM_{25}–$DMSP3_{0.4}$–MA_{25} hydrogels measured at λ = 1%.

Apart from mechanoresponsive property, we also investigate the photochromic and thermochromic properties of the AM_{25}–$DMSP3_{0.4}$–MA_{25} hydrogels. As shown in Figure 4A, the as-prepared AM_{25}–$DMSP3_{0.4}$–MA_{25} hydrogels were exposed to a UV light irradiation (365 nm) for 10 min and heated at 60 °C for 10 min. Optical images were taken before and after UV exposure and heating. Obviously, visual inspection of AM_{25}–$DMSP3_{0.4}$–MA_{25} hydrogels showed no color change after heating and UV light irradiation, indicating no SP-to-MC transition in the gel networks. This result confirmed that the DMSP3 mechanophore had no photochromism and thermochromism, which might be due to the absence of the electron-withdrawing nitro group at the 6-position of the benzopyran. Under external force stimuli (λ = 3), the as-prepared AM_{25}–$DMSP3_{0.4}$–MA_{25} hydrogels exhibited a color change from light yellow to blue-gray color. Moreover, the blue-gray color gradually faded and returned to the initial light yellow color in approximately 30 min after removing the external force, suggesting the reversible mechanoresponsive property of the AM_{25}–$DMSP3_{0.4}$–MA_{25} hydrogels. To quantitatively observe the color-changing degree, the gauge section of digital images of the gels were analyzed by the RGB (red, green, blue) values and located in the x, y chromaticity diagram (CIE 1931 color space). As shown in Figure 4B, during the deformation process, the gels showed an obvious color change in the pathways towards the blue-gray color under the stimuli of force and returned to the initial light yellow color without external force, whereas the colors of the hydrogels before and after UV light and temperature stimuli were located in similar light yellow areas. The results demonstrated the sole reversible mechanoresponsive property of the hydrogels.

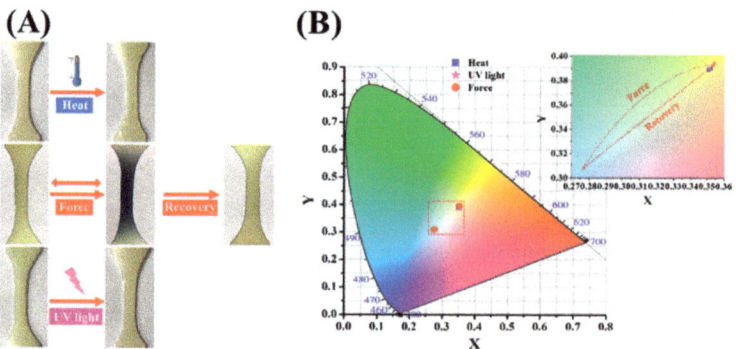

Figure 4. (**A**) Optical images of poly(AM–co–MA/DMSP3) hydrogels under the stimuli of heat, external force (λ = 3), and UV light, and (**B**) their corresponding CIE color coordinates.

In order to obtain poly(AM–co–MA/DMSP3) hydrogels with superior properties, the effect of MA:AM weight ratios and the concentration of DMSP3 crosslinker on the mechanical and mechanochromic performances were evaluated. To this end, a series of poly(AM–co–MA/DMSP3) hydrogels were firstly prepared under different weight ratios of MA:AM. The concentration of the DMSP3 crosslinker was kept at a constant value of 0.4% in proportion to the molar amount of MA monomer (0.4 mol%). The preparation conditions are summarized in Table 1. Figure 5A illustrates the typical stress–strain curves of poly(AM–co–MA/DMSP3) hydrogels with different MA:AM weight ratios. Obviously, at the MA:AM ratio of 1:4, the AM_{25}–$DMSP3_{0.4}$–MA_{40} hydrogels showed weak tensile stress (σ) of 0.32 MPa at a large tensile strain (λ) of ~900%. Generally, the tensile stress increased with the increasing of the weight ratio of MA:AM from 1:4 to 1:1. However, when the MA:AM ratio was beyond 1:1, the poly(AM–co–MA/DMSP3) hydrogels became brittle, accompanied by a significant decrease in tensile stress and tensile strain. Further, the effect of DMSP3 crosslinker concentration on the mechanical properties of the hydrogels at a fixed weight ratio of MA:AM at 1:1 was examined. Similarly, poly(AM–co–MA/DMSP3) hydrogel showed different mechanical properties with different DMSP3 concentrations (Figure 5B). As the DMSP3 concentrations increased from 0.1 mol% to 0.4 mol%, the hydrogels showed a monotonical increase in tensile stress from 0.49 to 1.91 MPa and elastic modulus from 0.27 to 0.67 kPa at similar fracture strains of ~800%. When the DMSP3 concentration further increased to 0.5 mol%, the poly(AM–co–MA/DMSP3) hydrogel exhibited enhanced stiffness, resulting in an inferior mechanical strength. Based on the above results, the AM_{25}–$DMSP3_{0.4}$–MA_{25} hydrogel prepared with the weight ratio of MA:AM at 1:1 and DMSP3 concentration at 0.4 mol% achieved the most remarkable mechanical properties (σ of 1.91 MPa, λ of 815%, and elastic modulus (E) of 0.67 kPa). In parallel, we also comparatively observed the external force-dependent color change of poly(AM–co–MA/DMSP3) hydrogels using tensile tests. Figure 5C,D summarized the color change of the hydrogels in response to an external force. As revealed, the AM_{25}–$DMSP3_{0.4}$–MA_{25} hydrogel strips under larger external force and the degree of discoloration in the blue-gray direction was greater. Thus, the AM_{25}–$DMSP3_{0.4}$–MA_{25} hydrogels were chosen as the research object in the following discussion, if specialty pointed out otherwise.

Due to the elastomer-like mechanical property and reversible transition of SP $\leftrightarrow$ MC in DMSP3 moiety, the AM_{25}–$DMSP3_{0.4}$–MA_{25} hydrogels were expected to have mechanical and mechanoresponsive self-recovery properties. To demonstrate the self-recovery properties of the hydrogels, loading and unloading experiments were performed on the AM_{25}–$DMSP3_{0.4}$–MA_{25} hydrogels with a maximum tensile strain of $\lambda = 3$. For comparison, the first two loading–unloading tests (i.e., first original and second no recovery) were carried out continuously without any rest period, while the third–fifth tests (i.e., third, fourth, and fifth-recovery) were conducted on the gels with 5, 10 and 30 min, respectively, for recovery during the unloading process. For each loading–unloading cycle, the hydrogel strip was spontaneously recovered to its original length without additional treatment. The mechanical recovery was estimated by cyclic stress–strain curves (Figure 6A), and the stiffness/toughness recovery ratios were summarized in Figure 6B. As shown in Figure 6A, the AM_{25}–$DMSP3_{0.4}$–MA_{25} hydrogels showed the largest hysteresis loop in the first original cycle, and the hysteresis loop became much smaller in the second no recovery cycle. Nevertheless, the hysteresis loop became larger with increasing the resting time in the third–fifth cycles. Quantitatively, the stiffness/toughness recovery of hydrogels after the second no recovery cycle was 82.0%/43.4%. After recovery for 5 min in the third cycle, the stiffness/toughness recovery of hydrogels increased to 88.0%/61.7%. For fourth and fifth loading–unloading cycles, with the resting time increased to 10 and 30 min, respectively, the stiffness/toughness recovery of hydrogels reached 91.0%/63.8% and 92.2%/69.5%, respectively. The results suggested that prolonged resting time benefited the mechanical recovery of the hydrogels. During the loading–uploading test, the mechanoresponsive self-recovery ability of the hydrogels was also evaluated by observing the color changes of the hydrogel strips at each loading–unloading cycle. As shown in Figure 6C, the color of the AM_{25}–$DMSP3_{0.4}$–MA_{25} hydrogel strip was changed from light yellow to blue-gray at the stress-concentrated area when the length of the hydrogel becomes three times the original length. Without any resting time, the hydrogels immediately self-recovered to their original length but still displayed blue-gray color at the beginning of the second cycle. Similarly, the hydrogels were also in blue-gray color at the beginning of the third and fourth cycles after a 5 or 10 min resting time, respectively. Notably, the hydrogels recovered to their original light yellow color after 30 min resting time at the fifth cycle, suggesting that prolonged resting time was favorable for the reversion of MC-to-SP.

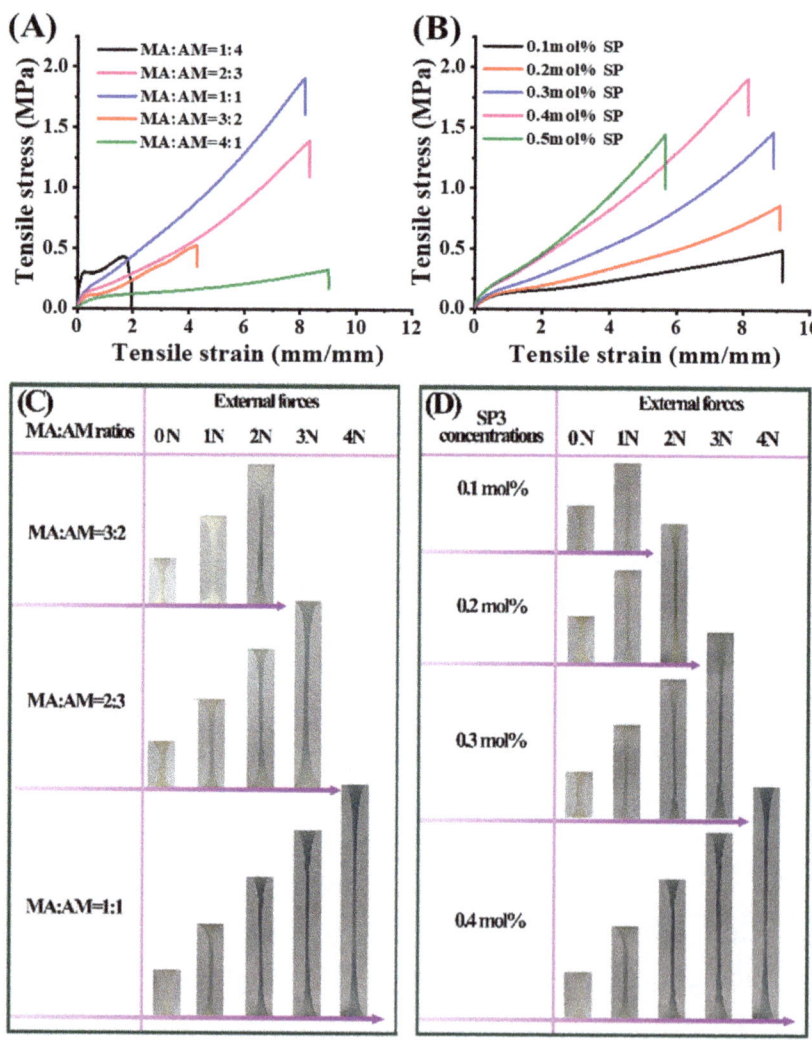

Figure 5. Mechanical and mechanoresponsive properties of the as-prepared poly(AM–*co*–MA/DMSP3) hydrogels with different MA:AM ratios and DMSP3 concentrations: (**A**,**B**) tensile stress–strain tests; (**C**,**D**) their corresponding color change in response to the external force.

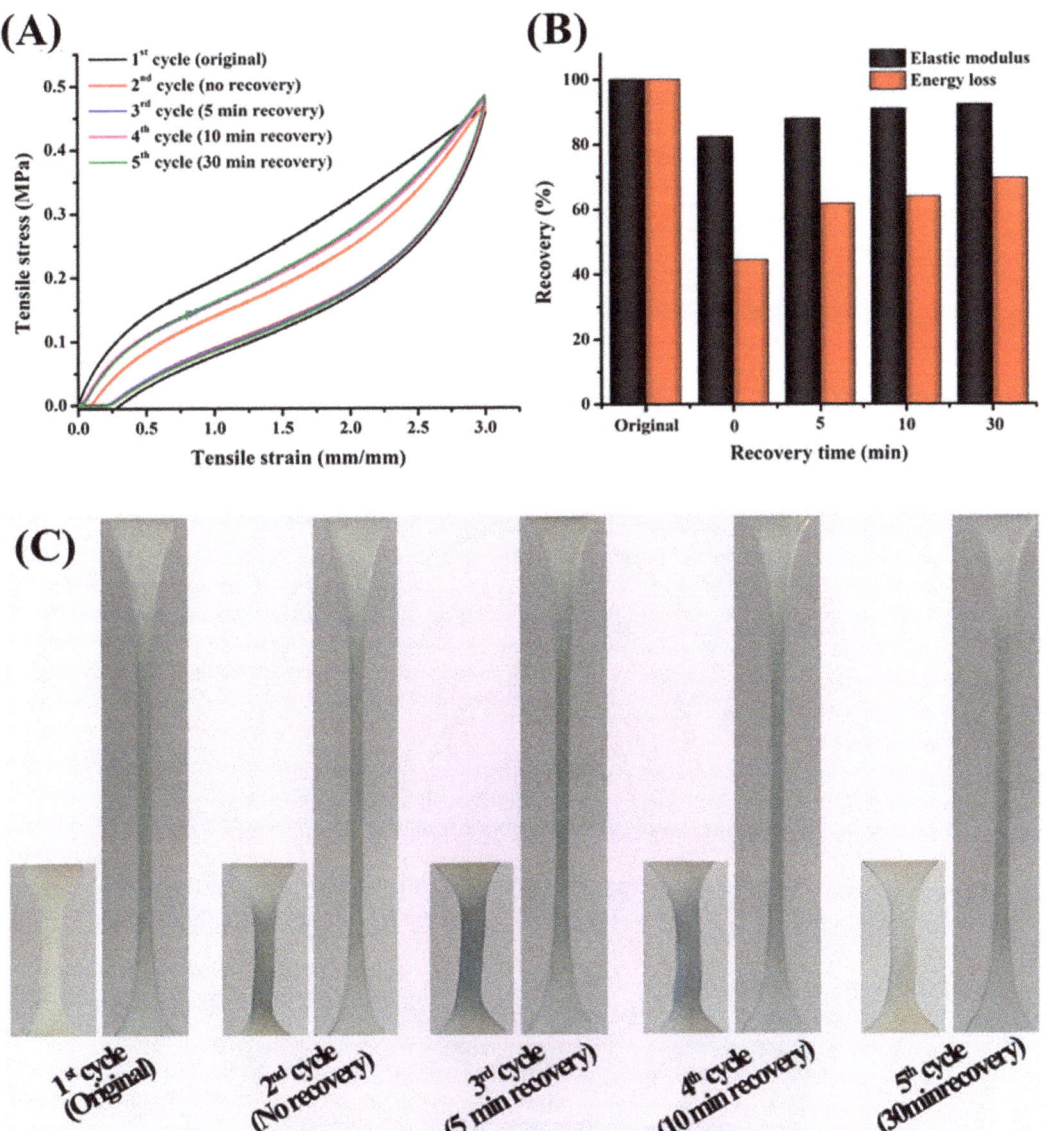

Figure 6. Force-induced color change and mechanical recovery of AM_{25}–$DMSP3_{0.4}$–MA_{25} hydrogels using cyclic loading–unloading tests, as demonstrated by (**A**) hysteresis loading–unloading tests; and (**B**) toughness (energy loss) and stiffness (elastic modulus) recovery; (**C**) visual inspection of color change and recovery.

In order to attain deep insight into the mechanochromic property, a quantitative relationship between the external force stimulation and color change was evaluated by successive loading–unloading tests conducted on AM_{25}–$DMSP3_{0.4}$–MA_{25} hydrogels at different strains, and the corresponding color changes were recorded by optical images. As shown in Figure 7A, the as-prepared gels did not show color changes at stretched or relaxed states when the tensile strain was less than 150%. This suggested that the small stretching force was dissipated to endure the deformation of the PAM network and was not large enough to be transferred from the PAM network to P(MA/DMSP3) polymer beads to trigger the SP to MC transformation or the transition degree of SP → MC was not large enough to induce the color change being observed by naked eyes. Obvious color changes can be observed as the tensile strain was larger than 200%, and the blue-gray coloration gradually deepened with the increase in tensile strains since a large tensile strain was beneficial for the conversion of SP to MC. Similarly, the relaxed gel can be recovered immediately to its original length immediately but still retain blue-gray color at each unloading state. To quantitatively analyze the color changes in the gels, the optical color change of the stress-concentrated area of the gels between the stretched and relaxed states were monitored using the RGB (red, green, blue) color channels and further located into the x, y chromaticity diagram. As shown in Figure 7B, a linear color change path of the stretched gels from light yellow to blue-gray with a linear fitting R^2 of 0.98 was observed as the tensile strain increased from 0% to 400%. In parallel, the relaxed gels showed a distinct pathway toward blue-gray color after unloading, with a linear fitting R^2 of 0.95. This result provided strong evidence that the gel indeed displayed strain-dependent color changes. The different color change pathways during the stretching and relaxing states may be due to the secondary color change by the isomerization and accumulation of MC. We also use UV–vis spectrum to quantitatively evaluate the mechanical activation of SP-to-MC transition degree of the hydrogels under different strains since the SP and MC moieties have distinguishable UV absorptions. Figure 7D showed the UV–vis spectra of the hydrogels at different strains of 0–700%. As shown, the hydrogel at a low strain of 50% showed almost identical UV–vis spectra to the virgin gel, suggesting that the SP → MC conversion of the gels cannot be detected under the low strains of 0% to 50%. When the gels were stretched to 100% or above, a new adsorption peak located at 587 nm corresponding to the MC moiety was observed, and the peak intensity increased with the increase of strains. The discoloration threshold of the hydrogels is about 100%. The peak intensity as a function of strains was summarized in Figure 7E. A linear relationship with an R^2 value of 0.99 was obtained, further confirming that the SP → MC conversion rate was related to the strains. Additionally, we also examined the recovery times on the same gel samples after being stretched at the different tensile strains from 200% to 400%. As shown in Figure 7C, the relaxed gels can be recovered to the original light yellow color gradually by prolonging the resting time. Moreover, the relaxed gels needed more time to recover from blue-gray to light yellow after stretching at larger tensile stress. For example, 10 min was needed for the gels to be recovered to their original light yellow color after stretching at λ = 2, 25 min was needed for λ = 2.5, 35 min was needed for λ = 3, and more than 60 min was needed for λ larger than 3.5. Based on the linear correlation between strain and chromaticity (x, y) as well as the self-recovery ability, it is expected that the hydrogels can be applied to strain sensors by monitoring the corresponding color change or color change path of the hydrogels due to different strains and vice versa. Furthermore, the reversible and mechanoresponsive property of poly(AM–*co*–MA/DMSP3) hydrogels can be instantly used for rewritable printing and rewritable data storage.

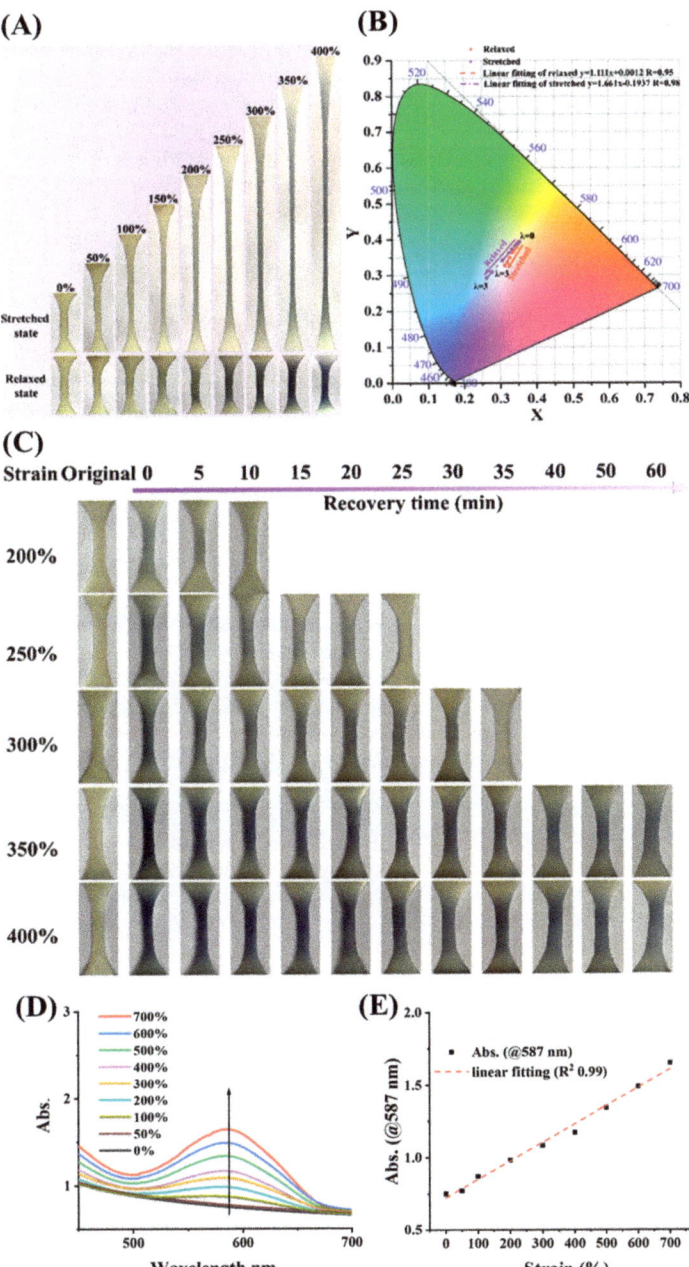

Figure 7. (**A**) Optical images of the corresponding color changes of AM_{25}–$DMSP3_{0.4}$–MA_{25} hydrogels under continuous loading and unloading tests with different strains, (**B**) their corresponding CIE color coordinates, (**C**) Color recovery of poly(AM–*co*–MA/DMSP3) hydrogel after as a function of recovery time under daylight and room temperature, (**D**) UV-vis spectrum of poly(AM–*co*–MA/DMSP3) hydrogels at different strains of 0–700%, (**E**) Data fitting to UV-vis absorbance at 587 nm for the gels at different strains of 0–700%.

3. Conclusions

In summary, a novel SP3 type crosslinker of dimethacrylate spiropyran (DMSP3) was synthesized and introduced into the copolymer of acrylamide (AM) and methyl acrylate (MA) by emulsion polymerization to prepare poly(AM–co–MA/DMSP3) mechanically responsive hydrogel. Under optimized conditions, the prepared poly(AM–co–MA/DMSP3) hydrogel presented as a rosary-like three-dimensional network structure with SP3 bearing polymer beads incorporated in PAM frameworks. Moreover, the hydrogels exhibited excellent mechanical properties (tensile stress of 1.91 MPa, a tensile strain of 815%, an elastic modulus of 0.67 MPa, and tear energy of 3920 J/m^2). In addition, the obtained poly(AM–co–MA/DMSP3) hydrogel only responded to a single mechanical force stimulus, and the response signal was not disturbed by the stimulation of heat and ultraviolet light, which ensured response sensitivity of the hydrogel-based sensors. The hydrogel showed a transition from light yellow to blue-gray under external stimulation, and when the external stimulation was removed, it could return to its original color in a short time, and the color change was reversible. More impressive, the poly(AM–co–MA/DMSP3) hydrogel showed a linear correlation between strain and chromaticity (x, y) as well as a stain and resting time-dependent color recovery rate. Based on this, poly(AM–co–MA/DMSP3) hydrogels are expected to serve as a mechanically responsive sensor for direct, simple, and visual detection of material damage/sensing/imaging.

4. Materials and Methods

4.1. Materials

Acrylamide (AM, 99%), methyl acrylate (MA, 99%), ethylene glycol dimethacrylate (98%), phenylbis(2,4,6-trimethylbenzoyl)phosphine oxide (PBPO), TWEEN 80, o-vanillin, boron tribromide (BBr$_3$), 2-iodoethanol, 2,3,3-trimethyl-3H-indole, and methacrylic anhydride were purchased from Shanghai Aladdin Chemistry Co., Ltd. (Shanghai, China). All chemical reagents were used directly as received without further purification. Water used in this work was purified by a DI-RO water purification system.

4.2. Synthesis of Dimethacrylate Spiropyran Mechanophore (DMSP3) Crosslinker

The DMSP3 crosslinker (compound **4**) was synthesized in four steps from compound **1** (2-hydroxyethyl-2,3,3-trimethyl-3H-indolium iodide) and compound **2** (2,3-dihydroxybenzaldehyde) in the presence of triethylamine and further functionalized with methylacryloyl ester group by reaction with methacrylic anhydride. Compound **1** was synthesized by the alkylation reactions of 2,3,3-trimethyl-3H-indole and 2-iodoethanol to give iodide salt (2-hydroxyethyl-2,3,3-trimethyl-3H-indolium iodide). Compound **2** was synthesized by replacing the methoxy group of o-vanillin with a hydroxyl group by hydrolysis reaction with BBr$_3$ to obtain 2,3-dihydroxybenzaldehyde. The reaction equations were schematically shown in Scheme 2. The chemical structure of all the intermediates and the final DMSP3 crosslinker was con-firmed by 1H NMR as shown in Figures S1–S4.

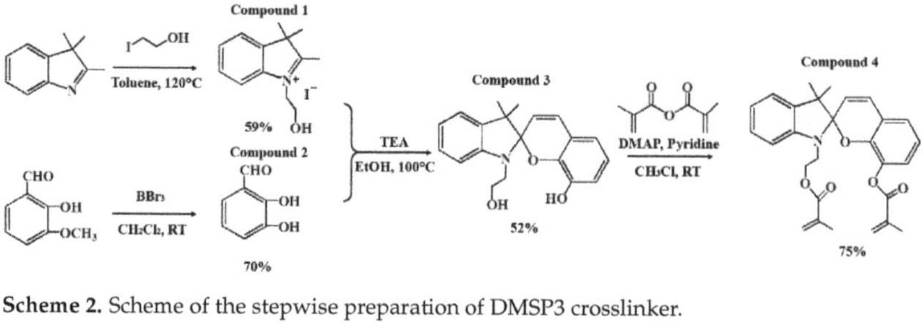

Scheme 2. Scheme of the stepwise preparation of DMSP3 crosslinker.

Compound **1** (2-hydroxyethyl-2,3,3-trimethyl-3H-indolium iodide). To a 500 mL round-bottom flask equipped with a reflux condenser, 150 mL acetonitrile, 2,3,3-trimethyl-

3H-indole (11.2 g, 70 mmol, 1 equiv), and 2-iodoethanol (6.5 mL, 84 mmol, 1.2 equiv) were sequentially added. Then, the reaction system was heated up to 85 °C and reacted for 12 h. After the completion of the reaction, the mixture was cooled down to ambient temperature. The solvent of acetonitrile was distilled off under reduced pressure. The resulting precipitate was washed with $CHCl_3$ three-time (3 × 50 mL) and dried under vacuum (0.1b kpa, 60 °C) overnight, finally obtaining a light purple solid powder compound **1** (13.7 g, 41 mmol 59% yield). ^{1}H NMR (300 MHz, DMSO-d_6) δ 7.96–7.93 (m, ^{1}H), 7.86–7.83 (m, 1H), 7.62–7.59 (m, ^{2}H), 4.60–4.57 (t, ^{2}H), 3.88–3.85 (t, ^{2}H), 2.81 (s, ^{3}H), 1.54 (s, ^{6}H).

Compound **2** (2,3-dihydroxybenzaldehyde). To a 500 mL round-bottom flask equipped with a dropping funnel, 250 mL CH_2Cl_2 and *o*-vanillin (15.0 g, 98.5 mmol, 1 equiv) were added. The reaction system was placed in an ice bath and cooled to 0 °C. After that, 30 mL CH_2Cl_2 contained BBr_3 (31.8 g, 126.9 mmol, 1.3 equivalent) in the dropping funnel was added dropwise to the reaction solution within 30 min under stirring. The mixture solution was stirred and reacted at room temperature for 19 h. After the reaction, 100 mL of water was added, and the mixture solution was further stirred for 1 h. The solution was then extracted with EtOAc (3 × 100 mL) and washed with saturated brine (3 × 100 mL) for further purification. The collected solution was dried with anhydrous Na_2SO_4 overnight and concentrated in vacuo to give a dark purple residue. The crude product was recrystallized by hot hexanes (50 °C) and acquired a yellow crystal compound **2** (10.9 g, 79 mmol, 70% yield). ^{1}H NMR (300 MHz, $CDCl_3$) δ 11.09 (s, ^{1}H), 9.90 (s, ^{1}H), 7.20–7.14 (m, ^{2}H), 6.97–6.92 (m, ^{1}H), 5.61(s, ^{1}H).

Compound **3** (dihydroxyl spiropyran). To a 500 mL round-bottom flask, compound **1** (13.9 g, 42 mmol, 1.05 equivalent), compound **2** (5.5 g, 40 mmol, 1 equivalent), and triethylamine (6.4 mL, 80 mmol, 2 equivalent) were soluted in 150 mL ethanol. The reaction mixture was brought to reflux under N_2 pressure and was stirred for 10 h. After cooling down to ambient temperature, the precipitate was filtered out and washed with cooled ethanol (3 × 20 mL) to yield compound **3** as a dark purple solid powder (6.9 g, 21 mmol, 52%). ^{1}H NMR (300 MHz, $CDCl_3$) δ 7.13–6.61 (m, 8H), 6.26 (s, ^{1}H), 6.11 (s, ^{1}H), 5.87–5.85 (d, ^{1}H), 5.57 (s, ^{1}H), 5.34 (s, ^{1}H), 4.29–4.26 (m, ^{2}H), 3.43–3.38 (m, ^{2}H), 1.97 (s, ^{3}H), 1.93 (s, ^{3}H), 1.31 (s, ^{3}H), 1.16 (s, ^{3}H).

Compound **4** (dimethacrylate spiropyran (DMSP3)). To a 250 mL round-bottom flask, compound **3** (6.4 g, 20 mmol, 1 equivalent) and *N*,*N*-dimethylaminopyridine (4.8 g, 40 mmol, 2 equivalent) was dissolved in 100 mL dry CH_2Cl_2. The reaction mixture was cooled to 0 °C and 20 mL CH_2Cl_2 contained methacrylic anhydride (8.9 mL, 60 mmol, 3 equivalent) were added dropwise to the cold solution under N_2 pressure within 15 min. The reaction system was stirred overnight under ambient temperature. After the reaction finished, the mixture solution was washed with 1 M HCl (2 × 100 mL) and saturated brine (2 × 100 mL) for further purification. The collected solution was dried by anhydrous $MgSO_4$ overnight and concentrated by rotary evaporation to give a gray solid. The crude product was purified by column chromatography eluting with 0.5% Et_3N/CH_2Cl_2 to yield DMSP3 as light purple oil (6.8 g, 15 mmol, 75%). ^{1}H NMR (300 MHz, $CDCl_3$) δ 7.13–6.61 (m, ^{8}H), 6.26 (s, ^{1}H), 6.11 (s, ^{1}H), 5.87–5.78 (m, ^{2}H), 5.57 (s, ^{1}H), 4.29–4.26 (m, ^{2}H), 3.43–3.38 (m, ^{2}H), 1.97 (s, ^{3}H), 1.93 (s, ^{3}H), 1.31 (s, ^{3}H), 1.16 (s, ^{3}H).

4.3. Synthesis of Poly(AM–co–MA/DMSP3) Hydrogels

Briefly, hydrophobic monomer of MA (2.5 g, about 25 wt% of total mass) and DMSP3 crosslinker (0.0532 g, 0.4 mol% of MA), and photo-initiator PBPO (0.0365 g, 0.3 mol% of MA) were added into 1 wt% TWEEN 80 aqueous solution (5 g, about 50 wt% of total mass) and vortexed for 5 min to form a uniform emulsion. Then, the hydrophilic monomer of AM (2.5 g, 25 wt% of total mass) monomer was added to the emulsion and vortexed for 5 min to dissolved into the aqueous phase of the emulsion. The mixture solution was injected into a glass mold with a 1 mm thick Teflon spacer and exposed to white light for 1 h to form poly(AM–*co*–MA/DMSP3) hydrogels. All poly(AM–*co*–MA/DMSP3) hydrogels were prepared in the same way just by tailoring the contents of MA, AM, and DMSP3 crosslinker in the gels. To further distinguish different poly(AM–co–MA/DMSP3) hydrogels were named as AM_X–SP_Y–MA_Z, where X, Y, and Z represented the contents of AM, DMSP3, and MA in the gels, respectively. For example, the above-prepared poly(AM–co–MA/DMSP3) hydrogels can be named AM_{25}–$DMSP3_{0.4}$–MA_{25}. The AM_{25}–MA_{25} hydrogels were prepared in the same way as the AM_{25}–$DMSP3_{0.4}$–MA_{25} hydrogels described above, except for the addition of the DMSP3 crosslinker.

4.4. Mechanical Tests

The hydrogels were prepared in a 1 mm thickness mold and cut into a standard dumbbell shape (ASTM-638-V) with 3.18 mm in width and 25 mm in gauge length before the tests. The tensile strain (λ), tensile stress (σ), elastic modulus (E), and dissipated energy (U_{hys}) were measured using a tensile tester (UTM 4304, SUNS) equipped with a 100 N load cell. The crosshead speed during the test was fixed at 100 mm·min^{-1}. The tensile strain was calculated by the elongation of the sample (Δl) to its initial length (l_0) ($\lambda = \Delta l/l_0$). The tensile stress (σ) was defined as the load force (F) applied per unit of the original cross-sectional area (A_0) of the sample ($\sigma = F/A_0$). The elastic modulus (E) was calculated by the slope of the initial linear regime of the stress-strain curve. The dissipated energy (U_{hys}) was estimated between the loading and unloading cycles. Before the tearing energy test, the specimen was cut into a trouser shape of 40 mm in length and 20 mm in width. The tearing energy (T) was calculated by the average force (F_{ave}) during steady-state tearing to the width (w) of the specimen ($T = 2F_{ave}/w$).

4.5. Rheological Measurement

The rheological properties of the prepared hydrogels were measured on a TA 2000ex rheometer using plate-and plate geometry (diameter 25 mm, gap 1000 µm), through two different modes:(i) the dynamic strain sweep from 0.1~1000% with a constant frequency of 10 rad/s was first performed at 25 °C, and the storage modulus was recorded to define the linear viscoelastic region in which the storage modulus is independent of the strain amplitude; (ii) the viscoelastic parameters, including shear storage modulus and loss modulus were measured over the ω range of 0.1–100 rad/s at strain 1% at 25 °C.

4.6. Optical Color Characterization

A white board was applied to the background during the test under ambient room light conditions. All-Optical images were taken by a Nikon D7000 camera. To further balance the white background, the background of all images was split into RGB channels and obtained the RGB value of pure white (RGB values of 220, 220, and 220) through Image-J software. After the background was white-balanced, the RGB values of the gel specimen center were obtained from the histogram of the area. To directly identify the color change of the gel specimen in response to force, the RGB values were converted to the (x, y) value and marked in the chromaticity diagram (CIE 1931 color space).

4.7. UV-Vis Spectrometer

The gel specimens of poly(AM–co–MA/DMSP3) hydrogel with the size of 26 mm × 9 mm × 1 mm were placed on the wall of 1 cm path length quartz cuvette. The UV-Vis spectra of gel specimen after being stretched to various strains were obtained using a UV-Visible spectrophotometer (TU-1810, PERSEE) over the 300−800 nm wavelength range.

Supplementary Materials: The following are available online at https://www.mdpi.com/article/10.3390/gels8040208/s1, Figure S1: 1H NMR spectrum of 2-hydroxyethyl-2,3,3-trimethyl-3H-indolium iodide. Figure S2: 1H NMR spectrum of 2,3-dihydroxybenzaldehyde. Figure S3: 1H NMR spectrum of dihydroxyl spiropyran. Figure S4: 1H NMR spectrum of DMSP3 crosslinker.

Author Contributions: Conceptualization, Y.L., Y.C., J.X. and L.X.; methodology, Z.G. and Y.Z.; data curation, S.X.; writing—original draft preparation, Y.L., S.X. and J.X.; writing—review and editing,

Y.L., Y.C., J.X. and L.X.; supervision, J.X. and L.X.; funding acquisition, J.X., N.L. and L.X. All authors have read and agreed to the published version of the manuscript.

Funding: This research was funded by the National Natural Science Foundation of China (51874129 and 52174247), the Natural Science Foundation of Hunan Province (2021JJ30212), the Scientific Research Fund of Hunan Provincial Education Department (19B153), the Graduate Innovation Research Foundation of Hunan Province (CX20211077). The APC was funded by the National Natural Science Foundation of China (51874129).

Institutional Review Board Statement: Not applicable.

Informed Consent Statement: Not applicable.

Data Availability Statement: Not applicable.

Conflicts of Interest: The authors declare no conflict of interest.

References

1. Nojoomi, A.; Arslan, H.; Lee, K.; Yum, K. Bioinspired 3D structures with programmable morphologies and motions. *Nat. Commun.* **2018**, *9*, 3705. [CrossRef] [PubMed]
2. Roy, D.; Brooks, W.L.A.; Sumerlin, B.S. New directions in thermoresponsive polymers. *Chem. Soc. Rev.* **2013**, *42*, 7214–7243. [CrossRef] [PubMed]
3. Meeks, A.; Mac, R.; Chathanat, S.; Aizenberg, J. Tunable long-range interactions between self-trapped beams driven by the thermal response of photoresponsive hydrogels. *Chem. Mater.* **2020**, *32*, 10594–10600. [CrossRef]
4. Zhang, F.; Xiong, L.; Ai, Y.; Liang, Z.; Liang, Q. Stretchable multi responsive hydrogel with actuatable, shape memory, and self-healing properties. *Adv. Sci.* **2018**, *5*, 1800450. [CrossRef] [PubMed]
5. Ikejiri, S.; Takashima, Y.; Osaki, M.; Yamaguch, H.; Harada, A. Solvent-free photoresponsive artificial muscles rapidly driven by molecular machines. *J. Am. Chem. Soc.* **2018**, *140*, 17308–17315. [CrossRef] [PubMed]
6. Xiao, S.; Zhang, Y.; Shen, M.; Chen, F.; Fan, P.; Zhong, M.; Ren, B.; Yang, J.; Zheng, J. Structural dependence of salt-responsive polyzwitterionic brushes with an anti-polyelectrolyte effect. *Langmuir* **2018**, *34*, 97–105. [CrossRef]
7. Xiao, S.; Zhang, M.; He, X.; Huang, L.; Zhang, Y.; Ren, B.; Zhong, M.; Chang, Y.; Yang, J.; Zheng, J. Dual salt-and thermoresponsive programmable bilayer hydrogel actuators with pseudo-interpenetrating double-network structures. *ACS Appl. Mater. Interfaces* **2018**, *10*, 21642–21653. [CrossRef]
8. Benson, P.J. Tunable materials respond to magnetic field. *Science* **2018**, *362*, 1124–1125.
9. Al Islam, S.; Jamil, Y.; Javaid, Z.; Hao, L.; Amin, N.; Javed, Y.; Rehman, M.; Anwar, H.A. Study on Thermal Response of Nanoparticles in External Magnetic Field. *J. Supercond. Nov. Magn.* **2021**, *34*, 3223–3228. [CrossRef]
10. Pham, S.H.; Choi, Y.; Choi, J. Stimuli-responsive nanomaterials for application in antitumor therapy and drug delivery. *Pharmaceutics* **2020**, *12*, 630. [CrossRef]
11. Municoy, S.; Álvarez Echazú, M.I.; Antezana, P.E.; Galdopórpora, J.M.; Olivetti, C.; Mebert, A.M.; Foglia, M.L.; Tuttolomondo, M.V.; Alvarez, G.S.; Hardy, J.G.; et al. Stimuli-responsive materials for tissue engineering and drug delivery. *Int. J. Mol. Sci.* **2020**, *21*, 4724. [CrossRef]
12. Liu, H.; Yao, J.; Guo, H.; Cai, X.; Jiang, Y.; Lin, M.; Jiang, X.; Leung, W.; Xu, C. Tumor microenvironment-responsive nanomaterials as targeted delivery carriers for photodynamic anticancer therapy. *Front. Chem.* **2020**, *8*, 758. [CrossRef]
13. Mamidi, N.; Delgadillo, R.M.V.; Castrejón, J.V. Unconventional and facile production of a stimuli-responsive multifunctional system for simultaneous drug delivery and environmental remediation. *Environ. Sci. Nano* **2021**, *8*, 2081–2097. [CrossRef]
14. Chen, Y.; Jiang, D.; Gong, Z.; Li, Q.; Shi, R.; Yang, Z.; Lei, Z.; Li, J.; Wang, L. Visible-light responsive organic nano-heterostructured photocatalysts for environmental remediation and H_2 generation. *J. Mater. Sci. Technol.* **2020**, *38*, 93–106. [CrossRef]
15. Wang, M.; Chen, M.; Niu, W.; Winston, D.D.; Cheng, W.; Lei, B. Injectable biodegradation-visual self-healing citrate hydrogel with high tissue penetration for microenvironment-responsive degradation and local tumor therapy. *Biomaterials* **2020**, *261*, 120301. [CrossRef]
16. Wang, D.; Chen, X.; Yuan, G.; Jia, Y.; Wang, Y.; Mumtaz, A.; Wang, J.; Liu, J. Toward artificial intelligent self-cooling electronic skins: Large electrocaloric effect in all-inorganic flexible thin films at room temperature. *J. Mater.* **2019**, *5*, 66–72. [CrossRef]
17. Zhang, X.; Chen, L.; Lim, K.H.; Gonuguntla, S.; Lim, K.W.; Pranantyo, D.; Yong, W.P.; Yam, W.J.T.; Low, Z.; Teo, W.J.; et al. The Pathway to Intelligence: Using Stimuli-Responsive Materials as Building Blocks for Constructing Smart and Functional Systems. *Adv. Mater.* **2019**, *31*, 1804540. [CrossRef] [PubMed]
18. Wang, B.; Shi, T.; Zhang, Y.; Chen, C.; Li, Q.; Fan, Y. Lignin-based highly sensitive flexible pressure sensor for wearable electronics. *J. Mater. Chem. C* **2018**, *6*, 6423–6428. [CrossRef]
19. Bai, C.; Wang, Z.; Yang, S.; Cui, X.; Li, X.; Yin, Y.; Zhang, M.; Wang, T.; Sang, S.; Zhang, W.; et al. Wearable electronics based on the gel thermogalvanic electrolyte for self-powered human health monitoring. *ACS Appl. Mater. Interfaces* **2021**, *13*, 37316–37322. [CrossRef]

20. Zhu, J.; Wang, X.; Xing, Y.; Li, J. Highly stretchable all-rubber-based thread-shaped wearable electronics for human motion energy-harvesting and self-powered biomechanical tracking. *Nanoscale Res. Lett.* **2019**, *14*, 247. [CrossRef]
21. Fallahi, H.; Taheri-Behrooz, F.; Asadi, A. Nonlinear mechanical response of polymer matrix composites: A review. *Polymer Rev.* **2020**, *60*, 42–85. [CrossRef]
22. Cho, Y.; Lee, J. Anisotropic mechanical responses of composites having water microchannels. *J. Ind. Eng. Chem.* **2018**, *60*, 498–504. [CrossRef]
23. Ishijima, Y.; Imai, H.; Oaki, Y. Tunable mechano-responsive color-change properties of organic layered material by intercalation. *Chem* **2017**, *3*, 509–521. [CrossRef]
24. Huang, H.; Zhou, Y.; Wang, Y.; Cao, X.; Han, C.; Liu, G.; Xu, Z.; Zhan, C.; Hu, H.; Peng, Y.; et al. Precise molecular design for BN-modified polycyclic aromatic hydrocarbons toward mechanochromic materials. *J. Mater. Chem. A* **2020**, *8*, 22023–22031. [CrossRef]
25. Wu, M.; Chen, Y. Developing real-time mechanochromic probes for polymeric materials. *Chem* **2021**, *7*, 838–840. [CrossRef]
26. Xie, S.; Chen, Y.; Guo, Z.; Luo, Y.; Tan, H.; Xu, L.; Xu, J.; Zheng, J. Agar/carbon dot crosslinked polyacrylamide double-network hydrogels with robustness, self-healing, and stimulus-response fluorescence for smart anti-counterfeiting. *Mater. Chem. Front.* **2021**, *5*, 5418–5428. [CrossRef]
27. Wang, H.; Xu, J.; Du, X.; Du, Z.; Cheng, X.; Wang, H. A self-healing polyurethane-based composite coating with high strength and anti-corrosion properties for metal protection. *Compos. Part B Eng.* **2021**, *225*, 109273. [CrossRef]
28. Klajn, R. Spiropyran-based dynamic materials. *Chem. Soc. Rev.* **2014**, *43*, 148–184. [CrossRef]
29. Xu, J.; Guo, Z.; Chen, Y.; Luo, Y.; Xie, S.; Zhang, Y.; Tan, H.; Xu, L.; Zheng, J. Tough, adhesive, self-healing, fully physical crosslinked κ-CG-K+/pHEAA double-network ionic conductive hydrogels for wearable sensors. *Polymer* **2021**, *236*, 124321. [CrossRef]
30. Tang, L.; Wu, S.; Xu, Y.; Cui, T.; Li, Y.; Wang, W.; Gong, L.; Tang, J. High toughness fully physical cross-linked double network organohydrogels for strain sensors with anti-freezing and anti-fatigue properties. *Mater. Adv.* **2021**, *2*, 6655–6664. [CrossRef]
31. Liu, A.; Xiong, C.; Ma, X.; Ma, W.; Sun, R. A multiresponsive hydrophobic associating hydrogel based on azobenzene and spiropyran. *Chin. J. Chem.* **2019**, *37*, 793–798. [CrossRef]
32. Francis, W.; Dunne, A.; Delaney, C.; Florea, L.; Diamond, D. Spiropyran based hydrogels actuators-Walking in the light. *Sensor. Actuat. B Chem.* **2017**, *250*, 608–616. [CrossRef]
33. Wang, W.; Hu, J.; Zheng, M.; Zheng, L.; Wang, H.; Zhang, Y. Multi-responsive supramolecular hydrogels based on merocyanine-peptide conjugates. *Org. Biomol. Chem.* **2015**, *13*, 11492–11498. [CrossRef]
34. Zhang, Y.; Ren, B.; Yang, F.; Cai, Y.; Chen, H.; Wang, T.; Feng, Z.; Tang, J.; Xu, J.; Zheng, J. Micellar-incorporated hydrogels with highly tough, mechanoresponsive, and self-recovery properties for strain-induced color sensors. *J. Mater. Chem. C* **2018**, *6*, 11536–11551. [CrossRef]
35. Chen, H.; Yang, F.; Chen, Q.; Zheng, J. A novel design of multi-mechanoresponsive and mechanically strong hydrogels. *Adv. Mater.* **2017**, *29*, 1606900. [CrossRef] [PubMed]
36. Li, M.; Zhang, Q.; Zhou, Y.; Zhu, S. Let spiropyran help polymers feel force. *Prog. Polym. Sci.* **2018**, *79*, 26–39. [CrossRef]
37. Cao, Z. Highly Stretchable Tough Elastomers Crosslinked by Spiropyran Mechanophores for Strain-Induced Colorimetric Sensing. *Macromol. Chem. Phys.* **2020**, *221*, 2000190. [CrossRef]
38. Xie, S.; Ren, B.; Gong, G.; Zhang, D.; Chen, Y.; Xu, L.; Zhang, C.; Xu, J.; Zheng, J. Lanthanide-doped upconversion nanoparticle-cross-linked double-network hydrogels with strong bulk/interfacial toughness and tunable full-color fluorescence for bioimaging and biosensing. *ACS Appl. Nano Mater.* **2020**, *3*, 2774–2786. [CrossRef]
39. Vidavsky, Y.; Yang, S.J.; Abel, B.A.; Agami, I.; Diesendruck, C.E.; Coates, G.W.; Silberstein, M.N. Enabling room-temperature mechanochromic activation in a glassy polymer: Synthesis and characterization of spiropyran polycarbonate. *J. Am. Chem. Soc.* **2019**, *141*, 10060–10067. [CrossRef]
40. Wang, K.; Deng, Y.; Wang, T.; Wang, Q.; Qian, C.; Zhang, X. Development of spiropyran bonded bio-based waterborne polyurethanes for mechanical-responsive color-variable films. *Polymer* **2020**, *210*, 123017. [CrossRef]
41. Wang, S.; Du, X.; Luo, Y.; Lin, S.; Zhou, M.; Du, Z.; Cheng, X.; Wang, H. Hierarchical design of waterproof, highly sensitive, and wearable sensing electronics based on MXene-reinforced durable cotton fabrics. *Chem. Eng. J.* **2021**, *408*, 127363. [CrossRef]
42. Zhang, Y.; Ren, B.; Xie, S.; Cai, Y.; Wang, T.; Feng, Z.; Tang, J.; Chen, Q.; Xu, J.; Xu, L.; et al. Multiple physical cross-linker strategy to achieve mechanically tough and reversible properties of double-network hydrogels in bulk and on surfaces. *ACS Appl. Polym. Mater.* **2019**, *1*, 701–713. [CrossRef]
43. Pinelli, F.; Magagnin, L.; Rossi, F. Progress in hydrogels for sensing applications: A review. *Mater. Today Chem.* **2020**, *17*, 100317. [CrossRef]
44. Zhang, J.; Jin, J.; Wan, J.; Jiang, S.; Wu, Y.; Wang, W.; Gong, X.; Wang, H. Quantum dots-based hydrogels for sensing applications. *Chem. Eng. J.* **2021**, *408*, 127351. [CrossRef]
45. Xu, J.; Wang, H.; Du, X.; Cheng, X.; Du, Z.; Wang, H. Self-healing, anti-freezing and highly stretchable polyurethane ionogel as ionic skin for wireless strain sensing. *Chem. Eng. J.* **2021**, *426*, 130724. [CrossRef]

Article

Formulation, Characterization, and In Vitro Drug Release Study of β-Cyclodextrin-Based Smart Hydrogels

Muhammad Suhail [1], Quoc Lam Vu [2] and Pao-Chu Wu [1,3,4,*]

1 School of Pharmacy, Kaohsiung Medical University, Kaohsiung 80708, Taiwan; u108830004@kmu.edu.tw
2 Department of Clinical Pharmacy, Thai Nguyen University of Medicine and Pharmacy, 284 Luong Ngoc Quyen Str., Thai Nguyen 24000, Vietnam; vuquoclam@tump.edu.vn
3 Department of Medical Research, Kaohsiung Medical University Hospital, Kaohsiung 80708, Taiwan
4 Drug Development and Value Creation Research Center, Kaohsiung Medical University, Kaohsiung 80708, Taiwan
* Correspondence: pachwu@kmu.edu.tw; Tel.: +886-7-3121101

Abstract: In this study, novel pH-responsive polymeric β-cyclodextrin-*graft*-poly(acrylic acid/itaconic acid) hydrogels were fabricated by the free radical polymerization technique. Various concentrations of β-cyclodextrin, acrylic acid, and itaconic acid were crosslinked by ethylene glycol dimethacrylate in the presence of ammonium persulfate. The crosslinked hydrogels were used for the controlled delivery of theophylline. Loading of theophylline was conducted by the absorption and diffusion method. The fabricated network of hydrogel was evaluated by Fourier transform infrared spectroscopy (FTIR), thermogravimetric analysis (TGA), X-ray diffractometry (XRD), and scanning electron microscopy (SEM). The crosslinking among hydrogel contents and drug loading by the fabricated hydrogel were confirmed by FTIR analysis, while TGA indicated a high thermal stability of the prepared hydrogel as compared to pure β-cyclodextrin and itaconic acid. The high thermal stability of the developed hydrogel indicated an increase in the thermal stability of β-cyclodextrin and itaconic acid after crosslinking. Similarly, a decrease in crystallinity of β-cyclodextrin and itaconic acid was observed after crosslinking, as evaluated by XRD analysis. SEM revealed an irregular and hard surface of the prepared hydrogel, which may be correlated with strong crosslinking among hydrogel contents. Crosslinked insoluble and uncrosslinked soluble fractions of hydrogel were evaluated by sol–gel analysis. An increase in gel fraction was seen with the increase in compositions of hydrogel contents, while a decrease in sol fraction was observed. Dynamic swelling and dissolution studies were performed in three various buffer solutions of pH 1.2, 4.6, and 7.4, respectively. Maximum swelling and drug release were observed at higher pH values as compared to the lower pH value due to the deprotonation and protonation of functional groups of the hydrogel contents; thus, the pH-sensitive nature of the fabricated hydrogel was demonstrated. Likewise, water penetration capability and polymer volume were evaluated by porosity and polymer volume studies. Increased incorporation of β-cyclodextrin, acrylic acid, and itaconic acid led to an increase in swelling, drug release, drug loading, and porosity of the fabricated hydrogel, whereas a decrease was detected with the increasing concentration of ethylene glycol dimethacrylate. Conclusively, the prepared hydrogel could be employed as a suitable and promising carrier for the controlled release of theophylline.

Keywords: hydrogel; β-cyclodextrin; sol–gel analysis; dissolution studies

1. Introduction

Theophylline (THP) is a bronchodilator drug used in the management of chronic obstructive pulmonary disease [1,2]. The biological source of THP is the leaves of Camellia sinensis. The intake of THP leads to relaxation of bronchioles and muscles of pulmonary sanguine vessels that determines the bronchodilator and relaxing behavior of THP upon the smooth muscles [3]. The reported half-life of THP is 8 h, which decreased to 5 h in smoking patients. Hence, in order to avoid fluctuations and maintain a constant plasma concentration, THP dose needs to be taken multiple times in a day [4], which leads to certain complications, including nausea, vomiting, abdominal pain, insomnia, jitteriness, and irregular or fast heartbeat. Furthermore, patient compliance is also decreased [5]. Hence, to overcome all these complications, different drug carrier systems were developed in order to prolong the release of THP for an extended period of time. Zhang et al. (2008) prepared chitosan/β-cyclodextrin-based microspheres by the spray drying method for sustained release of theophylline [6]. Likewise, Ahirrao and his coworkers fabricated sodium alginate based hydrogel beads for sustained delivery of theophylline [7]. Still, the challenges faced by rapid administration of THP are not overcome. Therefore, due to the unique properties of hydrogels such as high hydrophilicity, stability, biocompatibility, biodegradability, and low toxicity, they are considered the most suitable carrier agent for controlled delivery of therapeutic agents. Hence, the authors have prepared β-cyclodextrin-based hydrogels for controlled delivery of THP to avoid the adverse effects generated as a result of THP intake.

Hydrogels are crosslinked three-dimensional polymeric hydrophilic networks, which absorb a high quantity of water without losing their structural configuration. The use of hydrogels has been increased, especially in pharmaceutical and biomedical fields, due to their unique properties such as high stability, exceptional swelling index, drug loading, biocompatibility, degradability, biodegradability, and gelling capabilities [8,9]. Due to good response to external stimuli such as temperature, pH, magnetic fields, electric fields, and ionic strength [10,11], a special class of hydrogels which is known as stimuli-sensitive hydrogels has gained a great interest in drug delivery systems. pH-sensitive hydrogel, which is the most studied and important type of stimuli-sensitive hydrogels, plays an important role in controlled and targeted drug delivery systems [12].

β-Cyclodextrin (β-CD) is an acyclic oligosaccharide, having both hydrophobic and hydrophilic cavities. Both β-CD and γ-CD are considered safe chemicals by US Food and Drug Administration (FDA) and are employed for various purposes. A key role is played by β-CD on both industrial and pharmaceutical levels as it is easily available and economical. It has various cavity dimensions that are suitable for a number of drug candidates [13]. Acrylic acid (Aa) is a synthetic monomer that has the potential to swell highly in water due to the presence of –COOH groups. These functional groups of Aa bring changes in the ionic strength as the pH of the medium is changed. At low pH values, protonation of Aa occurs, whereas at high pH values, deprotonation of Aa occurs. Hence, protonation and deprotonation of functional groups of Aa influence the swelling of hydrogels as an increase in swelling is observed with the deprotonation of –COOH groups while almost a low swelling is perceived with the protonation of –COOH groups [14]. Itaconic acid (Ia), a synthetic monomer, plays an important role in controlled drug delivery systems [15]. The main capability of Ia is its ease of copolymerization, which generates –COOH side chains of the polymer chain. These functional groups have hydrophilic character and possess the capability to interact with other equivalent groups through hydrogen bonding. The hydrophilic nature of Ia is due to its two –COOH groups with altered pKa values. Therefore, a small quantity of IA is enough to show response to the external stimulus (pH) and enhance the swelling index of hydrogels [16,17]. Moreover, the stability, stiffness, and mechanical strength of hydrogels could be enhanced by the incorporation of comonomers, which could contribute further to H– bonding [18].

The current study is based on the preparation of pH-responsive hydrogels of β-cyclodextrin for controlled delivery of theophylline. The novelty of the presented study is the incorporation of two hydrophilic monomers, i.e., acrylic acid and itaconic acid, with β-cyclodextrin polymer. The sensitivity of the developed hydrogel was enhanced by both acrylic acid and itaconic acid, which enabled the developed hydrogel to swell highly at high pH values as compared to low pH values. The advantage of the current fabricated hydrogel is the protection of the drug from the acidity of stomach and also the protection of the stomach itself from the adverse effects of the drug. Hence, due to maximum swelling at pH 7.4, a high amount of drug was released at basic pH of 7.4 in a controlled way by the fabricated hydrogel. The prepared hydrogels were processed for further investigations.

2. Results and Discussion

2.1. Synthesis of β-CD-g-P(Aa/Ia) Hydrogels

Various concentrations of polymer (β-CD), monomers (Aa, Ia), and crosslinker (EGDMA) were employed for the development of β-cyclodextrin-*graft*-poly(acrylic acid/itaconic acid) (β-CD-g-P(Aa/Ia)) hydrogels by the free radical polymerization technique, as shown in Table 1. The polymerization of all formulations was carried out in a water bath at a temperature of 55 °C for an initial 2 h. The temperature was enhanced later up to 65 °C for the next 22 h. The prepared hydrogel discs were then placed in a vacuum oven at 40 °C for drying. After dehydration, the discs were subjected to a series of studies. The physical appearance of dried hydrogel is given in Figure 1.

Table 1. Feed ratio scheme for the formulation of β-CD-g-P(Aa/Ia) hydrogels.

F. Code	Polymer β-CD g/100 g	Monomer Aa g/100 g	Monomer Ia g/100 g	Initiator APS g/100 g	Crosslinker EGDMA g/100 g
BAIF-1	0.5	20	10	0.5	1.5
BAIF-2	1.0	20	10	0.5	1.5
BAIF-3	1.5	20	10	0.5	1.5
BAIF-4	0.5	20	10	0.5	1.5
BAIF-5	0.5	25	10	0.5	1.5
BAIF-6	0.5	30	10	0.5	1.5
BAIF-7	0.5	20	05	0.5	1.5
BAIF-8	0.5	20	10	0.5	1.5
BAIF-9	0.5	20	15	0.5	1.5
BAIF-10	0.5	20	10	0.5	1.0
BAIF-11	0.5	20	10	0.5	1.5
BAIF-12	0.5	20	10	0.5	2.0

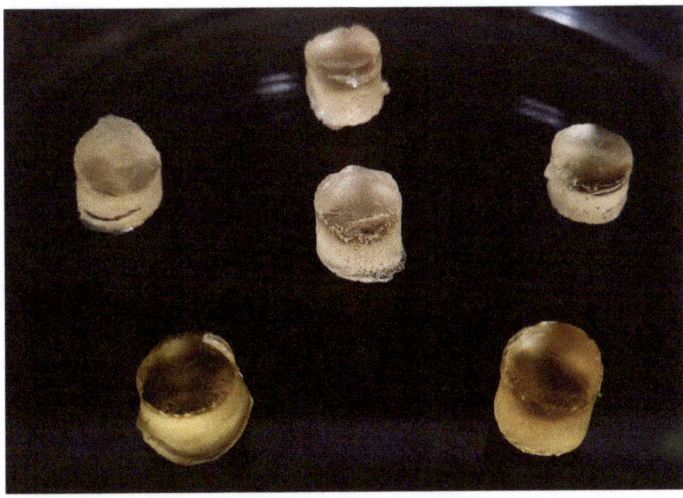

Figure 1. Physical appearance of prepared β-CD-g-P(Aa/Ia) hydrogel.

2.2. Fourier Transform Infrared Spectroscopy (FTIR)

The preparation of β-CD-g-P(Aa/Ia) hydrogel was confirmed by the evaluation of FTIR spectroscopy of β-CD, Aa, Ia, and fabricated hydrogels, as indicated in Figure 2. FTIR spectrum of β-CD (Figure 2A) presented stretching vibration of methylene C−H and ether C=O by peaks at 2972 and 1022 cm^{-1}, respectively, whereas a peak at 3308 cm^{-1} indicated stretching vibration of alcoholic O−H [13]. Similarly, Aa revealed FTIR spectra (Figure 2B) with peaks at 2887, 1673, and 1604 cm^{-1}, which were assigned to stretching vibration of –CH$_2$, C–O, and C–C. Likewise, the peak at 1201 cm^{-1} indicated stretching vibration of –C–O group [19,20]. The characteristic peaks of Ia were revealed by FTIR spectra (Figure 2C) at 1656, 1627, and 1480, which were assigned to stretching vibration of C=O, C=C, and C–O–H, respectively. Two peaks at 1228 and 1007 cm^{-1} indicated stretching vibration of the C–O group and out-of-plane banding of the OH group, respectively [21,22]. Certain peaks of polymer and monomers were changed due to the crosslinking and chemical interaction among hydrogel contents, as indicated in Figure 2D. The prominent peaks of β-CD were changed from 1022 and 2972 cm^{-1} to 1040 and 2950 cm^{-1} peaks of the developed hydrogel. Similarly, a few peaks of Aa and Ia were moved from 1201 and 2887 cm^{-1} and 1007 and 1228 cm^{-1} to the 1270 and 2950 cm^{-1} and 1030 and 1310 cm^{-1} peaks of the formulated polymeric network of the hydrogel. A few peaks of β-CD, Aa, and IA disappeared, while certain new peaks were formed. Hence, the shifting, disappearance, and formation of new peaks revealed the successful grafting of Aa and IA over the backbone of β-CD, which led to the development of a new polymeric network of the hydrogel. The FTIR spectral analysis of THP is indicated in Figure 2E and revealed stretching vibration of C=O, C=C, and C–O by peaks at 1566, 1537, and 1178 cm^{-1}, respectively. A fluctuation in certain peaks of THP was observed in drug-loaded hydrogels (Figure 2F). The peaks at 1178 and 1537 cm^{-1} were moved to 1205 and 1490 cm^{-1} peaks of the drug-loaded hydrogel. This change in peaks of THP was because of drug loading by the developed hydrogel, and hence no chemical interaction of drug with hydrogel contents was observed [23].

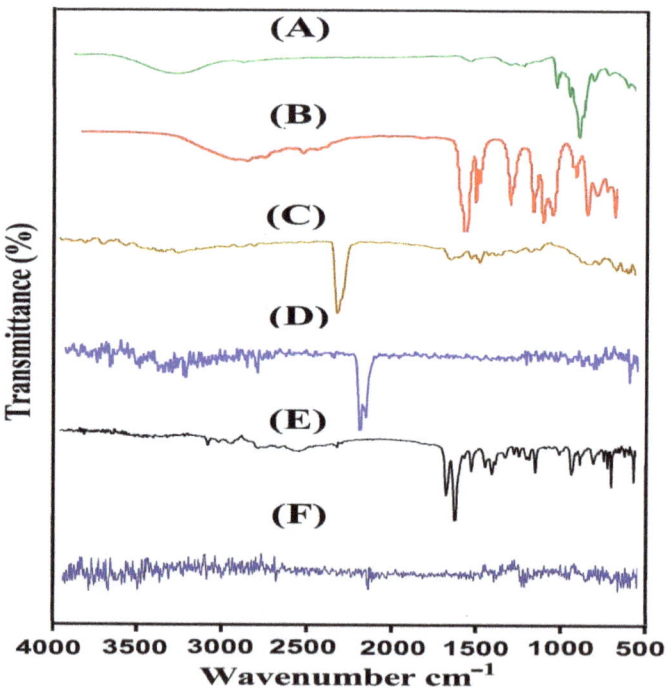

Figure 2. FTIR spectra of (**A**) β-CD, (**B**) Aa, (**C**) Ia, (**D**) unloaded β-CD-*g*-P(Aa/Ia) hydrogel, (**E**) THP, and (**F**) loaded β-CD-*g*-P(Aa/Ia) hydrogel.

2.3. Thermogravimetric Analysis (TGA)

TGA was performed in order to determine the thermal stability of pure polymer, monomer, and developed hydrogel, as shown in Figure 3. A weight reduction of 12%

was detected initially at 116 °C due to loss of bound water by TGA thermogram of β-CD (Figure 3A). Further decrease in weight was observed with the increase in temperature, and almost 63% weight reduction was perceived as the temperature approached 380 °C. Increased temperature led to degradation of β-CD [24]. Like β-CD, weight reduction of Ia occurred with the increase in temperature (Figure 3B). A 90% reduction in weight of Ia was detected as the temperature reached 223 °C due to the elimination of water and anhydride ring formation. A slight 4% weight reduction was observed at 278 °C. After that, due to the carbonization and decarboxylation process, degradation of IA started, which continued until complete paralysis [25]. Similarly, the TGA thermogram of fabricated hydrogel (Figure 3C) demonstrated a slight reduction of 8% in weight at 205 °C related to water loss of β-CD. After that, a reduction of 72% in weight was seen with the increase in temperature up to 488 °C. Further increase in temperature led to degradation of formulated hydrogel, which continued until entire paralysis. Hence, comparing the thermal stability of pure polymer and monomer with developed hydrogel, we can conclude that the thermal stability of both polymer and monomer was less than that of the developed hydrogel. An increase in thermal stability of β-CD and Ia occurred due to crosslinking and polymerization process among hydrogel contents, as indicated by the TGA thermogram of the fabricated hydrogel. The high thermal stability of the developed polymeric network of hydrogel presented strong intermolecular interaction of hydrogel contents, which occurred as a result of grafting, crosslinking, and polymerization reaction [26–28]. Nasir and her coworkers prepared Pluronic F127 based gels for controlled delivery of ivabradine hydrochloride and reported an increase in thermal stability of polymer after crosslinking with other gels' contents, which further supports our hypothesis [29].

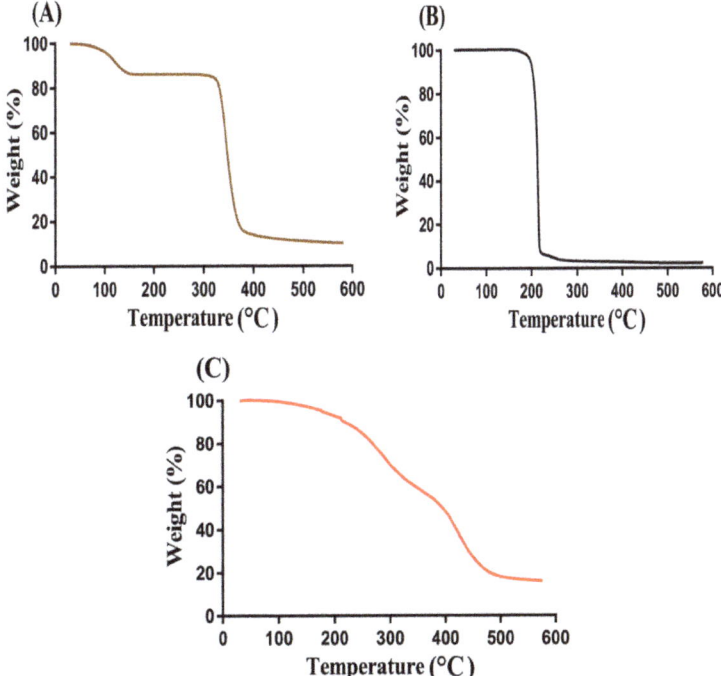

Figure 3. TGA of (**A**) β-CD, (**B**) Ia, and (**C**) β-CD-g-P(Aa/Ia) hydrogel.

2.4. X-ray Diffraction Studies (XRD)

The physical nature of β-CD, Ia, and developed hydrogel was evaluated by XRD analysis, as shown in Figure 4. The sharp and highly intense peaks of a compound indicate its crystalline nature, while diffuse peaks presented amorphous nature. The XRD pattern of β-CD (Figure 4A) exhibited sharp, highly intense, and prominent crystalline peaks at 2θ = 11.80°, 15.75°, 21.50°, and 27.90°. Similarly, XRD analysis of Ia (Figure 4B) indicated sharp and crystalline peaks at 2θ = 18.90°, 21.60°, 25.78°, 37.82°, and 54.45°. The sharp, crystalline, and highly intense peaks of β-CD and Ia were reduced or disappeared due to crosslinking and grafting, as indicated in the XRD analysis of the developed network of hydrogel (Figure 4C). The highly intense, crystalline, and sharp peaks of β-CD and Ia were replaced by low-intensity peaks, which demonstrated the successful crosslinking, grafting, and polymerization of hydrogel contents due to which crystallinity of pure reagent was reduced [30]. Abdullah et al. (2018) developed polyvinyl alcohol based hydrogel and demonstrated a decrease in crystallinity of polyvinyl alcohol by fabricated hydrogel [31].

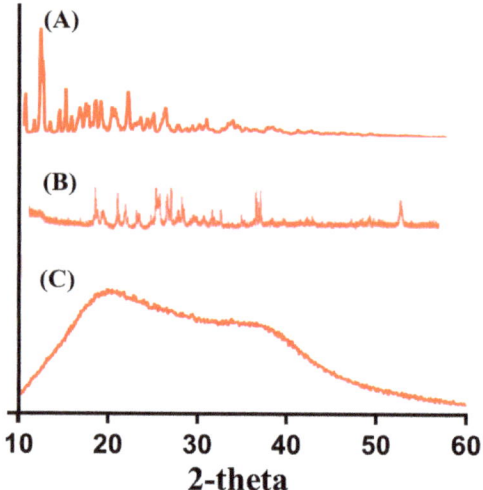

Figure 4. XRD of (**A**) β-CD, (**B**) Ia, and (**C**) β-CD-*g*-P(Aa/Ia) hydrogel.

2.5. Scanning Electron Microscopy (SEM)

The surface morphology of fabricated hydrogel was investigated by SEM, as shown in Figure 5. An irregular and hard surface was seen in the SEM examination of the formulated hydrogel. Large cracks and wrinkles could be seen, which may be correlated to the partial collapsing of gel during the process of dehydration. The hard surface of polymeric hydrogel revealed the strong intermolecular interaction that existed among the various contents of the fabricated hydrogel after the polymerization process [32].

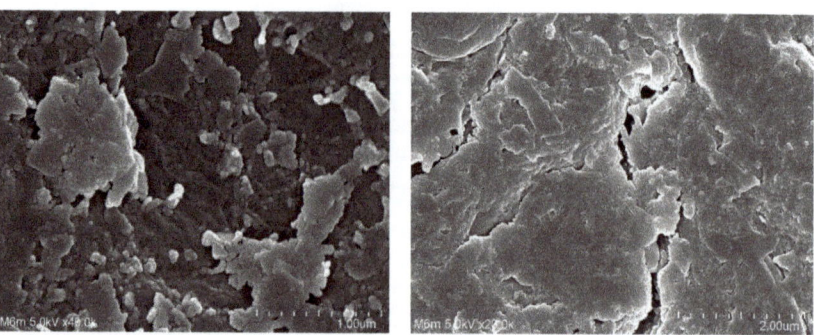

Figure 5. Scanning electron microscopy of β-CD-*g*-P(Aa/Ia) hydrogel.

2.6. Sol–Gel Fractions

Sol is the soluble uncrosslinked fraction, while gel is the insoluble crosslinked fraction of hydrogel. Due to the inverse relationship between the gel and sol fractions, an increase in one content leads to a decrease in other content. Both sol and gel fractions were influenced by the various feed ratios of hydrogel contents, i.e., β-CD, Aa, Ia, and EGDMA, as indicated in Figure 6A–D. An increase in the feed ratio of β-CD led to an increase in the gel fraction. The reason was the availability of a greater number of free radicals, which were generated highly with the increasing feed ratio of β-CD. Hence, a greater concentration of β-CD will increase the generation of free radicals for polymerization and crosslinking of hydrogel contents, thus increasing the gel fraction. Like β-CD, the gel fraction was increased with the increased incorporation of Aa and Ia contents. A high number of free radicals were generated with the increasing concentration of both Aa and Ia for the polymerization of β-CD content, which led to rapid polymerization of hydrogel contents; thus, as a result, an increase in gel fraction was detected. Similarly, bulk and crosslinking densities of the fabricated hydrogel were increased with the increase in feed ratio of EGDMA. The pore size of hydrogel was decreased due to high crosslinking, and thus an increase in gel fraction was observed. Khanum and her coworkers prepared polymeric HPMC-g-poly(AMPS) hydrogel and reported an increase in gel fraction with the increase in feed ratio of polymer, monomer, and crosslinker, which further supports our hypothesis [33]. Unlike the gel fraction, a decrease in the sol fraction was observed [34] as the feed ratio of β-CD, Aa, and Ia was increased and vice versa.

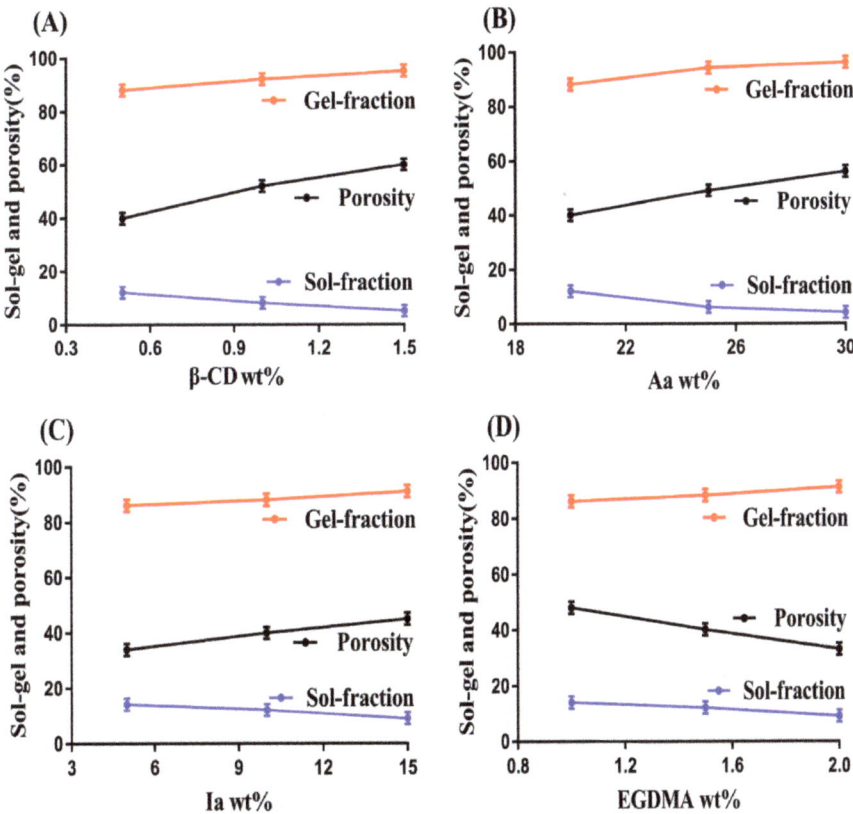

Figure 6. Effect of (**A**) β-CD, (**B**) Aa, (**C**) Ia, and (**D**) EGDMA on sol–gel and porosity of β-CD-*g*-P(Aa/Ia) hydrogel.

2.7. Porosity Study

A porosity study was conducted for all formulations of β-CD-g-P(Aa/Ia) hydrogel. The main purpose of this study was to evaluate the penetration capability of a fluid through the pores into the hydrogel network. Porosity was influenced by the various combinations of hydrogel contents as an increase in porosity was seen with the increasing composition of β-CD, Aa, and Ia, as indicated in Figure 6A–D. The reason may be the highly viscous nature of the reaction mixture, which was formed due to the polymerization of hydrogel contents. The increasing concentration of β-CD, Aa, and Ia content caused the formation of a highly viscous mixture, and as a result, evaporation of bubbles was restricted. This led to the generation of interconnected channels, and thus the porosity of the hydrogel was increased. Contrary to what was observed for the polymer and monomers, a decrease in porosity was observed as the feed ratio of EGDMA was increased. The decrease in porosity due to the increasing composition of EGDMA was due to the formation of a highly crosslinked bulk network of hydrogel, due to which the pore size of developed hydrogel was decreased. Thus, a limited number of channels were generated, and hence, a decrease in penetration and porosity was observed. Thus, we can conclude from the discussion that the higher the porosity, the greater the swelling and drug loading [35].

2.8. Swelling Study

A swelling study was performed for the β-CD-g-P(Aa/Ia) hydrogel in order to investigate the response of prepared hydrogel at pH 1.2, 4.6, and 7.4, as indicated in Figure 7. Almost low swelling was observed at pH 1.2 as compared to pH 4.6 and 7.4, which indicated the pH-sensitive nature of the fabricated hydrogel (Figure 7A). The low and high swelling of prepared hydrogel was due to the protonation and deprotonation of –COOH functional groups of both monomers, i.e., Aa and Ia. During protonation of –COOH groups, conjugates were formed with counterions through strong hydrogen bonding, due to which the charge density of –COOH groups was decreased, and as a result, low swelling was exhibited at pH 1.2. A change was seen in the swelling of the developed hydrogel as the pH changed from 1.2 to 4.6 and 7.4. Due to the deprotonation of –COOH groups at pH 4.6 and 7.4, the charge density of the same groups was increased, and as a result, strong electrostatic repulsive forces were generated. These forces counteracted each other, and thus maximum swelling at pH 7.4 was observed as compared to pH 4.6. Hence, we can conclude that swelling of fabricated hydrogel was perceived in an order of pH 7.4 > 4.6 > 1.2 [36–38].

The swelling of formulated hydrogel was affected by various combinations of β-CD, Aa, Ia, and EGDMA. The swelling was increased with the increased incorporation of polymer and monomer contents (Figure 7B). The –OH and CH_2OH functional groups of β-CD were produced highly with the increase in its composition, due to which charge density increased. Thus, an increase in swelling was observed [39]. Similarly, an increase in the swelling index of fabricated hydrogel was seen with the increasing composition of Aa and Ia (Figure 7C,D) due to the generation of a high number of –COOH groups. These functional groups of both monomers exerted strong electrostatic repulsive forces, which led to expansion in the volume of hydrogel, and thus an increase in swelling was perceived [40–42]. Unlike what was found for the polymer and monomers, a reduction in the swelling of hydrogel was found with the increasing composition of EGDMA (Figure 7E). The crosslinking and bulk densities of the formulated hydrogel were increased, due to which the pore size of the hydrogel was decreased; thus, as a result, water penetration into the polymeric hydrogel was decreased. This hard network and less penetration of water led to a decrease in swelling with the increasing composition of EGDMA [43–46].

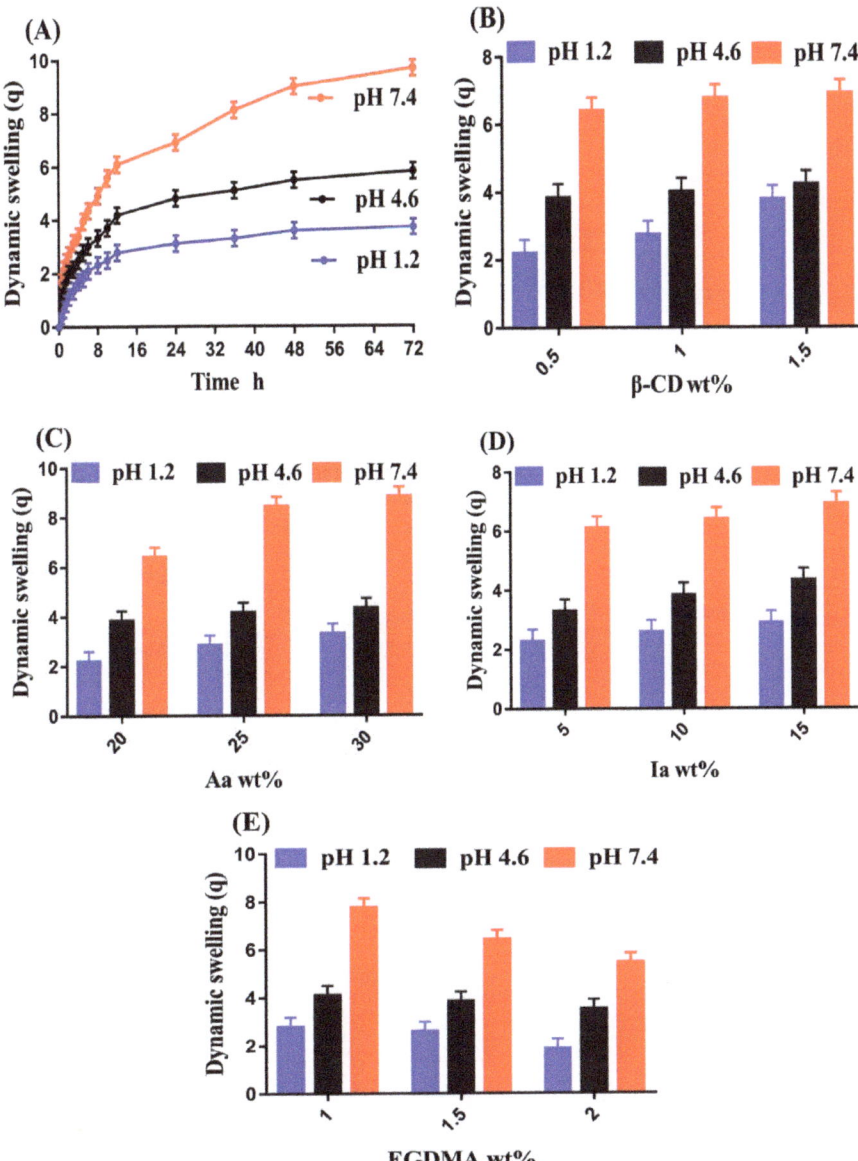

Figure 7. Effect of (**A**) pH, (**B**) β-CD, (**C**) Aa, (**D**) Ia, and (**E**) EGDMA on dynamic swelling of β-CD-*g*-P(Aa/Ia) hydrogel.

2.9. Polymer Volume Fraction

The determination of polymer volume fraction for developed hydrogel was carried out at three various pH values, i.e., pH 1.2, 4.6, and 7.4, as shown in Table 2. Polymer volume was found greater at pH 1.2 as compared to pH 4.6 and 7.4. Hydrogel contents, i.e., β-CD, Aa, Ia, and EGDMA, highly affected the polymer volume fraction at all pH values. A decrease was observed in polymer volume fraction with the increase in the composition of polymer and monomers. i.e., β-CD, Aa, and Ia. The main reason is attributed to the high swelling of hydrogel, which was achieved due to an increase in the concentration of polymer and monomers. Unlike what was observed for the polymer and monomers, an increase in polymer volume fraction was observed with the increasing composition of EGDMA, which is attributed to the low swelling of hydrogel. The low and high polymer volume values at pH 7.4, 4.6, and 1.2 indicated a high swelling index of the fabricated hydrogel at high pH values [47].

Table 2. Polymer volume fraction and drug loading of β-CD-g-P(Aa/Ia) hydrogels.

Formulation Code	Polymer Volume Fraction			Drug Loaded (mg)/500 mg of Dry Gel	
	pH 1.2	pH 4.6	pH 7.4	Weight Method	Extraction Method
BAIF-1	0.454	0.260	0.156	81.8 ± 0.8	80 ± 1
BAIF-2	0.363	0.250	0.150	104.4 ± 1	103.2 ± 1
BAIF-3	0.248	0.242	0.147	116.3 ± 1	115.6 ± 1
BAIF-4	0.454	0.260	0.156	81.8 ± 0.8	80 ± 1
BAIF-5	0.352	0.245	0.146	110.1 ± 1	109.5 ± 1
BAIF-6	0.240	0.236	0.138	121.2 ± 1	119.7 ± 1
BAIF-7	0.470	0.270	0.163	75.4 ± 1	73.6 ± 1
BAIF-8	0.454	0.260	0.156	81.8 ± 0.8	80 ± 1
BAIF-9	0.434	0.255	0.151	86.7 ± 0.9	85.3 ± 1
BAIF-10	0.416	0.245	0.128	88.2 ± 1	88 ± 1
BAIF-11	0.454	0.260	0.156	81.8 ± 0.8	80 ± 1
BAIF-12	0.532	0.280	0.184	70.3 ± 1	69.1 ± 1

2.10. Drug Loading

The amount of drug loaded by the formulated hydrogel was evaluated by the extraction and weight methods as indicated in Table 2. Drug loading depends on the porosity and swelling index of the hydrogel. A greater porosity allows a higher penetration capability of a medium into the hydrogel network and thus greater swelling and drug loading. Hence, we can conclude there is a direct relation between the swelling and drug loading [48]. Like porosity and swelling, the loading of a drug by the fabricated hydrogel was also influenced by the various compositions of β-CD, Aa, Ia, and EGDMA. An increase in drug loading was detected with the increasing composition of β-CD, Aa, and Ia. A high number of functional groups of polymer and monomers were generated with their high composition, which led to greater charge density and repulsive forces, and thus an increase in swelling and drug loading was observed. Contrary to what was observed for the polymer and monomers, drug loading was decreased as the concentration of EGDMA was enhanced. The reason is the low porosity and swelling, due to which a decrease in drug loading was perceived [49].

2.11. Drug Release Studies

In vitro drug release studies were performed for β-CD-g-P(Aa/Ia) hydrogel in order to evaluate the pH-responsive nature of the fabricated hydrogel at three pH values, i.e., pH 1.2, 4.6, and 7.4. Maximum drug release was achieved at pH 7.4 as compared to 4.6, while a very low drug release was perceived at pH 1.2, as shown in Figure 8A. Like swelling, protonation and deprotonation of COOH groups of both monomers occurred, which led to low and high release of drug from the fabricated hydrogel. Due to deprotonation, the charge density of COOH groups was increased very highly at pH 7.4 as compared to pH 4.6. The high charge density led to a decrease in attractive forces while enhancing the repulsive forces between COOH groups, and as a result, high swelling and drug release were perceived at pH 4.6 and 7.4. On other hand, due to protonation at pH 1.2, functional groups of monomers formed conjugates with counterions and thus strengthened the polymeric structure of hydrogel with strong hydrogen bonding. Due to strong hydrogen bonding, low swelling was observed, and thus low drug release was observed [50,51]. Similarly, drug release studies were performed at three pH values of 1.2, 4.6, and 7.4 for the commercially available tablets Theolin S.R (250 mg, PeiLi Pharmaceutical IND. Co., Ltd., Taichung, Taiwan), as shown in Figure 8B. More than 90% drug release was observed for Theolin at pH 7.4 (initial 6 h), 4.6 (initial 8 h), and 1.2 (initial 11 h).

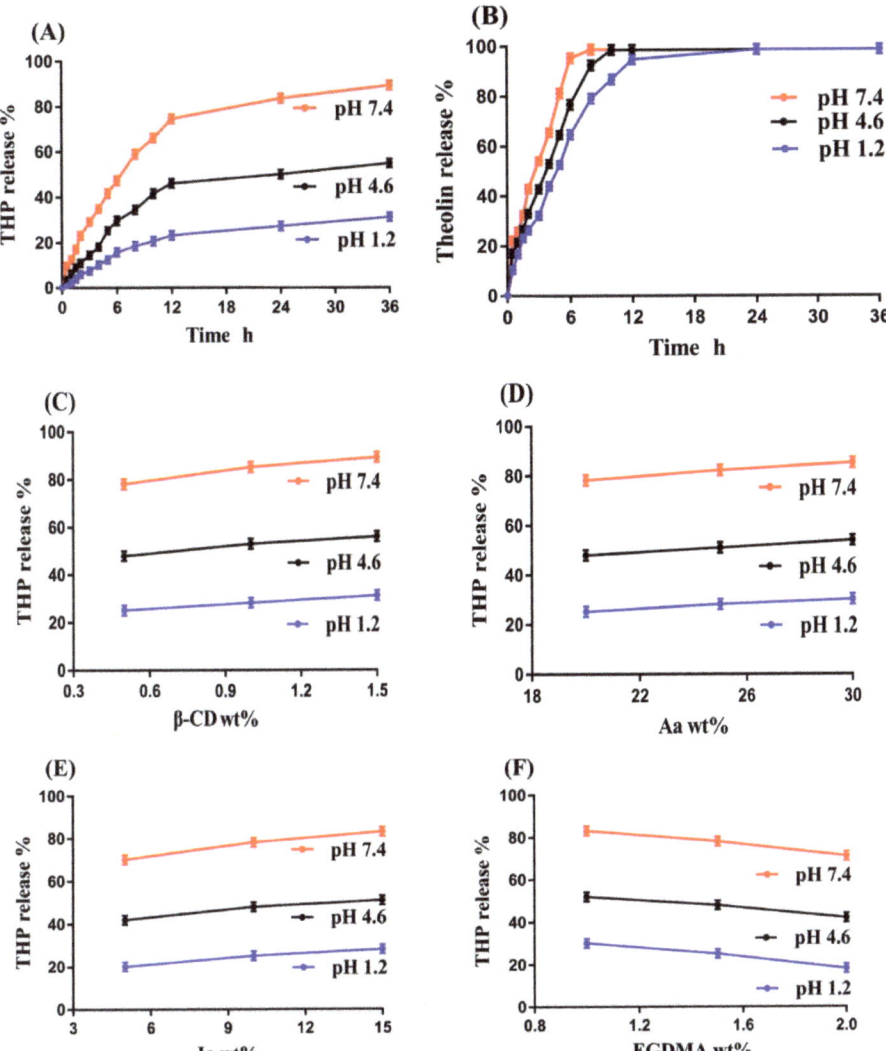

Figure 8. Effect of pH on (**A**) percent drug release from β-CD-g-P(Aa/Ia) hydrogel; (**B**) percent drug release from Theolin tablets; and effect of (**C**) β-CD, (**D**) Aa, (**E**) Ia, and (**F**) EGDMA on percent drug release from β-CD-g-P(Aa/Ia) hydrogel.

Like swelling, drug release was also influenced by the various combinations of polymer, monomers, and crosslinker. Drug release was increased with the incorporated compositions of β-CD, Aa, and Ia, as shown in Figure 8C–E. High composition of β-CD, Aa, and Ia led to an increase in generation of –OH, CH$_2$OH, and –COOH groups, due to which charge density was increased, and thus an increase in drug release was detected with the increasing composition of polymer and monomers [52,53]. A drop was detected in drug release with the increase in EGDMA composition (Figure 8F). The reason was the high crosslinking and hard network of hydrogel, which retard the penetration of water due to small pore size, and thus a decrease in drug release was observed [54].

Conclusively, we can demonstrate that at the low pH of 1.2, hydrogel remained protonated/un-ionized due to the pKa values of its reagents, and thus low swelling and drug release were observed. As the pH changed from 1.2 to 4.6 and 7.4, deprotonation/ionization of the functional groups of the polymer and monomers started and hydrogel networks expanded, and thus an increase in swelling and drug release was observed.

The literature indicates that different drug carrier systems have been developed for the sustained/controlled delivery of theophylline. Khan and his coworkers prepared gastroretentive floating tablets of theophylline using hydrophilic polymer METHOCEL K4M and reported release of theophylline for 8 h [55]. Bashir et al. (2016) prepared N-succinyl chitosan-g-poly(methacrylic acid) hydrogels and demonstrated swelling and in vitro release of theophylline for 10–11 h [56]. Similarly, Liu and his coworkers developed pH-sensitive hydroxypropyl methylcellulose acetate succinate based composite nanofibers for controlled delivery of theophylline for up to 12 h [57]. In the current study, the authors reported swelling and in vitro release of theophylline for 72 and 36 h, respectively. Comparing the drug released data of the commercial product Theolin and previously published research work with the present newly prepared hydrogels, we can conclude that release of theophylline was sustained for a long time by fabricated hydrogels in a controlled way. Thus, the newly prepared hydrogels can be considered as one of the most suitable and promising carriers for controlled drug delivery.

2.12. Kinetic Modeling

The best-fit release kinetic model and release rate of the drug from the fabricated hydrogel were determined from in vitro drug release data of all formulations by considering zero order, first order, Higuchi, and Korsmeyer–Peppas as kinetic models. The best-fit model and drug release mechanism were confirmed from the "r^2" value and "n" value individually. The "r^2" values indicate the regression coefficient. Table 3 indicates that all formulations of fabricated hydrogel followed Korsmeyer–Peppas model of kinetics because "r^2" values of Korsmeyer–Peppas model of all formulations were closer to 1 as compared to other kinetic models. Similarly, the "n" value determines the type of diffusion. If "n" > 0.45, then the diffusion is non-Fickian, whereas if "n" $\leq$ 0.45, then the diffusion is Fickian. The fabricated hydrogel exhibited non-Fickian diffusion because the "n" values of all formulations were within the range of 0.5316–0.7120 [58,59].

Table 3. Kinetic modeling release of THP from β-CD-g-P(Aa/Ia) hydrogels.

F. Code	Zero Order r^2	First Order r^2	Higuchi r^2	Korsmeyer–Peppas r^2	n
BAIF-1	0.9661	0.9283	0.9355	0.9843	0.5316
BAIF-2	0.9825	0.9894	0.9695	0.9963	0.6073
BAIF-3	0.9846	0.9952	0.9879	0.9986	0.6930
BAIF-4	0.9661	0.9283	0.9355	0.9843	0.5316
BAIF-5	0.9740	0.9795	0.9793	0.9893	0.6437
BAIF-6	0.9881	0.9649	0.9670	0.9980	0.7120
BAIF-7	0.9787	0.9081	0.9142	0.9822	0.6560
BAIF-8	0.9661	0.9283	0.9355	0.9843	0.5316
BAIF-9	0.9768	0.9763	0.9806	0.9810	0.6245
BAIF-10	0.9803	0.8426	0.9218	0.9874	0.6173
BAIF-11	0.9661	0.9283	0.9355	0.9843	0.5316
BAIF-12	0.9545	0.9141	0.9681	0.9794	0.6989

3. Conclusions

β-CD-g-P(Aa/Ia) hydrogels were prepared successfully by the free radical polymerization technique. The structural configuration of the prepared hydrogel and its contents was evaluated by FTIR analysis. The thermal stability of β-CD and Ia was found to be less than that of formulated hydrogel, as indicated by TGA thermogram. A decrease in crystalline and highly sharp peaks of β-CD and Ia was observed after crosslinking, as revealed by XRD analysis. Similarly, a hard surface was observed by the SEM. Swelling and drug release were found to be higher at pH 7.4 compared to pH 4.6 and 1.2 with an order of pH 7.4 > 4.6 > 1.2, presenting the pH-responsive nature of the fabricated hydrogel. Increased incorporation of β-CD, Aa, and Ia led to an increase in porosity, swelling, drug loading, and drug release, while a decrease was observed with the increased incorporation of EGDMA. A drop in sol fraction was seen with the increasing compositions of hydrogel contents, whereas gel fraction was increased. Polymer volume fractions were higher at low pH 1.2 and lower at high pH 7.4 and 4.6, indicating the maximum swelling capability of the developed hydrogel at high pH values. Thus, we can conclude from the discussion that prepared grafted hydrogel could be applied for controlled delivery of theophylline.

4. Materials and Methods

4.1. Materials

Theophylline and β-cyclodextrin were purchased from Sigma-Aldrich (St. Louis, MI, USA) and Alfa Aesar (Thermo Fisher Scientific, Ward Hill, MA, USA). Acrylic acid and itaconic acid were obtained from Acros (Carlsbad, CA, USA) and Acros Organics (Janssen Pharmaceuticalaan, Belgium). Similarly, ammonium persulfate was acquired from Showa (Tokyo, Japan), whereas ethylene glycol dimethacrylate was obtained from Alfa Aesar (Tewksbury, MA, USA).

4.2. Synthesis of β-CD-g-P(Aa/Ia) Hydrogels

Different formulations of β-cyclodextrin-*graft*-poly(acrylic acid/itaconic acid) (β-CD-g-P(Aa/Ia)) hydrogels were formulated by the free radical polymerization technique. An accurate amount of β-CD was dissolved in deionized distilled water. Similarly, ammonium persulfate (APS) was dissolved in deionized distilled water, while Aa was already available in liquid form. A mixture of water and ethanol was used for dissolving Ia at a temperature of 50 °C. APS solution was added into IA solution, and after proper mixing, the solution was poured into β-CD solution. The mixture was stirred for 20 min, and after that, Aa was added dropwise into the stirred mixture. Finally, ethylene glycol dimethacrylate (EGDMA) was added to the polymer and monomer mixture. After 5 min, a transparent solution was formed, which was purged by nitrogen gas in order to remove dissolved oxygen. The transparent solution was transferred into glass molds, which were kept in a water bath at 55 °C for the initial 2 h. The temperature was further increased up to 65 °C for the next 22 h. The prepared gel was sliced into 8 mm size discs. A mixture of water and ethanol was used for washing the gel discs in order to remove any impurity if attached to the surface of the gel discs. After that, the discs were placed at 40 °C in a vacuum oven for complete dehydration after exposing the gel discs to room temperature for 24 h. The dried discs of hydrogel were processed for further investigation.

4.3. Fourier Transform Infrared Spectroscopy (FTIR)

FTIR spectra of β-CD, Aa, Ia, THP, and the unloaded and loaded β-CD-g-P(Aa/Ia) hydrogel were obtained by using attenuated total reflectance FTIR (Nicolet 380 FTIR (Thermo Fisher Scientific, Ishioka, Japan)). Samples were ground thoroughly, and then the FTIR spectrum was analyzed within the range of 4000–500 cm^{-1} [60].

4.4. Thermogravimetric Analysis (TGA)

TGA (PerkinElmer Simultaneous Thermal Analyzer STA 8000 (PerkinElmer Ltd., Buckinghamshire, UK) was conducted in order to evaluate the thermal stability of β-CD, Ia, and fabricated β-CD-g-P(Aa/Ia) hydrogel. Samples of 0.3 to 5 mg were finely ground and placed in a platinum pan, which was attached with a microbalance. TGA thermogram was performed within the temperature range of 40–600 °C. Heat rate and nitrogen flow were maintained 20 °C/min and 20 mL/min, respectively, throughout the TGA analysis [61].

4.5. X-ray Diffraction Studies (XRD)

The crystalline or amorphous nature of β-CD, Ia, and fabricated β-CD-g-P(Aa/Ia) hydrogel was evaluated by X-ray diffraction (XRD-6000 Shimadzu, Tokyo, Japan) analysis. The crystallinity of a compound is identified by its sharp peaks, while diffuse peaks indicate the amorphous nature of the compound. The weighed crushed samples were placed in a sample holder and leveled with the help of a glass slide. The XRD analysis was performed within the range of 10–60° with an angle of 2θ 2°/min [62].

4.6. Scanning Electron Microscopy (SEM)

SEM (JSM-5300 model, (JEOL, Tokyo, Japan)) was carried out to determine the surface and structural characteristics of the developed hydrogels. Hence, the dried hydrogel disc was mounted on an aluminum point by sticky tape, covering the entire area. With the help of a gold splutter coater, a thin layer of gold was coated in an inert environment while using a vacuum evaporator. Photomicrographs were taken, and thus surface morphology of fabricated hydrogel was evaluated [63].

4.7. Sol–Gel Fractions

Sol–gel fractions of fabricated hydrogels were measured with the purpose of evaluating the amount of reactants consumed in their preparation by estimating the sol and gel content. Therefore, a dried hydrogel disc of known weight was placed in Soxhlet apparatus containing deionized distilled water. The extraction process was carried out at a temperature of 85 °C for 10 h. After that, the disc was removed and placed in a vacuum oven at 40 °C until dryness. The dried disc was weighed again [64]. Sol and gel fractions were estimated by using the following formulas:

$$\text{Sol fraction \%} = \frac{J_1 - J_2}{J_2} \times 100 \tag{1}$$

$$\text{Gel fraction} = 100 - \text{Sol fraction} \tag{2}$$

where J_1 is the initial weight of dried hydrogel disc before the extraction process and J_2 is the final weight after the extraction.

4.8. Porosity Study

A porosity study was conducted for all formulations of fabricated β-CD-g-P(Aa/Ia) hydrogels. Porosity was determined by the solvent displacement method. Absolute ethanol was employed as a displacement solvent. Thus, weighed dried hydrogel discs (M_1) were immersed for 72 h in absolute ethanol. After that, discs were taken out, blotted with filter paper to remove excess ethanol attached to the surface of the discs, and weighed again (M_2) [65]. The following formula was employed for the determination of (%) porosity:

$$(\%) \text{ Porosity} = \frac{M_2 - M_1}{\rho V} \times 100 \tag{3}$$

where ϱ indicates the density of absolute ethanol and V represents the swelling volume of hydrogel discs.

4.9. Swelling Study

A swelling study was performed for the developed hydrogels at pH 1.2, 4.6, and 7.4, at 37 °C in order to determine their pH-responsive nature. Therefore, a dried hydrogel disc of known weight was immersed in the respective 100 mL buffer solutions. After a regular interval of time, the disc was removed, blotted with filter paper, weighed again, and immersed back in the respective buffer medium. This process was continued until an equilibrium weight was achieved [66]. This study was performed in a triplicate. Dynamic swelling was estimated by using the following formula:

$$(q) = \frac{N_2}{N_1} \quad (4)$$

where q shows the dynamic swelling, N_1 represents the initial weight of the dried hydrogel disc before swelling, and N_2 indicates the final weight after swelling at time t.

4.10. Polymer Volume Fraction

Polymer volume fraction indicates the fraction of polymer in a completely swelled state and is represented by V2,s. Equilibrium volume swelling (Veq) data of the formulated hydrogels at three different pH values (1.2, 4.6, and 7.4) were employed for the estimation of polymer volume fraction [47]. Therefore, the following formula was used for the determination of polymer volume fraction:

$$V2,s = \frac{1}{V_{eq}} \quad (5)$$

4.11. Drug Loading

The diffusion and absorption method was used for drug loading of the developed hydrogels. Hence, a 1% drug (THP) solution was formed in a phosphate buffer solution of pH 7.4. Weighed dried hydrogel discs were immersed in the drug solution for 72 h. After achieving equilibrium swelling, discs were removed, washed with distilled water to remove any excess entrapped drug on the surface of hydrogel discs, and then placed in a vacuum oven at 40 °C for dryness. The dried hydrogel discs were weighed again. This experiment was performed in a triplicate.

Quantification of the drug loaded by the developed hydrogel was estimated by weight and extraction methods. In weight method, the weight of the unloaded dried hydrogel disc was subtracted from the drug-loaded dried hydrogel disc as indicated in the following formula:

$$\text{Drug loaded quantity} = X_L - X_{UL} \quad (6)$$

where X_L represents the weight of the drug-loaded hydrogel discs and X_{UL} indicates the weight of the unloaded hydrogel discs.

In the extraction method, a weighed dried disc of hydrogel was taken and immersed in a 25 mL phosphate buffer solution of pH 7.4. After a regular interval of time, samples were collected, and the buffer was replaced each time by a fresh medium of the same concentration. The process was continued until the entire drug was eliminated completely from the hydrogel disc. The collected samples were then evaluated on a UV-Vis spectrophotometer (U-5100,3J2-0014, Tokyo, Japan) at λ_{max} 272 nm, and hence drug content was determined [67].

4.12. Drug Release Studies

In vitro drug release studies were carried out to evaluate the drug release from commercially available tablets, Theolin S.R (250 mg, PeiLi Pharmaceutical IND. Co., Ltd.), and developed hydrogels at three various pH values, i.e., pH 1.2., 4.6 and 7.4. Hence, Theolin and loaded dried discs of hydrogel were immersed individually in 900 mL phosphate buffer solutions of pH 1.2, 4.6, and 7.4 in a USP dissolution apparatus type II (USP dissolution (Sr8plus Dissolution Test Station, Hanson Research, Chatsworth, CA, USA)) at 37 ± 0.5 °C and 50 rpm. An aliquot of 5 mL was taken periodically and replenished with fresh medium of the same concentration in order to keep the sink condition constant. The collected samples were then filtered and analyzed on a UV-Vis spectrophotometer (U-5100,3J2-0014, Tokyo, Japan) at λ_{max} 272 nm in a triplicate [36].

4.13. Kinetic Modeling

Various kinetic models such as zero order, first order, Higuchi, and Korsmeyer–Peppas were evaluated by fitting the achieved in vitro drug release data in order to determine the order and release mechanism of the drug from the fabricated hydrogels [68].

4.14. Statistical Analysis

SPSS Statistics software 22.0 (IBM Corp, Armonk, NY, USA) was employed for the statistical analysis. Student's t-test was used for the determination of variations between the tests, which were found statistically significant when the achieved p-value was less than 0.05.

Author Contributions: Conceptualization, P.-C.W.; data curation, M.S. and Q.L.V.; formal analysis, M.S.; funding acquisition, P.-C.W.; investigation M.S.; methodology, P.-C.W.; project administration, P.-C.W.; supervision, P.-C.W.; writing—original draft, M.S.; writing—review and editing, P.-C.W. All authors have read and agreed to the published version of the manuscript.

Funding: This research was funded by the National Science Council of Taiwan (MOST 110-2320-B-037-014-MY2).

Conflicts of Interest: The authors declare no conflict of interest.

References

1. Ogilvie, R.I. Clinical pharmacokinetics of theophylline. *Clin. Pharm.* **1978**, *3*, 267–293. [CrossRef] [PubMed]
2. Barnes, P.J. Theophylline: New perspectives for an old drug. *Am. J. Respir. Crit. Care Med.* **2003**, *167*, 813–818. [CrossRef] [PubMed]
3. Ram, F.S.; Jardin, J.R.; Atallah, A.; Castro, A.A.; Mazzini, R.; Goldstein, R.; Lacasse, Y.; Cendon, S. Efficacy of theophylline in people with stable chronic obstructive pulmonary disease: A systematic review and meta-analysis. *Respir. Med.* **2005**, *99*, 135–144. [CrossRef] [PubMed]
4. Mellstrand, T.; Svedmyr, N.; Fagerstrom, P.O. Absorption of theophylline from conventional and sustained-release tablets. *Eur. J. Respir. Dis. Suppl.* **1980**, *109*, 54–60. [PubMed]
5. Kumar Singh Yadav, H.; Shivakumar, H.G. In Vitro and In Vivo Evaluation of pH-Sensitive Hydrogels of Carboxymethyl Chitosan for Intestinal Delivery of Theophylline. *ISRN Pharm.* **2012**, *2012*, 763127. [CrossRef] [PubMed]
6. Zhang, W.F.; Chen, X.G.; Li, P.W.; He, Q.Z.; Zhou, H.Y. Preparation and characterization of theophylline loaded chitosan/β-cyclodextrin microspheres. *J. Mater. Sci. Mater. Med.* **2008**, *19*, 305–310. [CrossRef]
7. Ahirrao, S.; Gide, P.; Shrivastav, B.; Sharma, P. Extended release of theophylline through sodium alginate hydrogel beads: Effect of glycerol on entrapment efficiency, drug release. *Part. Sci. Technol.* **2014**, *32*, 105–111. [CrossRef]
8. Chai, Q.; Jiao, Y.; Yu, X. Hydrogels for biomedical applications: Their characteristics and the mechanisms behind them. *Gels* **2017**, *3*, 6. [CrossRef]
9. Suhail, M.; Rosenholm, J.M.; Minhas, M.U.; Badshah, S.F.; Naeem, A.; Khan, K.U.; Fahad, M. Nanogels as drug-delivery systems: A comprehensive overview. *Ther. Deliv.* **2019**, *10*, 697–717. [CrossRef]
10. Emam, H.E.; Shaheen, T.I. Design of a dual pH and temperature responsive hydrogel based on esterified cellulose nanocrystals for potential drug release. *Carbohydr. Polym.* **2022**, *278*, 118925. [CrossRef]
11. Emam, H.E.; Mohamed, A.L. Controllable Release of Povidone-Iodine from Networked Pectin@ Carboxymethyl Pullulan Hydrogel. *Polymers* **2021**, *13*, 3118. [CrossRef] [PubMed]
12. Mamidi, N.; Delgadillo, R.M.V. Design, fabrication and drug release potential of dual stimuli-responsive composite hydrogel nanoparticle interfaces. *Coll. Surf. B Biointerfaces* **2021**, *204*, 111819. [CrossRef] [PubMed]
13. Sarfraz, R.M.; Ahmad, M.; Mahmood, A.; Akram, M.R.; Abrar, A. Development of β-cyclodextrin-based hydrogel microparticles for solubility enhancement of rosuvastatin: An in vitro and in vivo evaluation. *Drug Des. Dev. Ther.* **2017**, *11*, 3083. [CrossRef] [PubMed]
14. Ali, L.; Ahmad, M.; Aamir, M.N.; Minhas, M.U.; Shah, H.H.; Shah, M.A. Cross-linked pH-sensitive pectin and acrylic acid based hydrogels for controlled delivery of metformin. *Pak. J. Pharm. Sci.* **2020**, *33*, 4.
15. Kirimura, K.; Sato, T.; Nakanishi, N.; Terada, M.; Usami, S. Breeding of starch-utilizing and itaconic-acid-producing koji molds by interspecific protoplast fusion between Aspergillus terreus and Aspergillus usamii. *Appl. Microbiol. Biotechnol.* **1997**, *47*, 127–131. [CrossRef]
16. Sen, M.; Yakar, A. Controlled release of antifungal drug terbinafine hydrochloride from poly(N-vinyl 2-pyrrolidone/itaconic acid) hydrogels. *Int. J. Pharm.* **2001**, *228*, 33–41. [CrossRef]

17. Tasdelen, B.; Kayaman-Apohan, N.; Guven, O.; Baysal, B.M. Preparation of poly (N-isopropylacrylamide/itaconic acid) copolymeric hydrogels and their drug release behavior. *Int. J. Pharm.* **2004**, *278*, 343–351. [CrossRef] [PubMed]
18. Peppas, N.; Torres-Lugo, M.; Pacheco-Gomez, J.; Foss, A.; Huang, Y.; Ichikawa, H.; Leobandung, W. Intelligent hydrogels and their biotechnological and separation applications. In *Radiation Synthesis of Intelligent Hydrogels and Membranes for Separation Purposes*; IAEA: Vienna, Austria, 2000; pp. 1–14.
19. Khan, M.Z.; Makreski, P.; Murtaza, G. Preparation, optimization, in vitro evaluation and ex vivo permeation studies of finasteride loaded gel formulations prepared by using response surface methodology. *Curr. Drug Deliv.* **2018**, *15*, 1312–1322. [CrossRef]
20. Gatiganti, D.L.; Srimathkandala, M.H.; Ananthula, M.B.; Bakshi, V. Formulation and evaluation of oral natural polysaccharide hydrogel microbeads of Irbesartan. *Anal. Chem. Lett.* **2016**, *6*, 334–344. [CrossRef]
21. Ge, H.C.; Hua, T.T.; Wang, J.C. Preparation and characterization of poly (itaconic acid)-grafted crosslinked chitosan nanoadsorbent for high uptake of Hg2+ and Pb2+. *Int. J. Biol. Macromol.* **2017**, *95*, 954–961. [CrossRef]
22. Betancourt, T.; Pardo, J.; Soo, K.; Peppas, N.A. Characterization of pH-responsive hydrogels of poly(itaconic acid-g-ethylene glycol) prepared by UV-initiated free radical polymerization as biomaterials for oral delivery of bioactive agents. *J. Biomed. Mater. Res. A* **2010**, *93*, 175–188. [CrossRef] [PubMed]
23. Khalid, I.; Ahmad, M.; Usman Minhas, M.; Barkat, K. Synthesis and evaluation of chondroitin sulfate based hydrogels of loxoprofen with adjustable properties as controlled release carriers. *Carbohydr. Polym.* **2018**, *181*, 1169–1179. [CrossRef] [PubMed]
24. Minhas, M.U.; Ahmad, M.; Khan, S.; Ali, L.; Sohail, M. Synthesis and characterization of β-cyclodextrin hydrogels: Crosslinked polymeric network for targeted delivery of 5-fluorouracil. *Drug Deliv.* **2016**, *9*, 10.
25. Coskun, R.; Soykan, C.; Delibas, A. Study of free-radical copolymerization of itaconic acid/2-acrylamido-2-methyl-1-propanesulfonic acid and their metal chelates. *Eur. Polym. J.* **2006**, *42*, 625–637. [CrossRef]
26. Wei, W.; Hu, X.; Qi, X.; Yu, H.; Liu, Y.; Li, J.; Zhang, J.; Dong, W. A novel thermo-responsive hydrogel based on salecan and poly (N-isopropylacrylamide): Synthesis and characterization. *Coll. Surf. B Biointerfaces* **2015**, *125*, 1–11. [CrossRef] [PubMed]
27. Hu, X.; Feng, L.; Wei, W.; Xie, A.; Wang, S.; Zhang, J.; Dong, W. Synthesis and characterization of a novel semi-IPN hydrogel based on Salecan and poly (N, N-dimethylacrylamide-co-2-hydroxyethyl methacrylate). *Carbohydr. Polym.* **2014**, *105*, 135–144. [CrossRef]
28. Ray, M.; Pal, K.; Anis, A.; Banthia, A. Development and characterization of chitosan-based polymeric hydrogel membranes. *Des. Monomers Polym.* **2010**, *13*, 193–206. [CrossRef]
29. Nasir, N.; Ahmad, M.; Minhas, M.U.; Barkat, K.; Khalid, M.F. pH-responsive smart gels of block copolymer [pluronic F127-co-poly (acrylic acid)] for controlled delivery of Ivabradine hydrochloride: Its toxicological evaluation. *J. Polym. Res.* **2019**, *26*, 1–15. [CrossRef]
30. Hu, X.; Wang, Y.; Zhang, L.; Xu, M.; Dong, W.; Zhang, J. Redox/pH dual stimuli-responsive degradable Salecan-g-SS-poly (IA-co-HEMA) hydrogel for release of doxorubicin. *Carbohydr. Polym.* **2017**, *155*, 242–251. [CrossRef]
31. Abdullah, O.; Usman Minhas, M.; Ahmad, M.; Ahmad, S.; Barkat, K.; Ahmad, A. Synthesis, optimization, and evaluation of polyvinyl alcohol-based hydrogels as controlled combinatorial drug delivery system for colon cancer. *Adv. Polym. Technol.* **2018**, *37*, 3348–3363. [CrossRef]
32. Sinha, P.; Ubaidulla, U.; Nayak, A.K. Okra (Hibiscus esculentus) gum-alginate blend mucoadhesive beads for controlled glibenclamide release. *Int. J. Biol. Macromol.* **2015**, *72*, 1069–1075. [CrossRef] [PubMed]
33. Khanum, H.; Ullah, K.; Murtaza, G.; Khan, S.A. Fabrication and in vitro characterization of HPMC-g-poly (AMPS) hydrogels loaded with loxoprofen sodium. *Int. J. Biol. Macromol.* **2018**, *120*, 1624–1631. [CrossRef] [PubMed]
34. Abbadessa, A.; Blokzijl, M.; Mouser, V.; Marica, P.; Malda, J.; Hennink, W.; Vermonden, T. A thermo-responsive and photopolymerizable chondroitin sulfate-based hydrogel for 3D printing applications. *Carbohydr. Polym.* **2016**, *149*, 163–174. [CrossRef] [PubMed]
35. Sarika, P.R.; James, N.R.; Kumar, P.R.A.; Raj, D.K. Preparation, characterization and biological evaluation of curcumin loaded alginate aldehyde-gelatin nanogels. *Mat. Sci. Eng. C Mater.* **2016**, *68*, 251–257. [CrossRef]
36. Hussain, A.; Khalid, S.H.; Qadir, M.I.; Massud, A.; Ali, M.; Khan, I.U.; Saleem, M.; Iqbal, M.S.; Asghar, S.; Gul, H. Water uptake and drug release behaviour of methyl methacrylate-co-itaconic acid [P(MMA/IA)] hydrogels cross-linked with methylene bis-acrylamide. *J. Drug Deliv. Sci. Technol.* **2011**, *21*, 249–255. [CrossRef]
37. Bukhari, S.M.H.; Khan, S.; Rehanullah, M.; Ranjha, N.M. Synthesis and characterization of chemically cross-linked acrylic acid/gelatin hydrogels: Effect of pH and composition on swelling and drug release. *Int. J. Polym. Sci.* **2015**, *2015*, 187961. [CrossRef]
38. Lim, S.L.; Tang, W.N.H.; Ooi, C.W.; Chan, E.S.; Tey, B.T. Rapid swelling and deswelling of semi-interpenetrating network poly (acrylic acid)/poly (aspartic acid) hydrogels prepared by freezing polymerization. *J. Appl. Polym. Sci.* **2016**, *133*, 1–9. [CrossRef]
39. Khan, S.A.; Azam, W.; Ashames, A.; Fahelelbom, K.M.; Ullah, K.; Mannan, A.; Murtaza, G. β-Cyclodextrin-based (IA-co-AMPS) Semi-IPNs as smart biomaterials for oral delivery of hydrophilic drugs: Synthesis, characterization, in-Vitro and in-Vivo evaluation. *J. Drug Deliv. Sci. Technol.* **2020**, *60*, 101970. [CrossRef]
40. Al-Tabakha, M.M.; Khan, S.A.; Ashames, A.; Ullah, H.; Ullah, K.; Murtaza, G.; Hassan, N. Synthesis, Characterization and Safety Evaluation of Sericin-Based Hydrogels for Controlled Delivery of Acyclovir. *Pharmaceuticals* **2021**, *14*, 234. [CrossRef]

41. Sullad, A.G.; Manjeshwar, L.S.; Aminabhavi, T.M. Novel pH-Sensitive Hydrogels Prepared from the Blends of Poly(vinyl alcohol) with Acrylic Acid-graft-Guar Gum Matrixes for Isoniazid Delivery. *Ind. Eng. Chem. Res.* **2010**, *49*, 7323–7329. [CrossRef]
42. Khalid, S.; Qadir, M.; Massud, A.; Ali, M.; Rasool, M. Effect of degree of cross-linking on swelling and drug release behaviour of poly (methyl methacrylate-co-itaconic acid)[P (MMA/IA)] hydrogels for site specific drug delivery. *J. Drug Deliv. Sci. Technol.* **2009**, *19*, 413–418. [CrossRef]
43. Caykara, T.; Turan, E. Effect of the amount and type of the crosslinker on the swelling behavior of temperature-sensitive poly(N-tert-butylacrylamide-co-acrylamide) hydrogels. *Colloid Polym. Sci.* **2006**, *284*, 1038–1048. [CrossRef]
44. Teijon, C.; Olmo, R.; Blanco, M.D.; Teijon, J.M.; Romero, A. Effect of the crosslinking degree and the nickel salt load on the thermal decomposition of poly (2-hydroxyethyl methacrylate) hydrogels and on the metal release from them. *J. Colloid Interface Sci.* **2006**, *295*, 393–400. [CrossRef] [PubMed]
45. Teijon, J.M.; Trigo, R.M.; Garcia, O.; Blanco, M.D. Cytarabine trapping in poly(2-hydroxyethyl methacrylate) hydrogels: Drug delivery studies. *Biomaterials* **1997**, *18*, 383–388. [CrossRef]
46. Vazquez, B.; Gurruchaga, M.; Goni, I. Hydrogels Based on Graft-Copolymerization of Hema Bma Mixtures onto Soluble Gelatin-Swelling Behavior. *Polymer* **1995**, *36*, 2311–2314. [CrossRef]
47. Badshah, S.F.; Akhtar, N.; Minhas, M.U.; Khan, K.U.; Khan, S.; Abdullah, O.; Naeem, A. Porous and highly responsive cross-linked β-cyclodextrin based nanomatrices for improvement in drug dissolution and absorption. *Life Sci.* **2021**, *267*, 118931. [CrossRef]
48. Majeed, A.; Pervaiz, F.; Shoukat, H.; Shabbir, K.; Noreen, S.; Anwar, M. Fabrication and evaluation of pH sensitive chemically cross-linked interpenetrating network [Gelatin/Polyvinylpyrrolidone-co-poly(acrylic acid)] for targeted release of 5-fluorouracil. *Polym. Bull.* **2022**, *79*, 1–20. [CrossRef]
49. Murthy, P.S.K.; Mohan, Y.M.; Sreeramulu, J.; Raju, K.M. Semi-IPNs of starch and poly(acrylamide-co-sodium methacrylate): Preparation, swelling and diffusion characteristics evaluation. *React. Funct. Polym.* **2006**, *66*, 1482–1493. [CrossRef]
50. Sohail, K.; Khan, I.U.; Shahzad, Y.; Hussain, T.; Ranjha, N.M. pH-sensitive polyvinylpyrrolidone-acrylic acid hydrogels: Impact of material parameters on swelling and drug release. *Braz. J. Pharm. Sci.* **2014**, *50*, 173–184. [CrossRef]
51. Bera, R.; Dey, A.; Chakrabarty, D. Synthesis, Characterization, and drug release study of acrylamide-co-itaconic acid based smart hydrogel. *Polym. Eng. Sci.* **2015**, *55*, 113–122. [CrossRef]
52. Sanli, O.; Ay, N.; Isiklan, N. Release characteristics of diclofenac sodium from poly(vinyl alcohol)/sodium alginate and poly(vinyl alcohol)-grafted-poly(acrylamide)/sodium alginate blend beads. *Eur. J. Pharm. BioPharm.* **2007**, *65*, 204–214. [CrossRef] [PubMed]
53. Khalid, Q.; Ahmad, M.; Usman Minhas, M. Hydroxypropyl-β-cyclodextrin hybrid nanogels as nano-drug delivery carriers to enhance the solubility of dexibuprofen: Characterization, in vitro release, and acute oral toxicity studies. *Adv. Polym. Technol.* **2018**, *37*, 2171–2185. [CrossRef]
54. Akhtar, M.F.; Ranjha, N.M.; Hanif, M. Effect of ethylene glycol dimethacrylate on swelling and on metformin hydrochloride release behavior of chemically crosslinked pH-sensitive acrylic acid-polyvinyl alcohol hydrogel. *Daru* **2015**, *23*, 41. [CrossRef] [PubMed]
55. Khan, F.; Razzak, S.M.I.; Khan, Z.R.; Azad, M.A.K.; Chowdhury, J.A.; Reza, S. Theophylline loaded gastroretentive floating tablets based on hydrophilic polymers: Preparation and in vitro evaluation. *Pak. J. Pharm. Sci.* **2009**, *22*, 155–161. [PubMed]
56. Bashir, S.; Teo, Y.Y.; Ramesh, S.; Ramesh, K. Synthesis, characterization, properties of N-succinyl chitosan-g-poly (methacrylic acid) hydrogels and in vitro release of theophylline. *Polymer* **2016**, *92*, 36–49. [CrossRef]
57. Liu, M.; Wang, X.; Wang, Y.; Jiang, Z. Controlled stimulation-burst targeted release by pH-sensitive HPMCAS/theophylline composite nanofibers fabricated through electrospinning. *J. Appl. Polym. Sci.* **2020**, *137*, 48383. [CrossRef]
58. Shoaib, M.H.; Tazeen, J.; Merchant, M.A.; Yousuf, R.I. Evaluation of drug release kinetics from ibuprofen matrix tablets using HPMC. *Pak. J. Pharm. Sci.* **2006**, *19*, 119–124.
59. Maziad, N.A.; EL-Hamouly, S.; Zied, E.; EL Kelani, T.A.; Nasef, N.R. Radiation preparation of smart hydrogel has antimicrobial properties for controlled release of ciprofloxacin in drug delivery systems. *Drug Deliv.* **2015**, *14*, 15.
60. Sohail, M.; Ahmad, M.; Minhas, M.U.; Ali, L.; Khalid, I.; Rashid, H. Controlled delivery of valsartan by cross-linked polymeric matrices: Synthesis, in vitro and in vivo evaluation. *Int. J. Pharm.* **2015**, *487*, 110–119. [CrossRef]
61. Ullah, K.; Sohail, M.; Buabeid, M.A.; Murtaza, G.; Ullah, A.; Rashid, H.; Khan, M.A.; Khan, S.A. Pectin-based (LA-co-MAA) semi-IPNS as a potential biomaterial for colonic delivery of oxaliplatin. *Int. J. Pharm.* **2019**, *569*, 118557. [CrossRef]
62. Ullah, K.; Sohail, M.; Mannan, A.; Rashid, H.; Shah, A.; Murtaza, G.; Khan, S.A. Facile synthesis of chitosan based-(AMPS-co-AA) semi-IPNs as a potential drug carrier: Enzymatic degradation, cytotoxicity, and preliminary safety evaluation. *Curr. Drug Deliv.* **2019**, *16*, 242–253. [CrossRef] [PubMed]
63. Sarfraz, R.; Khan, H.; Mahmood, A.; Ahmad, M.; Maheen, S.; Sher, M. Formulation and evaluation of mouth disintegrating tablets of atenolol and atorvastatin. *Indian J. Pharm. Sci.* **2015**, *77*, 83. [CrossRef] [PubMed]
64. Ullah, K.; Khan, S.A.; Murtaza, G.; Sohail, M.; Manan, A.; Afzal, A. Gelatin-based hydrogels as potential biomaterials for colonic delivery of oxaliplatin. *Int. J. Pharm.* **2019**, *556*, 236–245. [CrossRef] [PubMed]
65. Zia, M.A.; Sohail, M.; Minhas, M.U.; Sarfraz, R.M.; Khan, S.; de Matas, M.; Hussain, Z.; Abbasi, M.; Shah, S.A.; Kousar, M. HEMA based pH-sensitive semi IPN microgels for oral delivery; a rationale approach for ketoprofen. *Drug Dev. Ind. Pharm.* **2020**, *46*, 272–282. [CrossRef] [PubMed]

66. Ijaz, H.; Tulain, U.R.; Azam, F.; Qureshi, J. Thiolation of arabinoxylan and its application in the fabrication of pH-sensitive thiolated arabinoxylan grafted acrylic acid copolymer. *Drug Dev. Ind. Pharm.* **2019**, *45*, 754–766. [CrossRef] [PubMed]
67. Khan, S.; Ranjha, N.M. Effect of degree of cross-linking on swelling and on drug release of low viscous chitosan/poly (vinyl alcohol) hydrogels. *Polym. Bull.* **2014**, *71*, 2133–2158. [CrossRef]
68. Peppas, N.A.; Sahlin, J.J. A simple equation for the description of solute release. III. Coupling of diffusion and relaxation. *Int. J. Pharm.* **1989**, *57*, 169–172. [CrossRef]

Article

Preliminary Investigation of *Linum usitatissimum* Mucilage-Based Hydrogel as Possible Substitute to Synthetic Polymer-Based Hydrogels for Sustained Release Oral Drug Delivery

Arshad Mahmood [1,2], Alia Erum [3,*], Sophia Mumtaz [3], Ume Ruqia Tulain [3], Nadia Shamshad Malik [4] and Mohammed S. Alqahtani [5]

1. College of Pharmacy, Al Ain University, Abu Dhabi campus, Abu Dhabi 51133, United Arab Emirates; arshad.mahmood@aau.ac.ae
2. AAU Health and Biomedical Research Center, Al Ain University, Abu Dhabi 51133, United Arab Emirates
3. Faculty of Pharmacy, College of Pharmacy, University of Sargodha, Sargodha 40100, Pakistan; dr.sophia786@gmail.com (S.M.); umeruqia_tulain@yahoo.com (U.R.T.)
4. Faculty of Pharmacy, Capital University of Science and Technology, Islamabad 44000, Pakistan; nadia.malik@cust.edu.pk
5. Nanobiotechnology Unit, Department of Pharmaceutics, College of Pharmacy, King Saud University, Riyadh 11362, Saudi Arabia; msaalqahtani@ksu.edu.sa
* Correspondence: alia.erum@uos.edu.pk

Abstract: The aim of this study was to investigate the potential of *Linum usitatissimum* mucilage, a natural polymer, in developing a sustained release hydrogel for orally delivered drugs that require frequent dosing. For this purpose, nicorandil (a model drug)-loaded hydrogels with various feed ratios of *Linum usitatissimum* mucilage, acrylamide (monomer) and methylene bis-acrylamide (crosslinker) were prepared. The newly synthesized hydrogel formulations were probed fundamentally with respect to swelling behaviour, solvent penetration, and the release of the drug from the hydrogels. Later, the selected formulations were further characterized by Fourier-transform infrared spectroscopy, thermal analysis, X-ray diffraction analysis, and scanning electron microscopy. The swelling coefficient demonstrated a linear relation with the polymer ratio; however, an inverse behaviour in the case of monomer and crosslinker was observed. The drug release studies, performed at pH 1.2 and 4.5 and considering the dynamic environment of GIT, demonstrated that all formulations followed the Korsmeyer–Peppas model, displaying a slow drug release via diffusion and polymer erosion. FTIR analysis confirmed the successful grafting of acrylamide on linseed mucilage. Furthermore, scanning electron microscopy revealed a clear surface morphology with folds and pinholes in the hydrogel. Therefore, based upon the in-vitro outcomes, it can be concluded that a promising sustained release hydrogel can be prepared from natural polymer, *Linum usitatissimum* mucilage, offering many-fold benefits over the conventional synthetic polymers for oral delivery of drugs.

Keywords: *Linum usitatissimum* mucilage; hydrogel; nicorandil; copolymer; acrylamide

1. Introduction

The oral route of drug delivery is the most convenient way of administration but, at the same time, is associated with fluctuations in plasma drug concentrations that can be critical in certain diseases, for instance, cardiovascular problems [1]. In order to maintain a relatively uniform level of plasma concentrations, especially for the drugs with shorter half-life, and to avoid the problems of frequent administration, sustained release delivery systems have been the cream of the crop. Based on one mechanism or the other, a number of sustained release drug delivery systems have been established, including liposomes, microspheres, nano-emulsions and hydrogels, to name a few [2].

Hydrogels stand at the middle of the podium when it comes specifically to sustaining release delivery because of their ability to provide spatial and temporal control over the release of drugs. The development of tuneable hydrogels such as in-situ forming, stimuli (pH, temperature and enzymes) responsive [3–5], fatigue resistant [6,7], mechanical tune-ability with nano-particulate crosslinkers [8] and the degradation controlled hydrogel matrices are mainly based upon their physicochemical properties [9]. Over the past few decades, studies on novel drug delivery systems in general, and hydrogels in particular, have been focused on the so-called biocompatible synthetic polymers, which have dominated the era because of their consistency and considerable purity. However, when it comes to the biocompatibility, the biodegradability, the safety, the complications of the synthetic processes and, ultimately, the cost of production, there is no match to natural polymers. It has, further, been reported in the literature that natural polymers extemporized formulations via grafting and that improvised graft co-polymerization results in superabsorbent hydrogels with a swelling rate that ranges from a fraction of a minute to hours [10,11]. Considering the mentioned advantages, more and more natural polymers are being investigated in developing novel drug delivery systems under the umbrella of 'naturapolyceutics' for the delivery of all sorts of drugs.

Among the natural polymers, *Linum usitatissimum* (linseed) mucilage (polysaccharide) possesses the thickening, swelling and adhesive properties that render their potential for several applications in pharmaceutical preparations [12–14]. In the past decade or so, its initial utilization was reported as a gelling adhesive agent in combination with other polymers for buccal delivery [15,16] and colon targeting [17]. The swelling properties of the mucilage led to an investigation of their hydro-gelling potential; however, because of the lower mechanical strength, they were being used as an admixture with other polymers [18,19]. In our opinion, the graft copolymerization technique can help in fabricating harder and denser networks with enhanced mechanical strength of hydrogel with linseed mucilage.

Therefore, the aim of the study was to develop graft copolymeric hydrogel by free radical polymerization, using *Linum usitatissimum* (linseed) mucilage as a polymer with acrylamide monomer and methylene bis-acrylamide as a crosslinker for the gel frame. The composition based upon its intended characteristics was evaluated for influence on swelling, penetration of solvent and release of nicorandil, a model drug that has a very short half-life and that is an ideal candidate for the sustained release delivery.

2. Results & Discussion

2.1. Characterization of Linseed-Co-AAM Graft Copolymeric Hydrogels

The mucilage contains water-soluble non-starch polysaccharides, rhamnogalacturonan and arabinoxylans are the main polysaccharides and these components of linseed, having hydroxyl groups attached with its backbone. The mucilage has an insufficient amount of hydrolytic stability that can be improved by introducing acrylamides with groups such as alkyl and hydroxyl alkyl. The proposed chemical structure of graft copolymeric hydrogel is illustrated in Figure 1.

2.1.1. Swelling Studies

The swelling of naturally occurring polysaccharides depends upon the presence of hydrophilic/hydrophobic groups, degree of crosslinking, and elasticity of network [20]. Considering the physiochemical properties of the model drug, nicorandilnicorandil (weakly acidic drug with a pKa of 3.12), the swelling studies were performed at a pH of 1.2 and 4.5, which mimic the acidic environment of the gastrointestinal track, as per USP.

Effect of Varying Concentration of Monomer on Swelling Behaviour of Hydrogel

Comparative swelling ratios of different formulations of linseed-co-AAM hydrogel with varying monomer concentrations are shown in Figure 2. A slight decrease in swelling ratio was seen with increasing AAM concentration, as shown by swelling ratios 12.33 and 10.34 in the case of pH 1.2 and 11.82 and 10.56 in the case of pH 4.5 for F1 and F2, respectively.

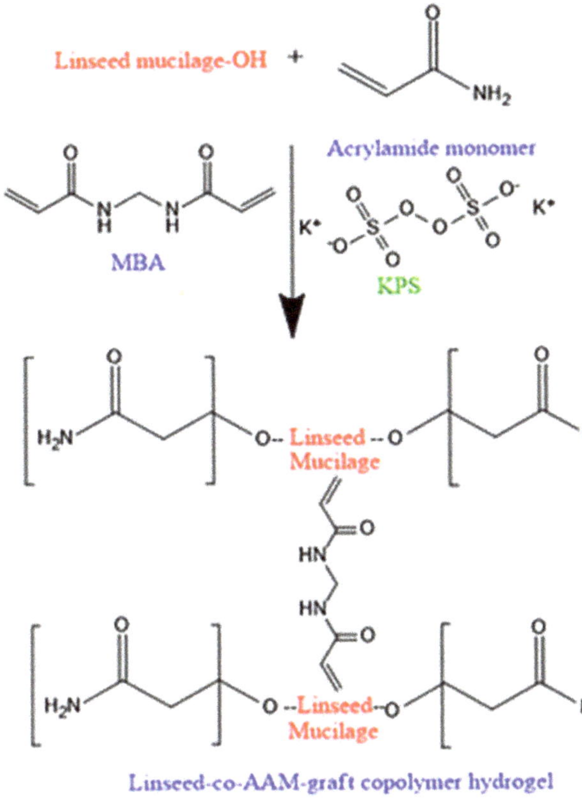

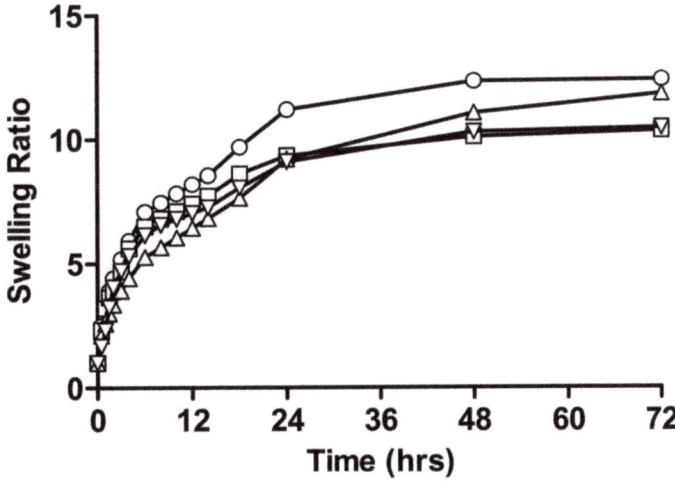

Figure 1. Schematic diagram for synthesis of linseed-co-AAM graft copolymeric hydrogels.

Figure 2. Relative swelling ratios of linseed-co-AAM graft copolymer hydrogel formulations with varying monomer concentration; (○) F1 at pH 1.2, (□) F2 at pH 1.2, (△) F1 at pH 4.5 and (▽) F2 at pH 4.5.

According to the literature, hydrogels formulated with lower total monomer concentration revealed a higher swelling, and the equilibrium mass swelling of the hydrogels decreased with increasing total monomer concentration. For the constant gel volume, the decrease in the total monomer concentration results in an increase in the degree of dilution of the matrix in the constant gel, which also results in an increase in the equilibrium water content of the gel [21–23].

Effect of Varying Concentration of Linseed Mucilage on Swelling Behaviour of Hydrogel

Comparative swelling ratios of different formulations of linseed-co-AAM hydrogel with varying mucilage concentration at pH 1.2 and 4.5 are specified in Figure 3. The swelling ratio increases with an increase in linseed mucilage concentration for F3, F4, and F5 as shown by swelling ratios 14.39, 16.92, 19.66, at pH 1.2 and 14.74, 17.54, and 20.27 at pH 4.5, respectively.

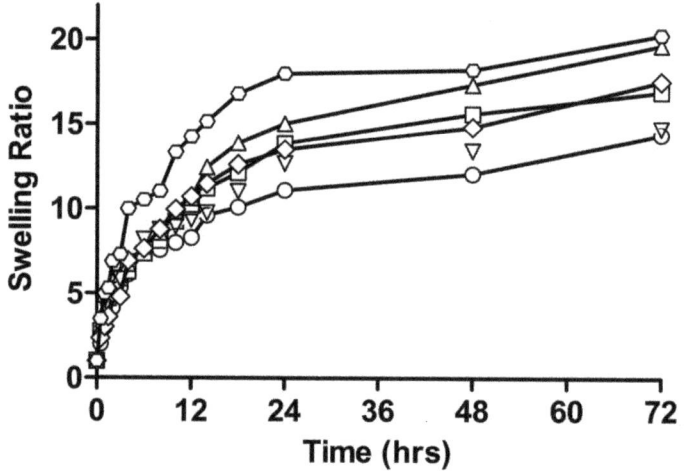

Figure 3. Relative swelling ratios of linseed-co-AAM graft copolymer hydrogels with varying linseed mucilage concentration; (○) F3 at pH 1.2, (□) F4 at pH 1.2, (△) F5 at pH 1.2, (▽) F3 at pH 4.5, (◇) F4 at pH 4.5 and (○) F5 at pH 4.5.

The swelling ratio of the hydrogels depends on the chemical structure of mucilage. Hydrogels that have hydrophilic groups swell greater then hydrogels containing hydrophobic groups because these groups break down in the presence of water, thus, minimizing their interaction with the water molecule [24,25]. The swelling of the polymer also depends on the number of hydrophilic groups on the polymeric network and the crosslinking density of the fine structure of the polymer. As the concentration of mucilage increased, the concentration of hydrophilic group increased having less crosslinking density which results in increased swelling of linseed mucilage [15,26].

Effect of Varying Concentration of Crosslinker

Comparative swelling ratios of different formulations of hydrogels with varying crosslinker concentrations are shown in Figure 4. Results showed that, as the concentration of crosslinker increased, a decrease in swelling ratio was observed, which was 10.09, 9.16 at pH 1.2 and 11.59, 10.22 swelling ratios at pH 4.5. The consequence of a high concentration of crosslinker (N,N-MBA) in formulation having a higher crosslink density profile, thus, decreasing water absorbency of hydrogel [10]. It was estimated that the concentration of MBA had a substantial effect on the permeability, absorbency, and swelling features of hydrogels. The swelling was decreased by increasing concentration of MBA in acidic and buffer medium due to tighter structure and hindered mobility. Mechanical strength was improved, but the porosity of hydrogels was decreased; due to this, rate of drug release by diffusion was also decreased [27].

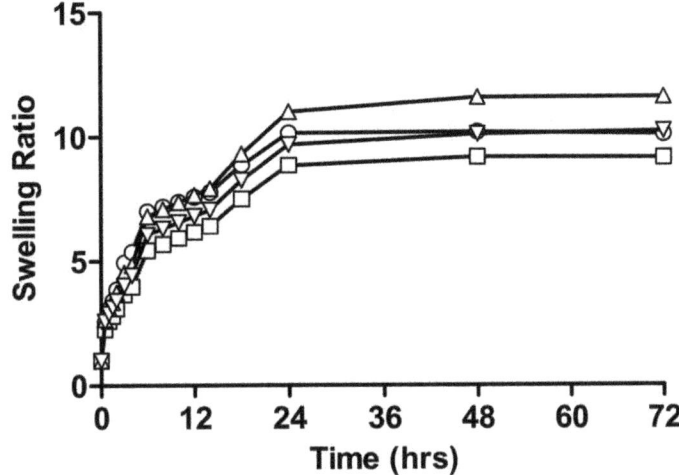

Figure 4. Relative swelling ratios of linseed-co-AAM graft copolymer hydrogels with varying crosslinker concentration; (○) F6 at pH 1.2, (□) F7 at pH 1.2, (△) F6 at pH 4.5 and (▽) F7 at pH 4.5.

The literature indicates that power-law performances of swelling that increase the crosslinker lower the water absorbency. Higher crosslinking results in an increased degree of crosslinking in the polymer network, which results in low swelling [28,29].

2.1.2. Percent Equilibrium Swelling (%ES) of All Formulations of Linseed-Co-AAM

The swelling index was carried out at pH 1.2 and 4.5 for all the above-mentioned formulations. Lower percentage of equilibrium swelling was obtained by increasing the crosslinker and monomer content in the hydrogel. With an increase in mucilage concentration, percentage of equilibrium swelling was increased [30].

Percentage of equilibrium swelling (% ES) of all formulations of hydrogels with varying mucilage, crosslinker and monomer concentrations at pH 1.2 and 7.4 are specified in Table 1.

Table 1. Percentage of equilibrium swelling of AAM-based hydrogels at pH 1.2 and 4.5.

Hydrogel Code	% Equilibrium Swelling (% ES)	
	pH 1.2	pH 4.5
F1	91.89	91.54
F2	90.26	90.53
F3	93.50	93.21
F4	93.72	93.26
F5	94.29	94.88
F6	90.09	91.37
F7	89.08	90.22

2.2. Determination of Drug Loading

Quantity of drug integrated in different formulations of linseed-co-AAM is given in Table 2. Results indicate that, as the concentration of crosslinker increased, drug loading decreased from 40 to 30 mg, while, in the case of increased linseed mucilage concentration, drug loading increased from 58 to 66 mg. With an increase in monomer concentration, drug loading also decreased from 69 to 65 mg. The literature also indicates that, as the concentration of crosslinker was increased, crosslinking density was high as a result of decreased drug loading [27]. By the increase in monomer concentration, drug loading was decreased because dilution of the polymeric network increased in the AAM-based hydrogels [23]. By increasing linseed mucilage concentration, low crosslinking density of the polymer would occur due to an increase in hydrophilic groups of the polymer, which results in increased drug loading.

Table 2. Loaded drug in different formulations of linseed-co-AAM.

Hydrogel Code	Nicorandil-Loaded mg $\pm$ S.E.M
F1	69 $\pm$ 1.4
F2	65 $\pm$ 1.7
F3	58 $\pm$ 1.1
F4	60 $\pm$ 1.3
F5	66 $\pm$ 1.5
F6	40 $\pm$ 2.1
F7	30 $\pm$ 1.9

2.3. Instrumental Analysis

2.3.1. Fourier Transforms Infrared (FTIR) Analysis

FTIR analysis was done to confirm AAM grafting on linseed mucilage. Secondly, to find out the loaded drug in the final formulation, discs revealed neither degradation nor reaction with the formulation. FTIR is based on the opinion that the basic components of a constituent that has chemical bonds usually show excitation and absorb infrared light at a specific peak frequency [31,32].

In regards to linseed mucilage FTIR, in Figure 5, the spectra showed 3433 cm^{-1} (OH stretching), 2870 cm^{-1} (aliphatic CH stretching) and 1722 cm^{-1} (carboxylic acid C=O stretching), which was probably due to acidic fractions present in linseed mucilage [33]. In the same overlay in the AAM spectrum, a sharp absorption peak at 1666.50 cm^{-1} indicates carbonyl C=O stretching [34]. In the IR spectrum of extra pure AAM, broadband at the region between 3300–3000 cm^{-1} indicates –NH stretching of the AAM unit. Carbonyl moiety of the AAM unit is found at the peak at 1615 cm^{-1} [35], while the spectrum of linseed-co-AAM with a clear and a sharp peak at 2938 cm^{-1} (C-H stretching) and a quite low-intensity peak at 3518 cm^{-1} (C-O stretching) is at 1042. The N-H intense peak at 1605 cm^{-1}, the presence of NH and C=O vibrations confirm the presence of grafting of AAM with linseed mucilage [36].

The FT-IR spectrum of nicorandil is shown in Figure 5 and the following characteristic peaks were observed: the peak at 3235 cm^{-1} represents NH bending, the absorption band at 1663 cm^{-1} for (C=O, CONH) bending and peak at 1391 cm^{-1} for CH2 and 1597 cm^{-1} for the Pyridinium ring [37].

In nicorandil, the loaded linseed-co-AAM spectra absorption peaks at 1661 cm^{-1} (aromatic C=C bending) and 1389 cm^{-1} (aromatic C-H bending) confirmed that the drug did not show any chemical interaction with the hydrogel preparation. By recording the FTIR, the spectra of the drug confirmed that the model drug was evenly distributed within hydrogels. The FTIR spectra of the crosslinker, the monomer, and the polymer were also recorded to determine the uniform distribution of all these ingredients in linseed-based hydrogel [38].

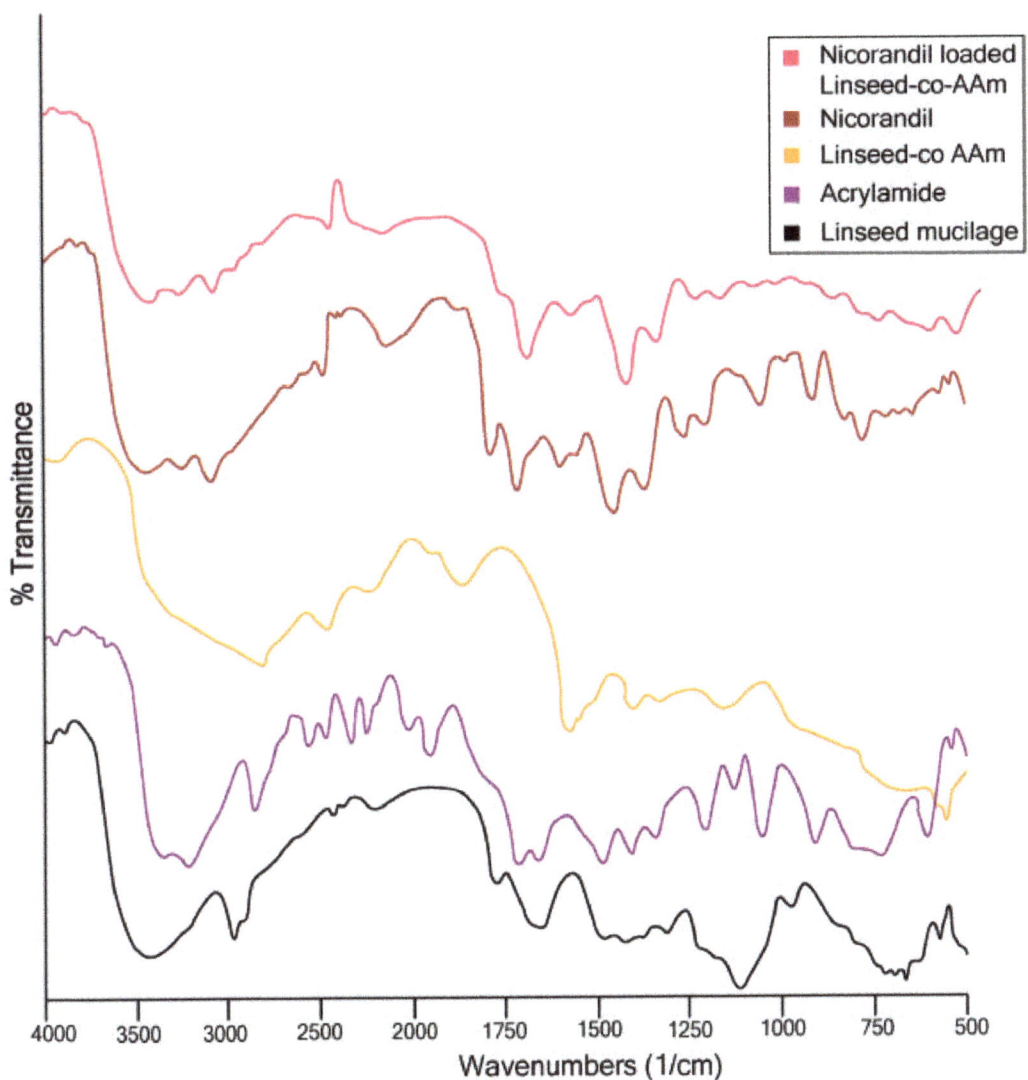

Figure 5. FTIR spectrum of linseed mucilage, AAM, linseed-co-AAM, Nicorandil and Nicorandil loaded Linseed-co-AAM formulation.

2.3.2. Scanning Electron Microscopy (SEM)

For evaluating morphological features of prepared hydrogels, scanning electron microscopy was performed on grafted linseed-co-AAM hydrogels. Clear surface morphology was detected with folds and pinholes. As the concentration of AAM increased, folds and holes were denser [35,36]. In formulations F1, F3 and F7, as the concentration of AAM increased, folds and holes were more compact, as shown in Figure 6.

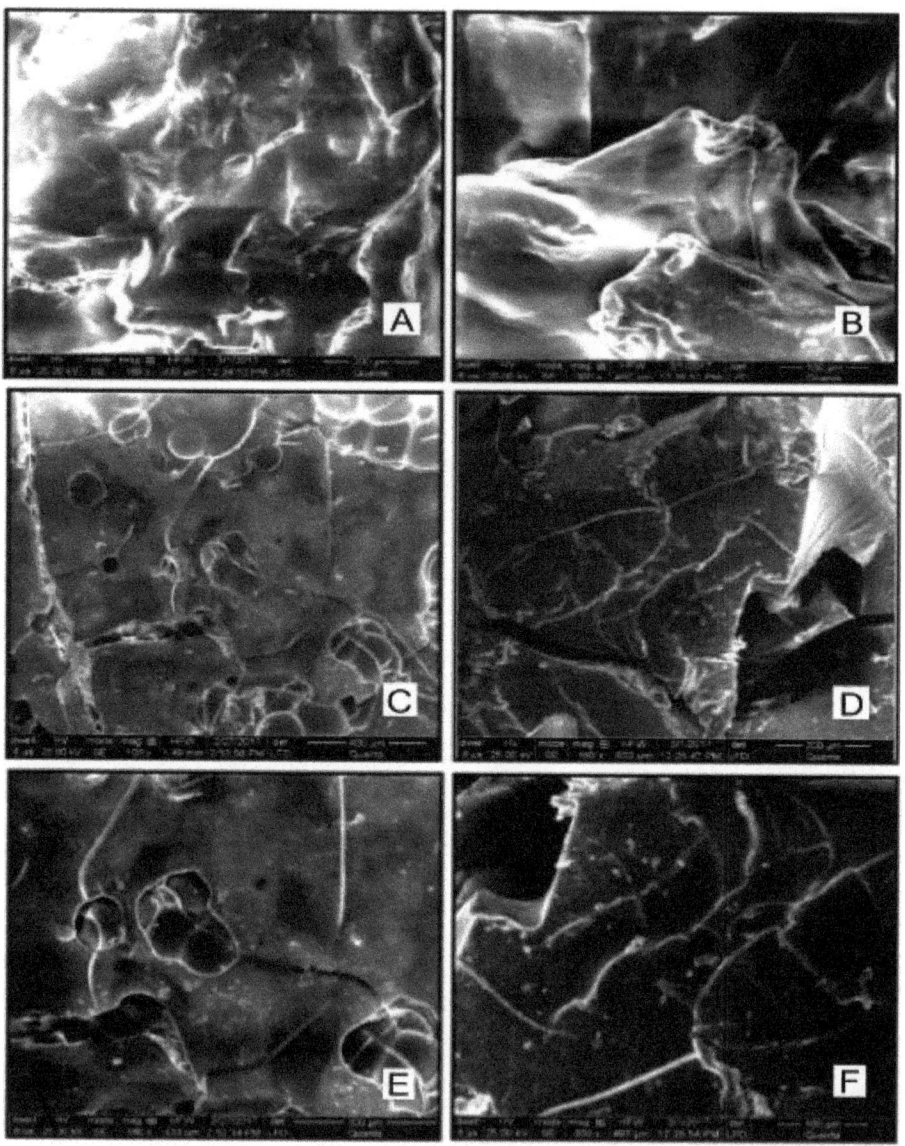

Figure 6. SEM image of F1 linseed-co-AAM formulation hydrogel with 100 μm (**A**) and 200 μm (**B**), F3 hydrogel with 300 μm (**C**) and 400 μm (**D**), F7 hydrogel with 100 μm (**E**) and 200 μm (**F**) scale.

2.4. In-Vitro Drug Release Measurement

2.4.1. In-Vitro Drug Release of Linseed-Co-AAM Formulations

The in-vitro drug release of linseed-co-AAM hydrogels with a varying monomer, linseed mucilage and crosslinker ratio is given in Figures 7–9. Drug release from a hydrogel depends on the swelling, the interaction between drug and polymer behaviour, and the solubility of the drug [39]. Formulations with varying AAM concentrations showed an increase in drug release with decreasing AAM content. As for F1 and F2, the percentage of drug release after 12 h was 80.1% and 90.3%, respectively [39].

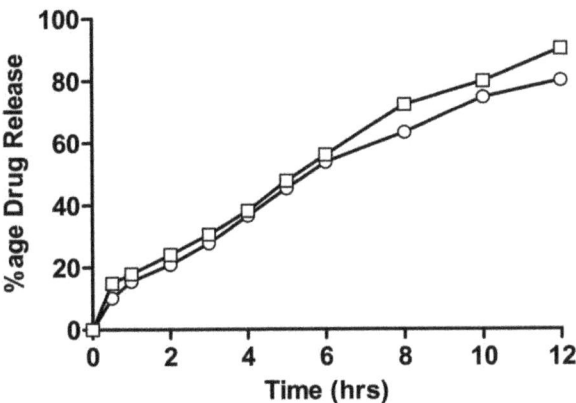

Figure 7. Percentage release of nicorandil from formulations having varying monomer concentrations; (○) F1 and (□) F2.

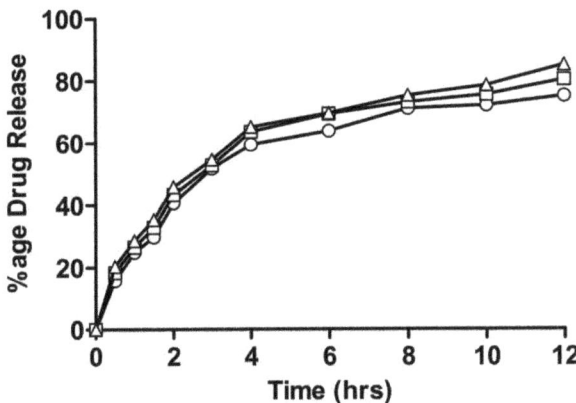

Figure 8. Percentage release of nicorandil from formulations having varying linseed mucilage concentration; (○) F1, (□) F2 and (△) F3.

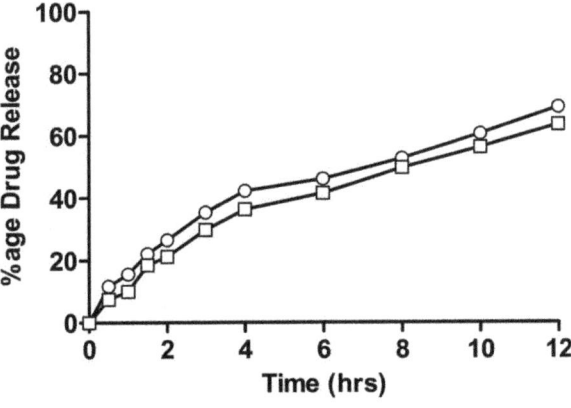

Figure 9. Percentage release of nicorandil with varying crosslinker concentration; (○) F6 and (□) F7.

Formulations with varying contents of linseed mucilage (F3, F4 and F5) show higher release of nicorandil, i.e., 75.1, 80.4 and 85.1, respectively, as shown in Figure 8. This significant difference in drug release from both polymeric networks is due to higher swelling ratios of F5, as drug release is directly proportional to the swelling of the polymeric network. When water penetrates the gel network, swelling occurs and drug dissolution starts inside the network, followed by its gradual diffusion out of the polymeric matrix. Linseed mucilage is a hydrophilic polymer. Hydrogel porosity was also enhanced by high polymer concentration, which is attributed to the reduced crosslinking density that resulted in improved swelling [40,41].

As for F6 and F7, the percentage of nicorandil release after 12 h was 69.1% and 63.5%, respectively. Formulations with varying crosslinker concentrations showed a decrease in drug release with increasing crosslinker content because swelling and the interaction of the drug molecule with the physiological medium decreased, so drug release also decreased.

2.4.2. Evaluation of Drug Release Kinetics

By using DD Solver software assessment of the drug release kinetics of numerous kinetic models, zero order, first order, Higuchi, Hixson–Crowell, and Korsmeyer–Peppas models were applied. The effects of different concentrations of crosslinker, monomer, and linseed mucilage concentrations on the kinetics of drug release from all linseed-co-AAM hydrogel formulations can be observed by the correlation coefficient (R^2) values and the release rates mentioned in Table 3.

Table 3. Impact of various constituents on drug release from linseed-co-AAM graft copolymer.

Hydrogel Code	Zero Order Model		First Order Model		Higuchi Model		Korsmeyer–Peppas Model			Hixson–Crowell Model	
	R^2	K_O	R^2	K_1	R^2	kH	R^2	kKP	N	R^2	kHC
F1	0.825	4.310	0.968	0.087	0.988	17.894	0.991	15.913	0.54	0.954	0.024
F2	0.824	3.928	0.976	0.074	0.992	16.329	0.995	14.541	0.54	0.952	0.020
F3	0.377	4.171	0.848	0.100	0.924	17.988	0.979	26.002	0.45	0.752	0.027
F4	0.284	4.288	0.828	0.111	0.898	18.578	0.975	28.427	0.45	0.724	0.030
F5	0.287	4.477	0.849	0.123	0.906	19.374	0.986	29.821	0.45	0.757	0.033
F6	0.707	3.385	0.899	0.058	0.994	14.243	0.996	15.641	0.46	0.852	0.016
F7	0.752	3.174	0.916	0.052	0.986	13.307	0.986	13.318	0.50	0.875	0.015

From the values of R^2 of the line graph, it can be seen that the Korsmeyer–Peppas model best fits all formulations. This model is applied to determine the release mechanism of drugs, i.e., Fickian diffusion or non-Fickian diffusion. If the value of (n) for cylindrical hydrogel discs equals 0.45, it corresponds to the Fickian diffusion, whereas, when the (n) value is between 0.45 and 1.0, it represents the non-Fickian diffusion. Values of (n) for F6, F7, F1 & F2 preparations in this study were less than 1 and greater than 0.45, showing non-Fickian diffusion, while F3, F4 & F5 shows Fickian diffusion, as the (n) value is equal to 0.45 [42].

Accordingly, formulation F5 exhibited comparatively higher drug release due to greater swelling ratios. A decrease in the crosslinking density provides a porous network that aids the influx of water, followed by swelling. Drug release characteristics were found to be directly proportional to the swelling capacity and inversely proportional to crosslinker and monomer concentration.

3. Conclusions

Within the current study, hydrogels based on *Linum ussitatissimum* mucilage were successfully designed, with the ability to demonstrate a sustaining mechanism for drug delivery. Through a thorough analysis, it was observed that an increase in the monomer and crosslinker concentration resulted in a decrease in the swelling, as well as the rate and extent of drug release. However, increasing the mucilage concentration had the opposite impact, i.e., swelling and an increase in the rate and extent of drug release. Moreover, linseed-Co-AAM hydrogel depicts pH-independent swelling and drug release behavior. With particular emphasis on the sustained release delivery systems, formulations F2 and F5 were considered to be superior to other formulations; a deeper analysis involving in-vivo investigation is highly recommended.

4. Materials and Methods

4.1. Materials

Seeds of *Linum usitatissimum* were purchased from the local food market of Sargodha, Pakistan. Acrylamide, potassium persulfate, sodium hydroxide, and potassium dihydrogen phosphate were from Sigma Aldrich, Germany. N,N-Methylene bis-acrylamide was obtained from Fluka, Switzerland. Hydrochloric acid, absolute ethanol, and n-hexane were gained by Riedel-de Haen, Germany. Nicorandil was obtained as a gift sample from GETZ Pharma Pakistan (Pvt) Ltd., Karachi, Pakistan.

4.2. Methods

4.2.1. Linseed Mucilage Extraction

Following the careful screening process, 200 g of *Linum usitatissimum* seeds were soaked in 600 mL of purified water with mild stirring at 25 °C. After a period of 24 h, the soaked seeds were incubated in a dry heat oven at 80 °C for 30 min. Mucilage was extracted via vacuum filtration and was washed with n-hexane using a cotton cloth, whereby n-hexane passed out very easily along with impurities and the filtrate was obtained in the form of mucilage. The purified mucilage was transferred to petri dishes and dried in an oven at 60 °C; the obtained scales were ground until a fine powder was obtained [14].

4.2.2. Preparation of Linseed-Based Hydrogels

As shown in Table 4, seven different formulations were articulated via a free radical polymerization method, with varying concentrations of polymer (linseed), crosslinker (methylene bis-AAM), monomer (AAM), and potassium persulfate as initiator [28,43].

Table 4. Composition of numerous formulations of linseed-co-AAM graft copolymer.

Formulation Code	Linseed Mucilage (g/100g)	Acrylamide (g/100g)	Initiator (g/100g)	Crosslinker (g/100g)
F1	1.0	12.5	0.2	0.2
F2	1.0	17.5	0.2	0.2
F3	1.0	15	0.2	0.2
F4	1.5	15	0.2	0.2
F5	2.0	15	0.2	0.2
F6	1.0	15	0.2	0.3
F7	1.0	15	0.2	0.4

Briefly, linseed mucilage powder was dissolved in distilled water with continuous stirring at 70 °C. Once dissolved, potassium persulfate solution (in water) was added to it and stirring was continued for another 10 min. Afterwards, crosslinker (N, N-Methylene bis-AAM) solution and monomer were added to polymer-initiator solution at room temperature; the final weight was adjusted using distilled water. The above mixture was incubated in a water bath at 50 °C for an hour, and then temperature was gradually increased up to 80 °C until transparent hydrogels were formed. Cylindrical hydrogels were cut into 0.5 cm diameter discs, washed in an ethanol-water solution to remove the unreacted contents and dried in an oven at 50 °C for 24 h. Dried discs were stored in airtight containers for further characterization of hydrogels [43].

4.2.3. Characterization of Linseed-Co-AAM Graft Copolymeric Hydrogels

Swelling Studies

Swelling studies were performed using the primly weighed dry hydrogel discs. The dry hydrogel discs were immersed in medium (having pH 1.2 and 4.5) and were allowed to swell until swelling equilibrium. At predetermined time intervals, swollen hydrogels were taken out of the medium; weight was noted after removing the excess water with filter paper that, ultimately, was placed back in same media. The following equation was used to calculate swelling behaviour:

$$\text{Swelling} = \frac{Ws - Wd}{Wd} \quad (1)$$

where Ws = weight of hydrogel in swollen form, W_d = weight of hydrogel in dry form [20,36,44].

Percentage of Equilibrium Swelling/Equilibrium Water Content

The swelling was continued until each gel achieved a constant weight. Percentage of equilibrium swelling (%ES) or equilibrium water content (EWC) was determined by the following equation:

$$\% \text{ES} = \frac{Meq - Wd}{Meq} \times 100 \quad (2)$$

where Meq is the weight of swollen gel at equilibrium and W_d is the weight of dried gel discs [45].

Drug Loading

Nicorandil was used as a model drug for the development of sustained-release formulations for the treatment of hypertension [46]. Drug loading into the hydrogel disc was performed by using the adsorption method. A 1% w/v drug solution was prepared in phosphate buffer having pH 4.5. One disc of each formulation was dipped in 100 mL of 1% drug solution until swelling equilibrium. Discs were removed from the solution and washed out with distilled water to remove an excess of the drug. After, they were allowed to air dry at room temperature first and then oven-dried at 40 °C. The amount of drug loaded in the discs was determined by the following formula given in equation 3.

$$\text{Total drug loaded} = W_L - W_U \quad (3)$$

where W_L = weight of dried drug loaded disc, W_U = weight of dried unloaded disc [46,47].

4.2.4. Instrumental Analysis

Fourier Transform Infrared (FTIR) Analysis

The Fourier transform infrared (FTIR) spectra were recorded on an FTIR (prestige-21 Shimadzu) spectrometer for the pure model drug, unloaded hydrogel and drug-loaded hydrogel in order to identify formation of any new bond [48]. For this purpose, samples of the hydrogels and drug were mixed with KBr solution, dried, crushed and kept under hydraulic pressure (150 kg/cm^2) to make the disc; spectra were recorded at a wavelength of 4000–500 cm^{-1}.

Scanning Electron Microscopy (SEM) Analysis

SEM is a widely applied technique for the evaluation of shape and surface morphology of hydrogels. Dried hydrogel discs were cut to specific sizes and were fixed on an aluminium stub. Hydrogels were freeze-dried and then coated with gold in a high vacuum evaporator. The coated samples were scanned and examined under an electron microscope to expose surface morphology [49,50].

4.2.5. In-Vitro Drug Release Study

In-vitro drug release studies were performed using USP-dissolution apparatus II at 37 ± 0.5 °C to evaluate the release behavior of all hydrogel formulations. Every disc was placed in dissolution medium, maintained at a temperature of 37 °C and stirred at a rate of 50 rpm to maintain a uniform drug concentration in the medium. An aliquot of 5 mL was withdrawn at specified time points, i.e., 0.5, 1, 2, 3, 4, 5, 6, 8, 10 and 12 h and absorbance of nicorandil was measured at a wavelength of 262 nm. In order to keep the dissolution medium volume constant, samples were replaced with an equal volume of fresh buffer maintained at 37 ± 0.5 °C. Standard calibration curves of nicorandil were obtained and absorbance was taken at 262 nm. Release kinetics of nicorandil from hydrogels was evaluated by dissolution data modeling by using DD Solver software [51].

Percentage of drug release in hydrogels was determined by using the following equation:

$$\text{In vitro percentage drug release} = F_t / F_{load} \times 100 \quad (4)$$

where F_t = release of drug at time t, F_{load} = amount of drug loaded in disc.

4.3. Mathematical Models of Drug Release Kinetics

To evaluate the release pattern of nicorandil, and zero-order, first order, Higuchi, Hixson–Crowell, Korsmeyer–Peppas kinetic models were applied.

Author Contributions: Conceptualization, A.M.; methodology, S.M.; formal analysis, A.E. and U.R.T.; investigation, A.M.; resources, N.S.M.; data curation, U.R.T. and M.S.A.; writing—original draft preparation, N.S.M.; writing—review and editing, A.M.; visualization, M.S.A.; supervision, A.E.; project administration, S.M.; funding acquisition, A.M. All authors have read and agreed to the published version of the manuscript.

Funding: This research received no external funding.

Institutional Review Board Statement: Not applicable.

Informed Consent Statement: Not applicable.

Data Availability Statement: Not applicable.

Conflicts of Interest: The authors declare no conflict of interest.

References

1. Nyholm, D.; Lennernäs, H.; toxicology. Irregular gastrointestinal drug absorption in Parkinson's disease. *Expert Opin. Drug Metab.* **2008**, *4*, 193–203. [CrossRef] [PubMed]
2. Feng, J.; Wu, Y.; Chen, W.; Li, J.; Wang, X.; Chen, Y.; Yu, Y.; Shen, Z.; Zhang, Y. Sustained release of bioactive IGF-1 from a silk fibroin microsphere-based injectable alginate hydrogel for the treatment of myocardial infarction. *J. Mater. Chem. B* **2020**, *8*, 308–315. [CrossRef] [PubMed]
3. Deen, G.R.; Loh, X.J. Stimuli-responsive cationic hydrogels in drug delivery applications. *Gels* **2018**, *4*, 13. [CrossRef] [PubMed]
4. Echeverria, C.; Fernandes, S.N.; Godinho, M.H.; Borges, J.P.; Soares, P.I. Functional stimuli-responsive gels: Hydrogels and microgels. *Gels* **2018**, *4*, 54. [CrossRef] [PubMed]
5. Xue, Y.; Zhang, J.; Chen, X.; Zhang, J.; Chen, G.; Zhang, K.; Lin, J.; Guo, C.; Liu, J. Trigger-Detachable Hydrogel Adhesives for Bioelectronic Interfaces. *Adv. Funct. Mater.* **2021**, *31*, 2106446. [CrossRef]
6. Liang, X.; Chen, G.; Lin, S.; Zhang, J.; Wang, L.; Zhang, P.; Wang, Z.; Wang, Z.; Lan, Y.; Ge, Q.A.M. Anisotropically Fatigue-Resistant Hydrogels. *Adv. Mater.* **2021**, *33*, 2102011. [CrossRef]
7. Liang, X.; Chen, G.; Lin, S.; Zhang, J.; Wang, L.; Zhang, P.; Lan, Y.; Liu, J. Bioinspired 2D Isotropically Fatigue-Resistant Hydrogels. *Adv. Mater.* **2022**, *34*, 2270064. [CrossRef]
8. Chen, W.; Kouwer, P.H. Combining Mechanical Tuneability with Function: Biomimetic Fibrous Hydrogels with Nanoparticle Crosslinkers. *Adv. Funct. Mater.* **2021**, *31*, 2105713. [CrossRef]
9. Caccavo, D.; Cascone, S.; Lamberti, G.; Barba, A.A. Controlled drug release from hydrogel-based matrices: Experiments and modeling. *Int. J. Pharm.* **2015**, *486*, 144–152. [CrossRef]
10. Kabiri, K.; Omidian, H.; Hashemi, S.; Zohuriaan-Mehr, M. Synthesis of fast-swelling superabsorbent hydrogels: Effect of crosslinker type and concentration on porosity and absorption rate. *Eur. Polym. J.* **2003**, *39*, 1341–1348. [CrossRef]
11. Omidian, H.; Rocca, J.G.; Park, K. Advances in superporous hydrogels. *J. Control. Release* **2005**, *102*, 3–12. [CrossRef] [PubMed]

12. Ghumman, S.A.; Noreen, S.; tul Muntaha, S. Linum usitatissimum seed mucilage-alginate mucoadhesive microspheres of metformin HCl: Fabrication, characterization and evaluation. *Int. J. Biol. Macromol.* **2020**, *155*, 358–368. [CrossRef] [PubMed]
13. Hasnain, M.S.; Rishishwar, P.; Rishishwar, S.; Ali, S.; Nayak, A.K. Isolation and characterization of Linum usitatissimum polysaccharide to prepare mucoadhesive beads of diclofenac sodium. *Int. J. Biol. Macromol.* **2018**, *116*, 162–172. [CrossRef] [PubMed]
14. Haseeb, M.T.; Hussain, M.A.; Yuk, S.H.; Bashir, S.; Nauman, M. Polysaccharides based superabsorbent hydrogel from Linseed: Dynamic swelling, stimuli responsive on–off switching and drug release. *Carbohydr. Polym.* **2016**, *136*, 750–756. [CrossRef]
15. Nerkar, P.P.; Gattani, S. In vivo, in vitro evaluation of linseed mucilage based buccal mucoadhesive microspheres of venlafaxine. *Drug Deliv.* **2011**, *18*, 111–121. [CrossRef]
16. Nerkar, P.P.; Gattani, S.G. Oromucosal delivery of venlafaxine by linseed mucilage based gel: In vitro and in vivo evaluation in rabbits. *Arch. Pharmacal Res.* **2013**, *36*, 846–853. [CrossRef]
17. Kurra, P.; Narra, K.; Puttugunta, S.B.; Kilaru, N.B.; Mandava, B.R. Development and optimization of sustained release mucoadhesive composite beads for colon targeting. *Int. J. Biol. Macromol.* **2019**, *139*, 320–331. [CrossRef]
18. Sheikh, F.A.; Hussain, M.A.; Ashraf, M.U.; Haseeb, M.T.; Farid-ul-Haq, M. Linseed hydrogel based floating drug delivery system for fluoroquinolone antibiotics: Design, in vitro drug release and in vivo real-time floating detection. *Saudi Pharm. J.* **2020**, *28*, 538–549. [CrossRef]
19. Haseeb, M.T.; Hussain, M.A.; Bashir, S.; Ashraf, M.U.; Ahmad, N. Evaluation of superabsorbent linseed-polysaccharides as a novel stimuli-responsive oral sustained release drug delivery system. *Drug Dev.* **2017**, *43*, 409–420. [CrossRef]
20. Liu, P.; Peng, J.; Li, J.; Wu, J. Radiation crosslinking of CMC-Na at low dose and its application as substitute for hydrogel. *Radiat. Phys. Chem.* **2005**, *72*, 635–638. [CrossRef]
21. Baker, J.P.; Hong, L.H.; Blanch, H.W.; Prausnitz, J.M. Effect of initial total monomer concentration on the swelling behavior of cationic acrylamide-based hydrogels. *Macromolecules* **1994**, *27*, 1446–1454. [CrossRef]
22. YÜRÜKSOY, B.I. Swelling behavior of acrylamide-2-hydroxyethyl methacrylate hydrogels. *Turk. J. Chem.* **2000**, *24*, 147–156.
23. Ganji, F.; Vasheghani, F.S.; Vasheghani, F.E. Theoretical description of hydrogel swelling: A review. *Iran. Polym. J.* **2010**, *19*, 375–398.
24. Peppas, N.; Bures, P.; Leobandung, W.; Ichikawa, H. Hydrogels in pharmaceutical formulations. *Eur. J. Pharm. Biopharm.* **2000**, *50*, 27–46. [CrossRef]
25. Chai, Q.; Jiao, Y.; Yu, X. Hydrogels for biomedical applications: Their characteristics and the mechanisms behind them. *Gels* **2017**, *3*, 6. [CrossRef]
26. Zhang, S.; Wang, W.; Wang, H.; Qi, W.; Yue, L.; Ye, Q. Synthesis and characterisation of starch grafted superabsorbent via 10 MeV electron-beam irradiation. *Carbohydr. Polym.* **2014**, *101*, 798–803. [CrossRef]
27. Chavda, H.; Patel, C. Effect of crosslinker concentration on characteristics of superporous hydrogel. *Int. J. Pharm. Investig.* **2011**, *1*, 17. [CrossRef]
28. Mahdavinia, G.; Pourjavadi, A.; Hosseinzadeh, H.; Zohuriaan, M. Modified chitosan 4. Superabsorbent hydrogels from poly (acrylic acid-co-acrylamide) grafted chitosan with salt-and pH-responsiveness properties. *Eur. Polym. J.* **2004**, *40*, 1399–1407. [CrossRef]
29. Kowalski, G.; Kijowska, K.; Witczak, M.; Kuterasiński, Ł.; Łukasiewicz, M. Synthesis and effect of structure on swelling properties of hydrogels based on high methylated pectin and acrylic polymers. *Polymers* **2019**, *11*, 114. [CrossRef]
30. Karadağ, E.; Saraydin, D.; Çetinkaya, S.; Güven, O. In vitro swelling studies and preliminary biocompatibility evaluation of acrylamide-based hydrogels. *Biomaterials* **1996**, *17*, 67–70. [CrossRef]
31. Mansur, H.S.; Oréfice, R.L.; Mansur, A.A. Characterization of poly (vinyl alcohol)/poly (ethylene glycol) hydrogels and PVA-derived hybrids by small-angle X-ray scattering and FTIR spectroscopy. *Polymer* **2004**, *45*, 7193–7202. [CrossRef]
32. Torres, R.; Usall, J.; Teixido, N.; Abadias, M.; Vinas, I. Liquid formulation of the biocontrol agent Candida sake by modifying water activity or adding protectants. *J. Appl. Microbiol.* **2003**, *94*, 330–339. [CrossRef] [PubMed]
33. Warrand, J.; Michaud, P.; Picton, L.; Muller, G.; Courtois, B.; Ralainirina, R.; Courtois, J. Flax (Linum usitatissimum) seed cake: A potential source of high molecular weight arabinoxylans? *J. Agric. Food Chem.* **2005**, *53*, 1449–1452. [CrossRef] [PubMed]
34. Durmaz, S.; Okay, O. Acrylamide/2-acrylamido-2-methylpropane sulfonic acid sodium salt-based hydrogels: Synthesis and characterization. *Polymer* **2000**, *41*, 3693–3704. [CrossRef]
35. Zheng, Y.; Li, P.; Zhang, J.; Wang, A. Study on superabsorbent composite XVI. Synthesis, characterization and swelling behaviors of poly (sodium acrylate)/vermiculite superabsorbent composites. *Eur. Pol. J.* **2007**, *43*, 1691–1698. [CrossRef]
36. Sorour, M.; El-Sayed, M.; Moneem, N.A.E.; Talaat, H.A.; Shalaan, H.; Marsafy, S.E. Characterization of hydrogel synthesized from natural polysaccharides blend grafted acrylamide using microwave (MW) and ultraviolet (UV) techniques. *Starch-Stärke* **2013**, *65*, 172–178. [CrossRef]
37. AHMED, A.B.; NATH, L.K. Drug-excipients compatibility studies of nicorandil in controlled release floating tablet. *Int. J. Pharm. Pharm. Sci* **2014**, *6*, 468–475.
38. Rashid, A.; Tulain, U.R.; Iqbal, F.M.; Malikd, N.S.; Erum, A. Synthesis, characterization and in vivo evaluation of ph sensitive hydroxypropyl methyl cellulose-graft-acrylic acid hydrogels for sustained drug release of model drug nicorandil. *Gomal J. Med. Sci.* **2020**, *18*, 99–106. [CrossRef]

39. Brazel, C.S.; Peppas, N.A. Mechanisms of solute and drug transport in relaxing, swellable, hydrophilic glassy polymers. *Polymer* **1999**, *40*, 3383–3398. [CrossRef]
40. Hu, Y.; Shim, Y.Y.; Reaney, M.J. Flaxseed gum solution functional properties. *Foods* **2020**, *9*, 681. [CrossRef]
41. Rocha, M.S.; Rocha, L.C.; da Silva Feijó, M.B.; dos Santos Marotta, P.L.L.; Mourao, S.C. Effect of pH on the flaxseed (Linum usitatissimum L. seed) mucilage extraction process. *Acta Scientiarum. Technol.* **2021**, *43*, e50457. [CrossRef]
42. Unagolla, J.M.; Jayasuriya, A.C. Drug transport mechanisms and in vitro release kinetics of vancomycin encapsulated chitosan-alginate polyelectrolyte microparticles as a controlled drug delivery system. *Eur. J. Pharm. Sci.* **2018**, *114*, 199–209. [CrossRef] [PubMed]
43. Katime, I.; Velada, J.; Novoa, R.; de Apodaca, E.D.; Puig, J.; Mendizabal, E. Swelling kinetics of poly (acrylamide)/poly (mono-n-alkyl itaconates) hydrogels. *Polym. Int.* **1996**, *40*, 281–286. [CrossRef]
44. Gulrez, S.K.; Al-Assaf, S.; Phillips, G.O. Hydrogels: Methods of preparation, characterisation and applications. In *Progress in Molecular and Environmental Bioengineering—From Analysis and Modeling to Technology Applications*; IntechOpen: London, UK, 2011; pp. 117–150.
45. Sharma, R.; Walker, R.B.; Pathak, K. Evaluation of the kinetics and mechanism of drug release from econazole nitrate nanosponge loaded carbapol hydrogel. *Indian J. Pharm. Educ. Res.* **2011**, *45*, 25–31.
46. Reddy, K.R.; Mutalik, S.; Reddy, S. Once-daily sustained-release matrix tablets of nicorandil: Formulation and in vitro evaluation. *AAPS Pharmscitech* **2003**, *4*, 480–488. [CrossRef] [PubMed]
47. Patel, H.; Panchal, D.R.; Patel, U.; Brahmbhatt, T.; Suthar, M. Matrix type drug delivery system: A review. *JPSBR* **2011**, *1*, 143–151.
48. Xu, Y.; Cui, B.; Ran, R.; Liu, Y.; Chen, H.; Kai, G.; Shi, J. Risk assessment, formation, and mitigation of dietary acrylamide: Current status and future prospects. *Food Chem. Toxicol.* **2014**, *69*, 1–12. [CrossRef]
49. Pourjavadi, A.; Kurdtabar, M. Collagen-based highly porous hydrogel without any porogen: Synthesis and characteristics. *Eur. Polym. J.* **2007**, *43*, 877–889. [CrossRef]
50. Minhas, M.U.; Ahmad, M.; Ali, L.; Sohail, M. Synthesis of chemically cross-linked polyvinyl alcohol-co-poly (methacrylic acid) hydrogels by copolymerization; a potential graft-polymeric carrier for oral delivery of 5-fluorouracil. *DARU J. Pharm. Sci.* **2013**, *21*, 44. [CrossRef]
51. Ahmed, A.B.; Nath, L.K. Fabrication and in vitro evaluation of floating matrix tablet of nicorandil using factorial design. *J. Pharm. Res.* **2011**, *4*, 1950–1954.

Article

Cleaning of Wastewater Using Crosslinked Poly(Acrylamide-*co*-Acrylic Acid) Hydrogels: Analysis of Rotatable Bonds, Binding Energy and Hydrogen Bonding

Salah Hamri [1,2], Tewfik Bouchaour [2], Djahida Lerari [1], Zohra Bouberka [3], Philippe Supiot [4] and Ulrich Maschke [4,*]

1. Center for Scientific and Technical Research in Physico-Chemical Analysis (CRAPC), BP 384, Industrial Zone, BouIsmaïl 42004, Algeria; salah_hamri@yahoo.fr (S.H.); lerari_zinai@yahoo.fr (D.L.)
2. Macromolecular Research Laboratory (LRM), Faculty of Sciences, Abou Bekr Belkaid University, BP 119, Tlemcen 13000, Algeria; bouchaour@yahoo.fr
3. Laboratoire Physico-Chimie des Matériaux-Catalyse et Environnement (LPCMCE), Université des Sciences et de la Technologie d'Oran Mohamed Boudiaf (USTOMB), Oran 31000, Algeria; bouberkazohra@yahoo.fr
4. CNRS, INRAE, Centrale Lille, UMR 8207—UMET—Unité Matériaux et Transformations, Université de Lille, 59000 Lille, France; philippe.supiot@univ-lille.fr
* Correspondence: ulrich.maschke@univ-lille.fr

Abstract: The discharge of untreated wastewater, often contaminated by harmful substances, such as industrially used dyes, can provoke environmental and health risks. Among various techniques, the adsorption of dyes, using three-dimensional (3D) networks consisting of hydrophilic polymers (hydrogels), represents a low-cost, clean, and efficient remediation method. Three industrially used dyes, Methylene Blue, Eosin, and Rose Bengal, were selected as models of pollutants. Poly(acrylamide) (poly(AM)) and poly(acrylamide-*co*-acrylic acid) (poly(AM-*co*-AA)) networks were chosen as adsorbent materials (hydrogels). These polymers were synthesized by crosslinking the photopolymerization of their respective monomer(s) in an aqueous medium under exposure to UV light. Experimental adsorption measurements revealed substantially higher dye uptakes for poly(AM-*co*-AA) compared to poly(AM) hydrogels. In this report, a theoretical model based on docking simulations was applied to analyze the conformation of polymers and pollutants in order to investigate some aspects of the adsorption process. In particular, hydrogen and halogen interactions were studied. The presence of strong hydrogen bonding plays a crucial role in the retention of dyes, whereas halogen bonding has a small or negligible effect on adsorption. An evaluation of binding energies allowed us to obtain information about the degree of affinity between polymers and dyes. The number of rotatable bonds in the copolymer exceeds those of poly(AM), meaning that poly(AM-*co*-AA) is revealed to be more suitable for obtaining a high retention rate for pollutants.

Keywords: wastewater; pollutant; dye; hydrogel; modeling; docking simulation

Citation: Hamri, S.; Bouchaour, T.; Lerari, D.; Bouberka, Z.; Supiot, P.; Maschke, U. Cleaning of Wastewater Using Crosslinked Poly(Acrylamide-*co*-Acrylic Acid) Hydrogels: Analysis of Rotatable Bonds, Binding Energy and Hydrogen Bonding. *Gels* 2022, 8, 156. https://doi.org/10.3390/gels8030156

Academic Editor: Wei Ji

Received: 4 February 2022
Accepted: 1 March 2022
Published: 3 March 2022

Publisher's Note: MDPI stays neutral with regard to jurisdictional claims in published maps and institutional affiliations.

Copyright: © 2022 by the authors. Licensee MDPI, Basel, Switzerland. This article is an open access article distributed under the terms and conditions of the Creative Commons Attribution (CC BY) license (https://creativecommons.org/licenses/by/4.0/).

1. Introduction

Water is an important liquid for human beings, the universe, and all life existing on earth [1,2]. This liquid can be easily polluted by different dyes [3,4]. Both water and dyes are still largely used in the textile industry, meaning that the wastewater after production is a mixture of dyes and water. Unfortunately, the elimination of the wastewater often occurs through discharge into rivers and other effluents [5–12].

This negative situation has motivated many researchers to publish many reports in the field of the treatment of water polluted with dyes [13–15]. Several physical and chemical techniques have been developed to purify water from these compounds, including photocatalysis, oxidation, filtration, coagulation/flocculation, and adsorption [16–20]. In particular, adsorption processes have been studied intensively because of their low cost, easy access, and effective dye removal, in which the dissolved dye compounds adsorb on the surface of suitable adsorbents [21].

Many biodegradable materials and effective adsorbents obtained from natural resources have been used to remove dyes from aqueous solutions. Hydrogels were frequently applied for this purpose, consisting of a three-dimensional polymeric material with the capacity to uptake an important amount of water due to the presence of hydrophilic groups in their structure, such as –OH, –CONH, and –SO$_3$H [22].

Copolymers based on acrylamide (AM) and acrylic acid (AA) have been applied to remove dyes. AM monomer is soluble in water, and linear poly(AM) finds many uses as water-soluble thickeners and flocculation agents [23–25], whereas AA represents the simplest unsaturated carboxylic acid. This colorless liquid is miscible with water, alcohols, ethers, and chloroform [26–28]. Solpan et al. [29] used (poly(AM-co-AA) hydrogels for the uptake of the cationic dyes, safranin-O and magenta. The diffusion of water and cationic dyes within hydrogels showed non-Fickian behavior. Corona-Rivera et al. [30] applied poly(AM-co-AA) crosslinked with N,N'-methylene bisacrylamide (NMBAM) for the removal of Remazol red dye from aqueous solutions, finding the maximum dye adsorption capacity for peculiar experimental conditions, with an adsorption mechanism well represented by the Langmuir model.

The diffusion of colored water inside the hydrogel depends on many factors, such as the dye structure and the functional groups on the polymer chains. The dye can generate an attraction through electrostatic interaction, which also represents an important key to removing dye from an aqueous medium, whereby a dye molecule and a receptor behave similarly to a ship and a harbor. In the field of biochemistry, the theory of docking was largely applied to study the interaction between ligand and protein, allowing us to explain the affinity between these components [31–37]. In this study, the interactions between the dyes and polymer networks were investigated. The docking method was applied to analyze different interactions, with the receptor and ligand representing the polymer matrix and dye, respectively. This simulation method has the advantage of enabling us to predict the preferred orientation of one molecule to a second one when bound to each other to form a stable complex [38–40]. Interestingly, this helps us to economize cost and time of research work.

In a previous paper [41], the interaction between a polymer based on HEMA monomer and Eosin Y (EY) as a pollutant was discussed. It was found that the theoretical prediction correlates well with experimental results. In the literature, some authors report on poly(AM-co-AA) crosslinked with NMBAM [42–46]. In this work, AM and AA were copolymerized and crosslinked with HDDA since it contributes to the high level of conversion of acrylic double bonds [47,48]. Under the UV-visible light exposure in the presence of a suitable photoinitiator (Darocur 1173), a chemically crosslinked three-dimensional copolymer was successfully obtained. In contrast to thermal polymerization, which often requires elevated temperatures, photopolymerization can be performed at room temperature [49]. In most reports on poly(AA), thermal polymerization was applied using a source of free radicals, together with a chemical stabilizer, such as ammonium persulfate and tetramethylethylenediamine [50–52]. This method has disadvantages such as long polymerization times, unstable and toxic reagents, and tedious preparation steps. On the other hand, photopolymerization using an initiator sensitive to light represents a quicker method that is less tedious and less toxic. The field of the exploitation of these polymeric materials is, thus, enlarged to applications in which the elevation of temperature is not advised. The final properties of UV-polymerized gels depend on the UV-visible spectrum of the source, light intensity and uniformity, and exposure times [53].

The model dyes studied in this report were Rose Bengal (RB), EY, and Methylene Blue (MB), presenting anionic and cationic natures. These dyes are widely used in many applications, thus increasing the probability that they contribute to water pollution, since even small quantities can easily affect the water quality [54–56].

To understand the different interatomic interactions between dyes and polymers, the docking simulation method was exploited. Two model systems, crosslinked poly(AM)/dye and poly(AM-*co*-AA)/dye, were considered using Avogadro software. These model systems were all energy-minimized, and the conformation of polymer/dye systems was simulated using Auto-Dock Vina software [57,58], and then visualized and analyzed using UCSF Chimera.

2. Results and Discussion
2.1. Effect of Crosslinker Content on Equilibrium Swelling

To find out the optimal dye concentration for the retention study, the UV-visible spectra of the dyes were screened in the concentration range from 32×10^{-3} mg·mL^{-1} to 64×10^{-3} mg·mL^{-1}. According to the obtained results (Figure S1), the spectra corresponding to 64×10^{-3} mg·mL^{-1} reveal saturation effects for all absorbance bands except those from RB. Electronic spectra associated with the lower concentration of 32×10^{-3} mg·mL^{-1} were acceptable; therefore, this concentration was chosen for the retention study.

Figure 1 presents the evolution of poly(AM) swelling equilibrium versus the composition of a crosslinking agent (HDDA) for each dye solution. Equilibrium swelling data were remarkably increased by decreasing the crosslinker content. The best results of maximum equilibrium swelling were obtained with 1 wt% of HDDA. In this case, swelling values in solutions of RB, BM, and EY were found at around 870%, 850%, and 900%, respectively. The crosslinking density essentially governs the diffusion of the dyes in the polymer networks, as well as the swelling behavior.

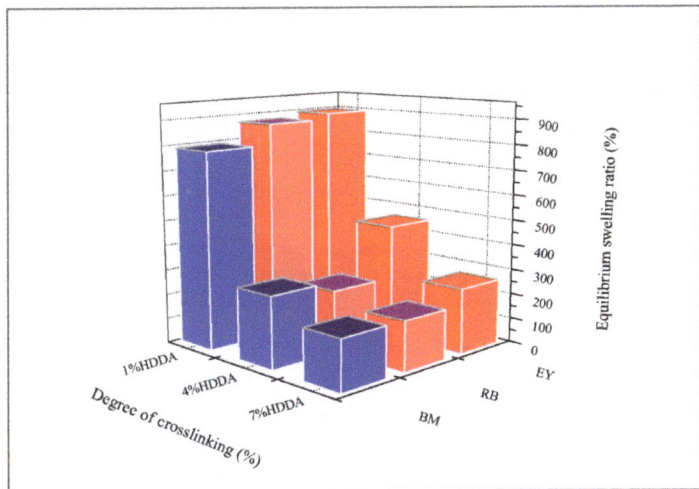

Figure 1. Effect of the composition of HDDA (wt%) on equilibrium swelling of poly(AM) hydrogel in dye solutions.

Figure 2 presents the evolution of the swelling equilibria of poly(AM-*co*-AA) versus the composition of HDDA for each dye solution. The copolymer prepared with 1 wt% HDDA shows the highest equilibrium swelling with 70%, 72%, and 71% in solutions of BM, RB, and EY, respectively. In comparison with poly(AM), poly(AM-*co*-AA) presents a much lower equilibrium swelling due to the addition of AA units, thus increasing the crosslinking density. Swelling equilibrium values of 34%, 35%, and 29% were obtained for 4 wt% HDDA in solutions of BM, RB, and EY, respectively. For 7 wt% HDDA, the corresponding swelling data were 18%, 19%, and 19%.

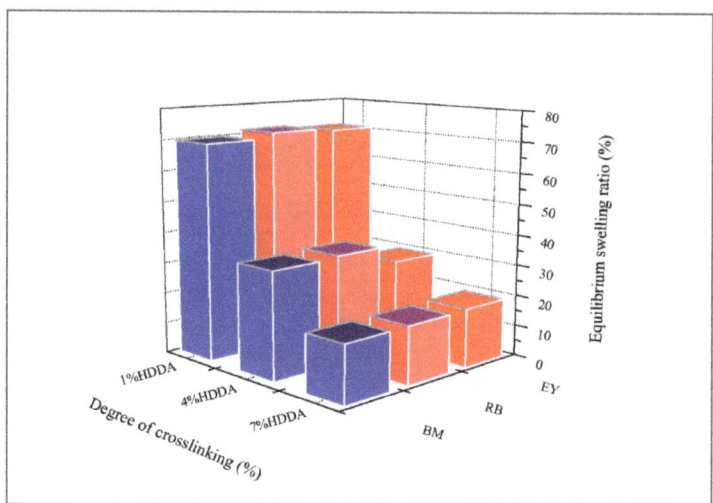

Figure 2. Effect of the composition of HDDA (wt%) on equilibrium swelling of poly(AM-*co*-AA) hydrogel in dye solutions.

2.2. Effect of Crosslinker Content on Absorbance
2.2.1. Rose Bengal Dye

Figure 3 illustrates the retention behavior of RB using poly(AM) and poly(AM-*co*-AA). A retention rate of about 7% for RB by poly(AM) was obtained, whereas 97% of RB was removed in the case of poly (AM-*co*-AA) (Figure S2).

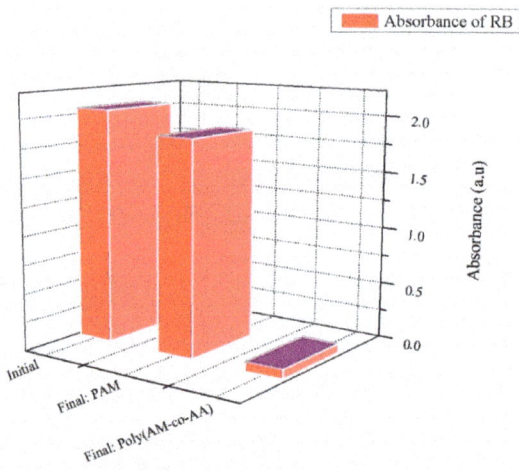

Figure 3. Retention behavior of RB in presence of crosslinked poly(AM) and poly(AM-*co*-AA) (1 wt% HDDA, after 24 h contact time).

The hydrogen bonding between chains of the neutral poly(AM) provokes physical crosslinking effects; the poly(AM) network is brittle and its glass transition was reported to be around 450 K [59,60].

From the Morse curve [61], considering large distances, the energy is zero (no interaction). This means that two atoms placed infinitely far away do not interact with each other, or they are not bonded to each other. At inter-nuclear distances in the order of the atomic diameter, attractive forces dominate. At smaller distances between two atoms, the force is repulsive and the energy of the two atoms is high.

The distance between atoms has, thus, an important effect on the interactions of the system. It was found that interactions can be classified as strong and medium for a distance interval of [2.5, 3.1] Å and [3.1, 3.55] Å, respectively, whereas distances between selected atoms greater than 3.55Å correspond to weak or non-existing interactions [41].

In the case of crosslinked poly(AM) hydrogel, the distances between chlorine and oxygen atoms are greater than 3.55 Å, which shows that the interactions are weak. The hydrogen bond with 2.48 Å represents a strong interaction, but the neutrality of the hydrogel cannot allow this bond to be constructed (Table 1 and Figure 4).

Table 1. Interatomic distances obtained from interactions of the two polymers with RB using the docking simulation method.

System	Bonds	Distance (Å)
Poly(AM)/RB	I...O	4.721
	C...O	6.197
	O...H	2.486
Poly(AM-co-AA)/RB	I...O	3.823
	Cl...O	3.752
	O...H	2.246

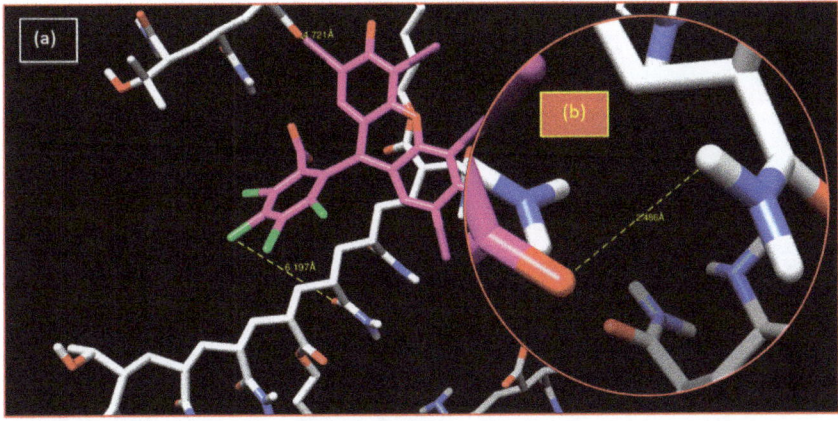

Figure 4. Crosslinked poly(AM)/RB system: (**a**) 3-D representation of results of the interaction; (**b**) enlargement of the hydrogen bonding interaction.

In the crosslinked poly(AM-co-AA)/RB system, there are weak and average electrostatic interactions between the AA fraction and RB. The interaction between chlorine and iodine with oxygen is of a halogen type. The corresponding interatomic distance is higher than 3.55 Å, thus resulting in a weak interaction. Furthermore, the hydrogen bond with 2.24 Å represents a strong interaction, because the copolymer is charged in the aqueous medium and becomes a polyelectrolyte. Repulsion occurs between the negative charges of the AA parts and, consequently, the dye was retained by the strong hydrogen bond (O...H). The AA fraction of poly(AM-co-AA) has increased the retention percentage from 7% to 97%; therefore, it can be concluded that this copolymer effectively retains RB in an aqueous medium (Table 1 and Figure 5).

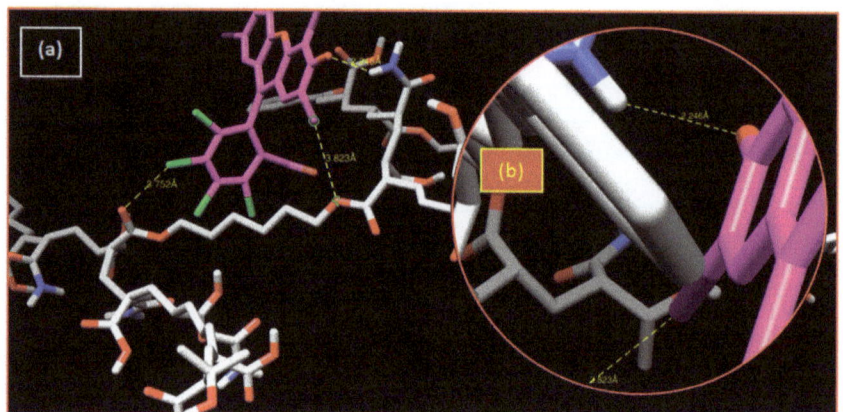

Figure 5. Crosslinked poly(AM-co-AA)/RB system: (**a**) 3-D representation of results of the interaction; (**b**) enlargement of the hydrogen bonding interaction.

2.2.2. Methylene Blue Dye

Figure 6 shows that crosslinked poly(AM) presents a negligible retention of MB (about 1%), whereas poly(AM-*co*-AA) removes MB at rates of 45% for a contact time of 24 h.

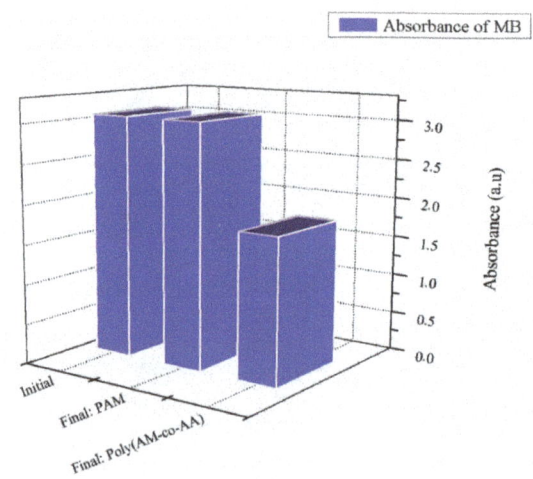

Figure 6. Retention of MB in presence of crosslinked poly(AM) and poly(AM-co-AA) hydrogels (1% wt HDDA, after 24 h contact time).

In the neutral hydrogel, the interatomic distance between nitrogen and oxygen is 5.37 Å, which means that there is a weak attraction between these two atoms. A similar situation occurs between sulfur and hydrogen atoms (Table 2 and Figure 7). Initial and final spectra of the dye are shown in Figure S3.

Table 2. Interatomic distances obtained from interactions of the two polymers with MB using the docking simulation method.

System	Bonds	Distance (Å)
Poly(AM)/MB	S...H	4.229
	N...O	5.373
Poly(AM-co-AA)/MB	S...H	2.670
	N...O	3.924

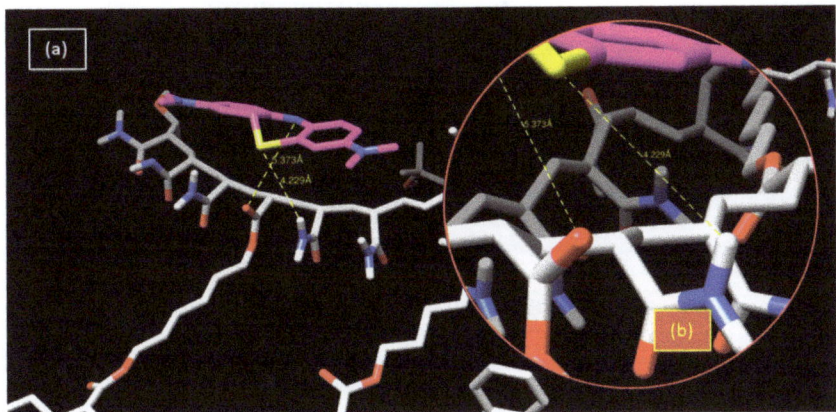

Figure 7. Crosslinked poly(AM)/MB system: (**a**) 3-D representation of results of the interaction; (**b**) enlargement of the hydrogen bonding interaction.

Sulfur atoms have been known to participate in hydrogen bonds. It has been shown that the sulfur atom is a poor H-bond acceptor, but a moderately good H-bond donor [62]. In the copolymeric hydrogel, there is a hydrogen bond with an interatomic distance of 2.67 Å, which is considered to be a strong bond, facilitating an increase in the retention percentage from 1% to 45% (Table 2 and Figure 8).

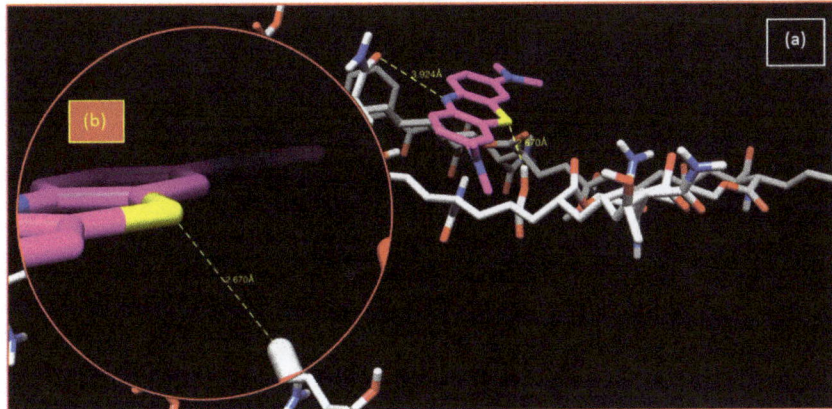

Figure 8. Crosslinked poly(AM-co-AA)/MB system: (**a**) 3-D representation of results of the interaction; (**b**) enlargement of the hydrogen bonding interaction.

2.2.3. Eosin Y Dye

Figure 9 reveals a very small adsorption effect of EY by poly(AM) (0.4%). Halogens participating in the halogen bonding of the investigated dyes include iodine (I) (present in RB), bromine (Br) (present in EY), and chlorine (Cl) (present in RB). These halogens are able to act as donors and follow the general trend of Cl < Br < I, with iodine normally forming the strongest interactions [63,64].

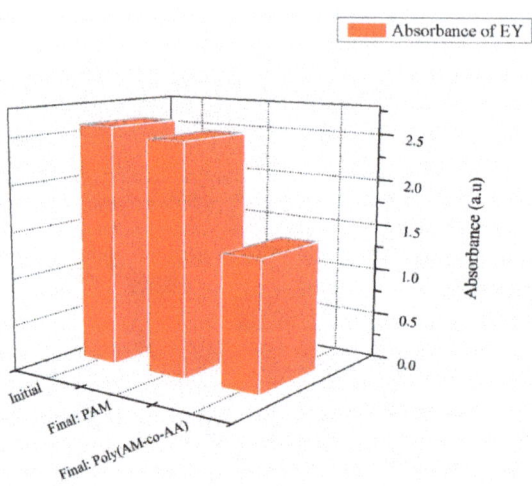

Figure 9. Retention of EY in the presence of crosslinked poly(AM) and poly(AM-*co*-AA) hydrogels (1 % wt HDDA, after 24 h contact time).

For this neutral system (poly(AM)), there is a Br...O bond with an interatomic distance higher than 3.07 Å and a hydrogen bond with 6.53 Å (Table 3 and Figure 10) that qualify these bonds as weak bonds. Initial and final spectra are presented in Figure S4. A strong hydrogen bonding of poly(AM-*co*-AA) exists with 1.872 Å, which improves the retention of the MB dye (Figure 11).

Table 3. Interatomic distances obtained from interactions of the two polymers with EY using the docking simulation method.

System	Bonds	Distance (Å)
Poly(AM)/EY	O...H	6.536
	Br...O	3.072
Poly(AM-*co*-AA)/EY	O...H	1.872
	Br...O	3.618

Figure 12 represents a summary of the adsorption results for poly(AM-*co*-AA) hydrogel, showing retention percentages of 45%, 50%, and 97% for MB, EY, and RB, respectively. Poly(AM-*co*-AA) functions, thus, with considerable efficiency, removing a high percentage of RB, though it is less effective for EY and MB. This difference can be explained by the different molecular structures and architectures of these dyes. Moreover, RB possesses more functional groups compared to EY and MB.

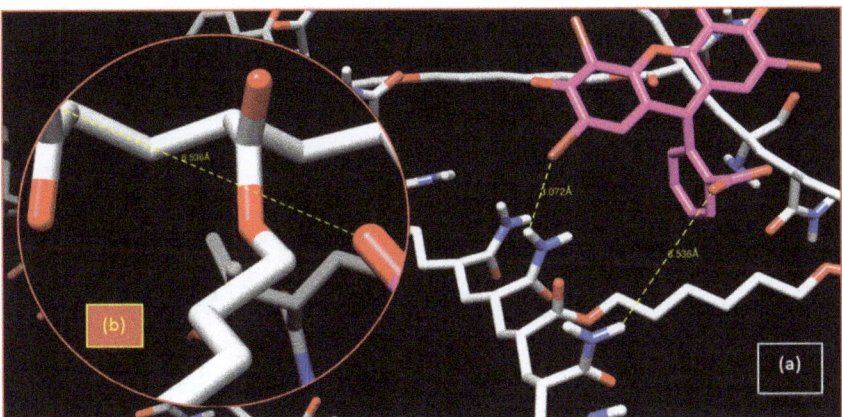

Figure 10. Crosslinked poly(AM)/EY system: (**a**) 3-D representation of results of the interaction; (**b**) enlargement of the hydrogen bonding interaction.

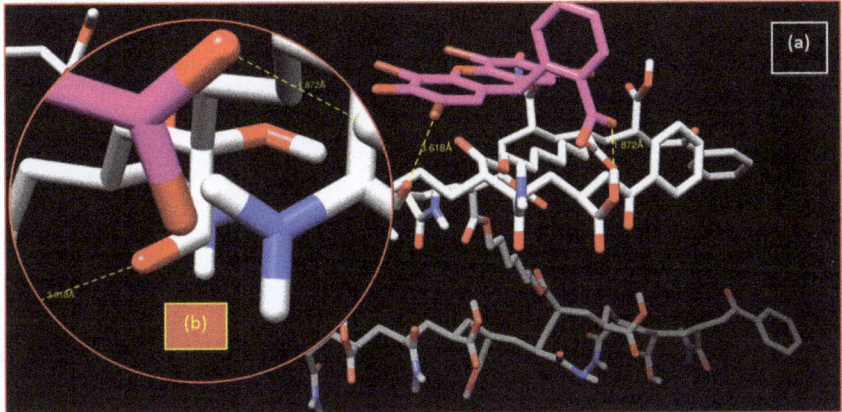

Figure 11. Crosslinked poly(AM-co-AA)/EY system: (**a**) 3-D representation of results of the interaction; (**b**) enlargement of the hydrogen bonding interaction.

2.3. Binding Energy and Number of Rotatable Bonds Analysis

AutodockVina software allows us to determine the binding energies, which were used to evaluate if dyes could have stable complex interactions with polymeric hydrogels. The negative sign of the binding energy means that the dye was bound spontaneously without consuming energy. If the sign is positive, the binding occurs only if the required energy is available. Lower values of binding affinity correspond to a higher stability of polymer/dye complexes. Consequently, hydrogel/dye systems with the highest and lowest stability in Table 4 were poly(AM-*co*-AA)/RB and poly(AM)/MB, respectively.

A rotatable bond is defined as any single non-ring bond attached to a non-terminal, non-hydrogen atom. In Figure 13a, presenting the crosslinked poly(AM) model using Autodock tools, most bonds were nonrotatable. When two HDDA units were very close, the crosslinker creates rigidity in the polymer network. The presence of a single crosslinking unit leads to more rotatable bonds. In Figure 13b, representing the AM repetition unit, we can see that C214–N215 bonds are nonrotatable; the same situation applies for C219–N220: amide C-N bonds present a high energy barrier for rotation [65,66]. C212–C214 and C223–C219 bonds are rotatable, i.e., the AM repetition unit possesses one rotatable bond.

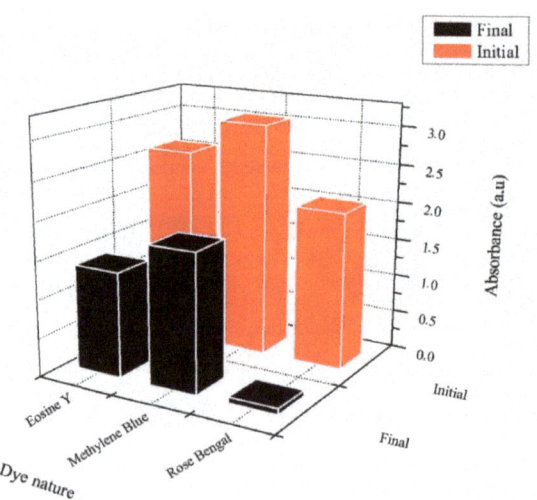

Figure 12. Illustration of the uptake efficiency for all dyes in presence of poly(AM-*co*-AA) hydrogel.

Table 4. Binding energies of polymer/dye systems.

Polymer/Dye	Binding Energy (kcal/mol)
Poly(AM)/RB	−7.0
Poly(AM)/EY	−5.2
Poly(AM)/MB	−4.1
Poly(AM-*co*-AA)/RB	−7.7
Poly(AM-*co*-AA)/EY	−5.1
Poly(AM-*co*-AA)/MB	−4.4

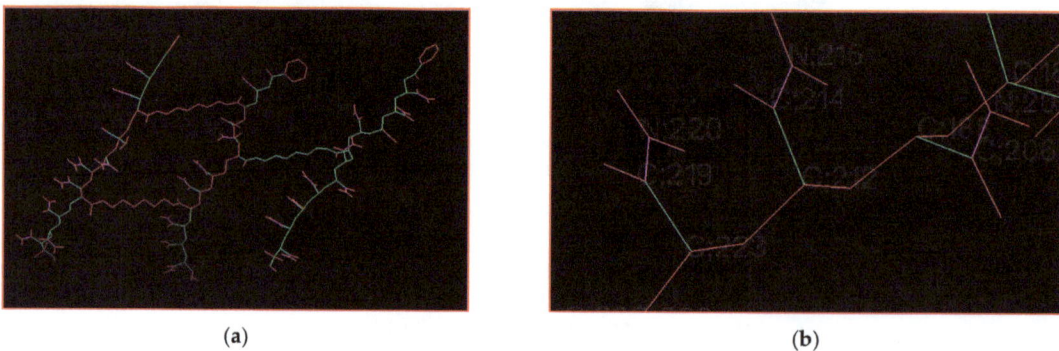

(a)　　　　　　　　　　　　　　　　(b)

Figure 13. Rotatable bonds of the crosslinked poly(AM) model using Autodock tools: (**a**) 92 rotatable bonds, (**b**) one rotatable bond of the AM repetition unit. Green: rotatable, magenta: nonrotatable, red: unrotatable bond.

Figure 14a presents 104 rotatable bonds of the crosslinked poly(AM-*co*-AA) model, using Autodock tools. Figure 14b shows the AA repetition unit of poly(AM-*co*-AA), exhibiting two rotatable bonds, C201–C203 and C203–C204.

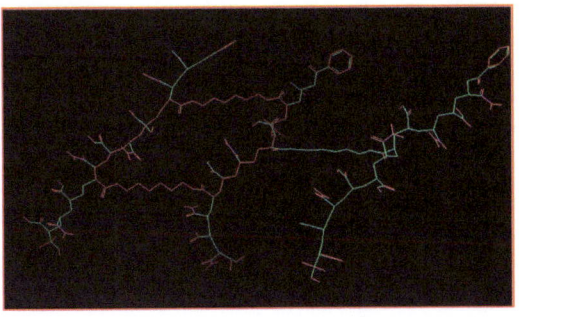 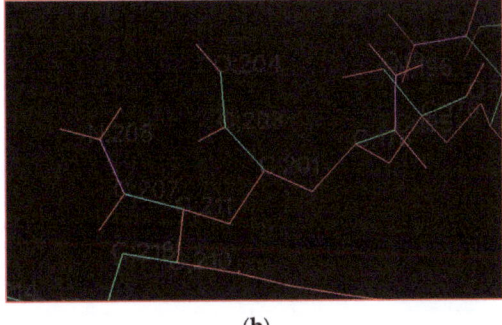

(a) (b)

Figure 14. Rotatable bonds of the crosslinked poly(AM-*co*-AA) model using Autodock tools: (**a**) 104 rotatable bonds, (**b**) two rotatable bonds of the AA repetition unit. Green: rotatable, magenta: nonrotatable, red: unrotatable bond.

The crosslinked poly(AM) model presents 92 rotatable bonds whereas the crosslinked poly(AM-*co*-AA) presents 104 rotatable bonds (Table 5). This difference creates more conformations for the copolymer compared to the homopolymer, so that polymer–dye interactions are favored for crosslinked poly(AM-*co*-AA).

Table 5. Number of rotatable bonds of poly(AM)/HDDA, poly(AM-co-AA)/HDDA and dyes, obtained by Autodock software.

Product	Number of Rotatable Bonds
poly(AM/HDDA)	92
poly(AM-*co*-AA)/HDDA	104
AM repeat unit	01
AA repeat unit	02
RB	2
EY	2
MB	2

3. Conclusions

The UV photopolymerization technique in an aqueous medium was chosen to elaborate chemically crosslinked poly(AM) and poly(AM-*co*-AA) as dye adsorbent hydrogels. Experimental parameters, such as the optimal percentage of HDDA as a crosslinking agent, as well as the suitable dye concentration for analysis, were found to be1 wt% and 32×10^{-3} mg·mL^{-1}, respectively. All dyes show negligible retention effects using the neutral poly(AM), and significant adsorption for the polyelectrolyte poly(AM-*co*-AA). It was found that 97% of RB was removed efficiently by the copolymer (MB: 45%, EY: 50%), which can be related to the presence of one more functional group compared to the other dyes, and also due to strong hydrogen bonding (O...H) with an interatomic distance of 2.24 Å, which plays a key role in interaction. As a consequence, this copolymer could be considered to be an efficient hydrogel with which to remove the considered dyes from a water medium.

The conformation of polymers and pollutants were analyzed via a docking simulation. Interestingly, halogen bonding could be neglected, whereas hydrogen bonding plays a key role for dye retention. The system composed of poly(AM-*co*-AA)/RB has a binding energy of −7.7 kcal/mol, which means that this system has the highest stability compared to the other investigated polymer/dye systems. An analysis of rotatable bonds shows that the AA repetition unit presents two rotatable bonds, whereas that of AM has one; therefore, poly(AM-*co*-AA) has more conformations than poly(AM), thus increasing dye retention.

4. Materials and Methods

4.1. Materials

The monomers used in this study were AM and AA (both from Sigma-Aldrich (Saint-Quentin-Fallavier, France), purity: 99%), the crosslinking agent was HDDA (from Cray Valley, Courbevoie, France), purity: 98%), and the photoinitiator was 2-hydroxy-2-methyl-1-phenyl-propane-1 (commercial designation: Darocur 1173) (from Ciba-Geigy, purity: 97%). RB (purity: 95%), EY (purity: 99%), and MB (purity: 70%) (all from Sigma-Aldrich) were applied as dyes. All products were used as received without purification. The chemical structures of the reagents are illustrated in Table 6. Abbreviations are given in Table S1.

Table 6. Chemical structure of monomers, crosslinker, and dyes.

Name	Chemical Structure
Acrylamide (AM)	
Acrylic acid (AA)	
1,6-Hexanedioldiacrylate (HDDA)	
Rose Bengal sodium salt (RB)	
Eosin Y (EY)	
Methylene Blue (MB)	

4.2. Hydrogel Synthesis

First, the AM monomer was dissolved in distilled water. Then, 0.5 wt% of Darocur 1173 was added. To underline the crosslinker effect on the water uptake capacity of the obtained hydrogels, a set of three solutions with different percentages of HDDA (1 wt%, 4 wt%, and 7 wt%) was prepared. After steering for 24 h, the solutions were exposed to UV irradiation for 30 min, using a TL08 UV lamp, with a characteristic wavelength of λ = 365 nm and an intensity of 1.5 mW/cm^2. For the sake of comparison, a hydrogel copolymer was generated. A stock solution of 50 wt%/50 wt% AM/AA was firstly prepared. In the second step, three solutions were prepared with different percentages of HDDA: 1 wt%, 4 wt%, and 7 wt%, with 98.5 wt%, 95.5 wt%, and 92.5 wt% of the AM/AA solution, respectively. Finally, Darocur 1173 as a photoinitiator was added to each of these solutions (0.5 wt%). All solutions were prepared at room temperature.

After the polymerization/crosslinking process (see also Figure S5), all samples were obtained in pellet form (sample thickness: 3 mm, diameter: 2.5 cm) and washed in distilled water to remove all remaining residue.

4.3. Dye Retention Experiments

The selected cylindrical hydrogel (1.5 g), as an adsorbent, was immersed in 32×10^{-3} mg·mL^{-1} of dye solution at room temperature (T = 23 °C) for 24 h. Then, the hydrogel was separated by filtration, and the residual concentration of the considered dye solution was introduced in a glass flask and evaluated using a dual-beam ultraviolet–visible spectrometer (Specord 200 plus, Analytik Jena, Jena, Germany).

4.4. Equilibrium Swelling Measurements

In order to underline the equilibrium swelling of the elaborated hydrogels, the sample was weighed in the dry state and then immersed in a dye solution for 24 h under stirring. Then, the sample was wiped with a filter paper to remove free liquid on the surface before being weighed. The degree of swelling was calculated according to Equation (1).

$$\tau(\%) = 100 \left(\frac{m_t - m_0}{m_0} \right) \tag{1}$$

where $\tau(\%)$ represents the degree of swelling, m_t stands for the weight of the swollen network at time t, and m_0 is the weight of the initially dried network.

4.5. Model Proposition

Two model systems were proposed. The first one represents the crosslinked poly(AM)/HDDA system, based on three chains of poly(AM) containing ten units each. Chains were connected by three HDDA crosslinking nodes. The second model system, the crosslinked poly(AM-*co*-AA), was created similar to the first model. RB, EY, and MB were all presented in 3-D. All models were energy-minimized using auto-optimization with force field UFF and the steepest descent algorithm of the Avogadro software. The output simulation implies eight conformations; the best conformation of each hydrogel/dye system was illustrated based on their energy.

4.6. Software

Avogadro version 4.8.6 was used to visualize and optimize the model systems. The files were saved in the molecule file format pdb. AutoDock version 1.5.6, a molecular modeling simulation software, represents a suite of automated docking tools. It is designed to predict how dyes bind to polymer networks; the grid box allows the user to limit the space of interaction analysis (Figure 15).

The simulation was conducted in dimensions of grid box points (x = y = z = 126 Å), and the grid box center dimensions were set as mentioned in Table S2. The other parameters were maintained and were used as defaults. Finally, the output file (log.txt) was analyzed, and the best docking results regarding binding energies were selected and investigated with other software programs. The Chimera calculation software UCSF version 1.5.3 was used to analyze interatomic distances.

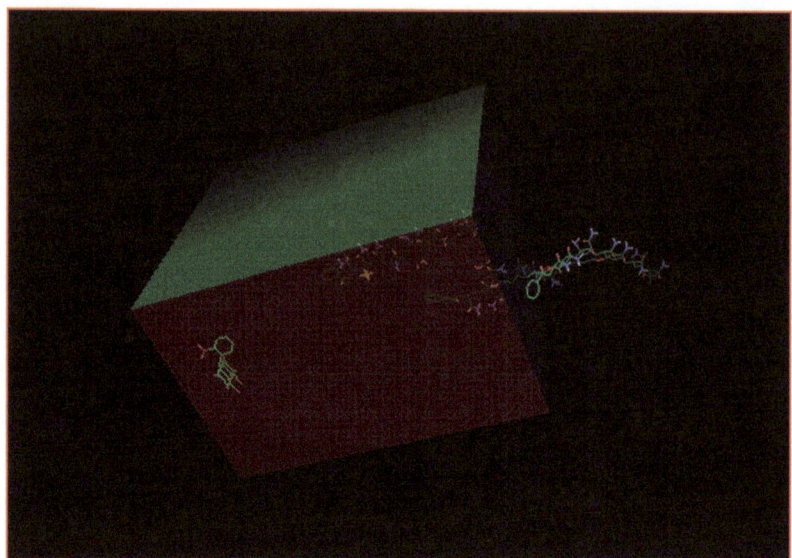

Figure 15. Grid box containing dye and polymer network using Autodock tools.

Supplementary Materials: The following are available online at https://www.mdpi.com/article/10.3390/gels8030156/s1, Figure S1: UV-visible absorption spectra of the three dyes at different concentrations (solid line: C = 0.064 mg/mL, dashed line: C = 0.032 mg/mL); Figure S2: UV-visible absorption spectra of RB solutions with a contact time of 24 h; Figure S3: UV absorption spectra of MB solutions with a contact time of 24 h; Figure S4: UV-visible absorption spectra of EY solutions with a contact time of 24 h; Figure S5: Representation of crosslinked poly(acrylamide-*co*-acrylic acid)/HDDA; Table S1: Definition of Abbreviation; Table S2: Grid box center dimension for all systems.

Author Contributions: S.H.: conceptualization, methodology, writing—original draft preparation. T.B.: software, validation, simulation calculations. D.L.: supervision, reviewing, editing. Z.B.: reviewing, editing. P.S.: reviewing, editing. U.M.: editing, reviewing, submission. All authors have read and agreed to the published version of the manuscript.

Funding: This research received no external funding.

Institutional Review Board Statement: Not applicable.

Informed Consent Statement: Not applicable.

Data Availability Statement: Not applicable.

Acknowledgments: The authors gratefully acknowledge the support of the Research Center CRAPC Tipaza-Algeria and LRM Laboratory of Tlemcen University-Algeria.

Conflicts of Interest: The authors declare no conflict of interest.

References

1. Popkin, B.M.; D'Anci, K.E.; Rosenberg, I.H. Water, Hydration and Health. *Nutr. Rev.* **2010**, *68*, 439–458. [CrossRef] [PubMed]
2. Bayu, T.; Kim, H.; Oki, T. Water Governance Contribution to Water and Sanitation Access Equality in Developing Countries. *Water Resour. Res.* **2020**, *56*, e2019WR025330. [CrossRef]
3. Aminul, I.M.; Imran, A.; Karim, S.M.A.; Firoz, M.S.H.; Al-Nakib, C.; Morton, D.W.; Angove, M.J. Removal of dye from polluted water using novel nano manganese oxide-based materials. *J. Water Process. Eng.* **2019**, *32*, 100911.
4. Piaskowski, K.; Świderska-Dąbrowska, R.; Zarzycki, P.K. Dye removal from water and wastewater using various physical, chemical, and biological processes. *J. AOAC Int.* **2018**, *101*, 1371–1384. [CrossRef]
5. Ranganathan, K.; Jeyapaul, S.; Sharma, D.C. Assessment of water pollution in different bleaching based paper manufacturing and textile dyeing industries in India. *Environ. Monit. Assess.* **2007**, *134*, 363. [CrossRef]

6. Pang, Y.L.; Abdullah, A.Z. Current status of textile industry wastewater management and research progress in Malaysia: A review. *Clean–Soil Air Water* **2013**, *41*, 751–764. [CrossRef]
7. Deng, H.; Wei, R.; Luo, W.; Hu, L.; Li, B.; Di, Y.; Shi, H. Microplastic pollution in water and sediment in a textile industrial area. *Environ. Pollut.* **2019**, *258*, 113658. [CrossRef]
8. Mia, R.; Selim, M.; Shamim, A.; Chowdhury, M.; Sultana, S.; Armin, M.; Hossain, M.; Akter, R.; Dey, S.; Naznin, H. Review on various types of pollution problem in textile dyeing & printing industries of Bangladesh and recommandation for mitigation. *J. Text. Eng. Fash. Technol.* **2019**, *5*, 220–226.
9. Wang, Z.; Shao, D.; Westerhoff, P. Wastewater discharge impact on drinking water sources along the Yangtze River (China). *Sci. Total Environ.* **2017**, *599*, 1399–1407. [CrossRef]
10. Adar, E. Removal of acid yellow 17 from textile wastewater by adsorption and heterogeneous persulfate oxidation. *Int. J. Environ. Sci. Technol.* **2021**, *18*, 483–498. [CrossRef]
11. Kong, Y.; Zhuang, Y.; Han, Z.; Yu, J.; Shi, B.; Han, K.; Hao, H. Dye removal by eco-friendly physically cross-linked double network polymer hydrogel beads and their functionalized composites. *J. Environ. Sci.* **2019**, *78*, 81–91. [CrossRef] [PubMed]
12. Lellis, B.; Fávaro-Polonio, C.Z.; Pamphile, J.A.; Polonio, J.C. Effects of textile dyes on health and the environment and bioremediation potential of living organisms. *Biotechnol. Res. Innov.* **2019**, *3*, 275–290. [CrossRef]
13. Wang, J.; Wang, Z.; Vieira, C.L.Z.; Wolfson, J.M.; Pingtian, G.; Huang, S. Review on the treatment of organic pollutants in water by ultrasonic technology. *Ultrason. Sonochem.* **2019**, *55*, 273–278. [CrossRef]
14. Ribeiro, A.R.; Nunes, O.C.; Pereira, M.F.R.; Silva, A.M.T. An overview on the advanced oxidation processes applied for the treatment of water pollutants defined in the recently launched Directive 2013/39/EU. *Environ. Int.* **2015**, *75*, 33–51. [CrossRef] [PubMed]
15. Hamri, S.; Bouchaour, T.; Maschke, U. Erythrosine/triethanolamine system to elaborate crosslinked poly(2-hydroxyethylmethacrylate): UV-photopolymerization and swelling studies. *Macromol. Symp.* **2014**, *336*, 75–81. [CrossRef]
16. Adewuyi, Y.G. Sonochemistry in environmental remediation. 2. Heterogeneous sonophotocatalytic oxidation processes for the treatment of pollutants in water. *Environ. Sci. Technol.* **2005**, *39*, 8557–8570. [CrossRef] [PubMed]
17. Wang, A.M.; Qu, J.H.; Liu, H.J.; Lei, P.J. Dyes wastewater treatment by reduction-oxidation process in an electrochemical reactor packed with natural manganese mineral. *J. Environ. Sci.* **2006**, *18*, 17–22.
18. Orimolade, B.O.; Arotiba, O.A. Towards visible light driven photoelectrocatalysis for water treatment: Application of a FTO/BiVO$_4$/Ag$_2$S heterojunction anode for the removal of emerging pharmaceutical pollutants. *Sci. Rep.* **2020**, *10*, 5348. [CrossRef]
19. Amin, M.T.; Alazba, A.A.; Manzoor, U. A review of removal of pollutants from water/wastewater using different types of nanomaterials. *Adv. Mater. Sci. Eng.* **2014**, *24*, 825910. [CrossRef]
20. Poulopoulos, S.G.; Yerkinova, A.; Ulykbanova, G.; Inglezakis, V.J. Photocatalytic treatment of organic pollutants in a synthetic wastewater using UV light and combinations of TiO$_2$, H$_2$O$_2$ and Fe(III). *PLoS ONE* **2019**, *14*, e0216745. [CrossRef]
21. Ali, I.; Gupta, V. Advances in water treatment by adsorption technology. *Nat. Protoc.* **2006**, *1*, 2661–2667. [CrossRef] [PubMed]
22. Basak, S.; Nandi, N.; Paul, S.; Banerjee, I.W.H.A.A. A tripeptide-based self-shrinking hydrogel for waste-water treatment: Removal of toxic organic dyes and lead (Pb^{2+}) ions. *Chem. Commun.* **2017**, *53*, 5910–5913. [CrossRef] [PubMed]
23. Craciun, G.; Manaila, E.; Stelescu, M.D. Flocculation efficiency of poly(acrylamide-*co*-acrylic acid) obtained by electron beam irradiation. *J. Mater.* **2013**, *7*, 297123. [CrossRef]
24. Sun, W.; Zhang, G.; Pan, L.; Li, H.; Shi, A. Synthesis, characterization, and flocculation properties of branched cationic polyacrylamide. *Int. J. Polym. Sci.* **2013**, *10*, 397027. [CrossRef]
25. Munishwar, N.G. Electrophoresis. Gel electrophoresis: Polyacrylamide gels. In *Encyclopedia of Analytical Science*, 3rd ed.; Worsfold, P., Poole, C., Townshend, A., Miró, M., Eds.; Elsevier: Amsterdam, The Netherlands, 2019; pp. 447–456.
26. Parikh, P.; Sina, M.; Banerjee, A.; Wang, X.; D'Souza, M.S.; Doux, J.M.; Wu, E.A.; Trieu, O.Y.; Gong, Y.; Zhou, Q.; et al. Role of polyacrylic acid (paa) binder on the solid electrolyte interphase in silicon anodes. *Chem. Mater.* **2019**, *31*, 2535–2544. [CrossRef]
27. Wiśniewska, M.; Urban, T.; Grządka, E.; Zarko, V.I.; Gun'ko, V.M. Comparison of adsorption affinity of polyacrylic acid for surfaces of mixed silica-alumina. *Colloid. Polym. Sci.* **2014**, *292*, 699–705. [CrossRef] [PubMed]
28. Tomar, R.S.; Gupta, I.; Singhal, R.; Nagpal, A.K. Synthesis of poly (acrylamide-*co*-acrylic acid) based superabsorbent hydrogels: Study of network parameters and swelling behavior. *Polym. Plast. Technol. Eng.* **2007**, *46*, 481–488. [CrossRef]
29. Şolpan, D.; Duran, S.; Torun, M. Removal of cationic dyes by poly(acrylamide-*co*-acrylic acid) hydrogels in aqueous solutions. *Radiat. Phys. Chem.* **2008**, *77*, 447–452. [CrossRef]
30. Corona-Rivera, M.A.; Ovando-Medina, V.M.; Bernal-Jacome, L.A.; Cervantes-González, E.; Antonio-Carmona, I.D.; Dávila-Guzmán, N.E. Remazol red dye removal using poly(acrylamide-*co*-acrylic acid) hydrogels and water absorbency studies. *Colloid Polym. Sci.* **2017**, *295*, 227–236. [CrossRef]
31. Loulidi, I.; Boukhlifi, F.; Ouchabi, M.; Amar, A.; Jabri, M.; Kali, A.; Chraibi, S.; Hadey, C.; Aziz, F. Adsorption of crystal violet onto an agricultural waste residue: Kinetics, isotherm, thermodynamics, and mechanism of adsorption. *Sci. World J.* **2020**, *9*, 5873521. [CrossRef]
32. Labena, A.; Abdelhamid, A.E.; Amin, A.S.; Husien, S.; Hamid, L.; Safwat, G.; Diab, A.; Gobouri, A.A.; Azab, E. Removal of methylene blue and congo red using adsorptive membrane impregnated with dried ulvafasciata and sargassumdentifolium. *Plants* **2021**, *10*, 384. [CrossRef] [PubMed]

33. Jabar, J.M.; Odusote, Y.A.; Alabi, K.A.; Ahmed, I.B. Kinetics and mechanisms of congo-red dye removal from aqueous solution using activated Moringa oleifera seed coat as adsorbent. *Appl. Water Sci.* **2020**, *10*, 136. [CrossRef]
34. Weill, N.; Therrien, E.; Campagna-Slater, V.; Moitessier, N. Methods for docking small molecules to macromolecules: A user's perspective. 1. The theory. *Curr. Pharm. Des.* **2014**, *20*, 3338–3359. [CrossRef] [PubMed]
35. Meng, X.Y.; Zhang, H.X.; Mezei, M.; Cui, M. Molecular docking: A powerful approach for structure-based drug discovery. *Curr. Comput. Aided Drug Des.* **2011**, *7*, 146–157. [CrossRef] [PubMed]
36. Takahashi, O.; Masuda, Y.; Muroya, A.; Furuya, T. Theory of docking scores and its application to a customizable scoring function. *SAR QSAR Environ. Res.* **2010**, *21*, 5–6. [CrossRef] [PubMed]
37. Kumalo, H.M.; Bhakat, S.; Soliman, M.E.S. Theory and applications of covalent docking in drug discovery: Merits and pitfalls. *Molecules* **2015**, *20*, 1984–2000. [CrossRef] [PubMed]
38. Rudden, L.S.P.; Degiacomi, M.T. Protein docking using a single representation for protein surface, electrostatics, and local dynamics. *J. Chem. Theory Comput.* **2019**, *15*, 5135–5143. [CrossRef] [PubMed]
39. Balius, T.E.; Fischer, M.; Stein, R.M.; Adler, T.B.; Nguyen, C.N.; Cruz, A.; Gilson, M.K.; Kurtzman, T.; Shoichet, B.K. Testing inhomogeneous solvation theory in docking. *Proc. Natl. Acad. Sci. USA* **2017**, *114*, E6839–E6846. [CrossRef]
40. Sarfaraz, S.; Muneer, I.; Liu, H. Combining fragment docking with graph theory to improve ligand docking for homology model structures. *J. Comput. Aided Mol. Des.* **2020**, *34*, 1237–1259. [CrossRef]
41. Bendahma, Y.H.; Hamri, S.; Merad, M.; Bouchaour, T.; Maschke, U. Conformational modeling of the system pollutant/three-dimensional poly (2-hydroxyethyl methacrylate) (PHEMA) in aqueous medium: A new approach. *Polym. Bull.* **2019**, *76*, 1517–1530. [CrossRef]
42. Fuxman, A.M.; McAuley, K.B.; Schreiner, L.J. Modeling of free-radical crosslinking copolymerization of acrylamide and n,n'-methylenebis(acrylamide) for radiation dosimetry. *Macromol. Theory Simul.* **2003**, *12*, 647–662. [CrossRef]
43. Paljevac, M.; Jeřabek, K.; Krajnc, P. Crosslinked poly(2-hydroxyethyl methacrylate) by emulsion templating: Influence of crosslinker on microcellular structure. *J. Polym. Environ.* **2012**, *20*, 1095–1102. [CrossRef]
44. Chavda, H.; Patel, C. Effect of crosslinker concentration on characteristics of superporous hydrogel. *Int. J. Pharm. Investig.* **2011**, *1*, 17–21. [CrossRef] [PubMed]
45. Zhao, X.F.; Li, Z.J.; Wang, L.; Lai, X.J. Synthesis, characterization, and adsorption capacity of crosslinked starch microspheres with N,N'-methylene bisacrylamide. *J. Appl. Polym. Sci.* **2008**, *109*, 2571–2575. [CrossRef]
46. Abraham, J.; Pillai, V.N.R. N,N'-methylene bisacrylamide-crosslinked polyacrylamide for controlled release urea fertilizer formulations. *Commun. Soil Sci. Plant Anal.* **1995**, *26*, 3231–3241. [CrossRef]
47. Hamri, S.; Lerari, D.; Sehailia, M.; Dali-Youcef, B.; Bouchaour, T.; Bachari, K. Prediction of equilibrium swelling ratio on synthesized polyacrylamide hydrogel using central composite design modeling. *Int. J. Plast. Technol.* **2018**, *22*, 247–261. [CrossRef]
48. Gölgelioğlu, C.; Tuncel, A. Butyl methacrylate based monoliths with different cross-linking agents using DMF-aqueous buffer as porogen. *Electrophoresis* **2013**, *34*, 331–342. [CrossRef]
49. Hamri, S.; Bouchaour, T. pH-dependent swelling behaviour of interpenetrating polymer network hydrogels based on poly(hydroxybutyl methacrylate) and poly(2-hydroxyethyl methacrylate). *Int. J. Plast. Technol.* **2016**, *20*, 279–293. [CrossRef]
50. Zhu, L.; Guan, C.; Zhou, B.; Zhang, Z.; Yang, R.; Tang, Y.; Yang, J. Adsorption of dyes onto sodium alginate graft poly(acrylic acid-co-2-acrylamide-2-methyl propane sulfonic acid)/ kaolin hydrogel composite. *Polym. Polym. Compos.* **2017**, *25*, 627. [CrossRef]
51. Saraydin, D.; Solpan, D.; Işıkver, Y.; Ekici, S.; Güven, O. Radiation crosslinked poly(acrylamide/2-hydroxypropyl methacrylate/maleic acid) and their usability in the uptake of uranium. *J. Macromol. Sci. A* **2002**, *39*, 969–990. [CrossRef]
52. Heidari, S.; Esmaeilzadeh, F.; Mowla, D.; Ghasemi, S. Synthesis of an efficient copolymer of acrylamide and acrylic acid and determination of its swelling behavior. *J. Petrol. Explor. Prod. Technol.* **2018**, *8*, 1331–1340. [CrossRef]
53. Sheth, S.; Jain, E.; Karadaghy, A.; Syed, S.; Stevenson, H.; Zustiak, S.P. UV dose governs uv-polymerized polyacrylamide hydrogel modulus. *Int. J. Polym. Sci.* **2017**, *9*, 5147482. [CrossRef]
54. Lum, C.H.; Marshall, W.J.; Kozoll, D.D.; Meyer, K.A. The use of radioactive (I 131-labeled) rose bengal in the study of human liver disease: Its correlation with liver function tests. *Ann. Surg.* **1959**, *149*, 353–367. [CrossRef]
55. Dukhopelnykov, E.; Bereznyak, E.; Gladkovskaya, N.; Skuratovska, A.; Krivonos, D. Studies of eosin Y—DNA interaction using a competitive binding assay. *Spectrochim. Acta A Mol. Biomol. Spectrosc.* **2021**, *15*, 119114. [CrossRef]
56. Kellner-Rogers, J.S.; Taylor, J.K.; Masud, A.M.; Aich, N.; Pinto, A.H. Kinetic and thermodynamic study of methylene blue adsorption onto chitosan: Insights about metachromasy occurrence on wastewater remediation. *Energ. Ecol. Environ.* **2019**, *4*, 85–102. [CrossRef]
57. Trott, O.; Olson, A.J. AutoDock Vina: Improving the speed and accuracy of docking with a new scoring function, efficient optimization and multithreading. *J. Comput. Chem.* **2010**, *31*, 455–461. [CrossRef] [PubMed]
58. Murcko, M.A. Computational methods to predict binding free energy in ligand-receptor complexes. *J. Med. Chem.* **1995**, *38*, 4953–4967. [CrossRef]
59. Zhou, Q.H.; Li, M.; Yang, P.; Gu, Y. Effect of hydrogen bonds on structures and glass transition temperatures of maleimide–isobutene alternating copolymers: Molecular dynamics simulation study. *Macromol. Theory Simul.* **2013**, *22*, 107–114. [CrossRef]
60. Maurer, J.J.; Schulz, D.N.; Siano, D.B.; Bock, J. Thermal analysis of acrylamide-based polymers. In *Analytical Calorimetry*; Johnson, J.F., Gill, P.S., Eds.; Springer: Boston, MA, USA, 1984; pp. 43–55.

61. Ackland, G.J.; Bonny, G. Interatomic potential development. In *Comprehensive Nuclear Materials*, 2nd ed.; Konings, R.J.M., Stoller, R.E., Eds.; Elsevier: Amsterdam, The Netherlands, 2020; pp. 544–572.
62. Zhou, P.; Tian, F.; Lv, F.; Shang, Z. Geometric characteristics of hydrogen bonds involving sulfur atoms in proteins. *Proteins* **2009**, *76*, 151–163. [CrossRef]
63. Metrangolo, P.; Resnati, G. Halogen bonding: A paradigm in supramolecular chemistry. *Chem. Eur. J.* **2001**, *7*, 2511–2519. [CrossRef]
64. Politzer, P.; Lane, P.; Concha, M.C.; Ma, Y.; Murray, J.S. An overview of halogen bonding. *J. Mol. Model.* **2007**, *13*, 305–311. [CrossRef] [PubMed]
65. Veber, D.F.; Johnson, R.; Cheng, Y.; Smith, R.; Ward, W.; Kopple, D. Molecular properties that influence the oral bioavailability of drug candidates. *J. Med. Chem.* **2002**, *12*, 2615–2623. [CrossRef] [PubMed]
66. Djamaa, Z.; Lerari, D.; Mesli, A.; Bachari, K. Poly(acrylic acid-*co*-styrene)/clay nanocomposites: Efficient adsorbent for methylene blue dye pollutant. *Int. J. Plast. Technol.* **2019**, *23*, 110–121. [CrossRef]

Article

Evaluation of the Physical Stability of Starch-Based Hydrogels Produced by High-Pressure Processing (HPP)

Dominique Larrea-Wachtendorff [1], Vittoria Del Grosso [1] and Giovanna Ferrari [1,2,*]

[1] Department of Industrial Engineering, University of Salerno, 84084 Fisciano, Italy; dlarrea-wachtendorff@unisa.it (D.L.-W.); vdelgrosso@unisa.it (V.D.G.)
[2] ProdAl Scarl, c/o University of Salerno, 84084 Fisciano, Italy
* Correspondence: gferrari@unisa.it; Tel.: +39-089-964028

Abstract: Starch-based hydrogels are natural polymeric structures with high potential interest for food, cosmeceutical, and pharmaceutical applications. In this study, the physical stability of starch-based hydrogels produced via high-pressure processing (HPP) was evaluated using conventional and accelerated methods. For this purpose, conventional stability measurements, namely swelling power, water activity, texture, and organoleptic properties, as well as microbiological analysis of rice, corn, wheat, and tapioca starch hydrogels, were determined at different time intervals during storage at 20 °C. Additionally, to assess the stability of these structures, accelerated tests based on temperature sweep tests and oscillatory rheological measurements, as well as temperature cycling tests, were performed. The experimental results demonstrated that the physical stability of starch-based HPP hydrogels was interdependently affected by the microorganisms' action and starch retrogradation, leading to both organoleptic and texture modifications with marked reductions in swelling stability and firmness. It was concluded that tapioca starch hydrogels showed the lowest stability upon storage due to higher incidence of microbial spoilage. Accelerated tests allowed the good stability of HPP hydrogels to be predicted, evidencing good network strength and the ability to withstand temperature changes. Modifications of the rheological properties of corn, rice, and wheat hydrogels were only observed above 39 °C and at stress values 3 to 10 times higher than those necessary to modify commercial hydrogels. Moreover, structural changes to hydrogels after cycling tests were similar to those observed after 90 days of conventional storage. Data obtained in this work can be utilized to design specific storage conditions and product improvements. Moreover, the accelerated methods used in this study provided useful information, allowing the physical stability of starch-based hydrogels to be predicted.

Keywords: starch-based; high-pressure processing; stability

Citation: Larrea-Wachtendorff, D.; Del Grosso, V.; Ferrari, G. Evaluation of the Physical Stability of Starch-Based Hydrogels Produced by High-Pressure Processing (HPP). *Gels* **2022**, *8*, 152. https://doi.org/10.3390/gels8030152

Academic Editors: Yi Cao and Wei Ji

Received: 31 January 2022
Accepted: 25 February 2022
Published: 1 March 2022

Publisher's Note: MDPI stays neutral with regard to jurisdictional claims in published maps and institutional affiliations.

Copyright: © 2022 by the authors. Licensee MDPI, Basel, Switzerland. This article is an open access article distributed under the terms and conditions of the Creative Commons Attribution (CC BY) license (https://creativecommons.org/licenses/by/4.0/).

1. Introduction

Hydrogels are a commercially widespread group of polymeric materials, consisting of three-dimensional crosslinked networks of hydrophilic or hydrophobic biopolymers capable of absorbing and retaining a significant amount of water [1]. In recent years, hydrogels have been recognized as "smart structures" with tailor-made characteristics conferring different functional attributes of the utmost importance for the design, synthesis, and self-assembly of novel biomaterials, including drug delivery systems [2,3]. Currently, natural hydrogels produced from renewable sources, as alternatives used to replace or reduce the use of synthetic materials, are receiving significant attention in the scientific community due to their compatibility with the human body [4]. Among this trending group of hydrogels, starch-based hydrogels are among the most promising alternatives for producing polymeric biomaterials [5,6]. Their biocompatibility, hydrophilicity, and biodegradability have been highlighted as remarkable characteristics of these structures, encouraging their extensive use in several applications [2,3,5,7–11]. Moreover, given the wide range of applications of starch-based hydrogels, their physical–mechanical properties and durability can be fine-tuned by changing the processing methods and operation conditions, the type of biopolymer, and the composition of the liquid phase [4].

Recently, high-pressure processing technology (HPP) was proposed to produce natural-starch-based hydrogels [12–14]. HPP is a non-thermal technology that is widely used in the food industry for pasteurization, since it allows extended shelf life, minimizing nutritional and sensorial property losses in processed products [15]. Moreover, it is well known that HPP promotes sol–gel transitions in proteins and other food components [16]. For this reason, this technology has also been proposed as a physical method for modifying or gelatinizing different types of starch suspensions, allowing the limitations of gelation processes that are currently utilized to be overcome, such as the long operation times, high energy consumption levels, and use of hazardous materials [13,17–25].

Starch-based hydrogels form thanks to the physical entanglement of polymer chains or other non-covalent interactions, giving a reversible nature to these structures, which represents an advantage in biomedical and food applications for these materials. However, due to their non-permanent bonds, HPP hydrogels are commonly considered weak structures with reduced rheological properties compared to chemical hydrogels. This hinders their utilization in applications where good mechanical properties and stability are desired [4]. It is interesting to note that starch-based HPP hydrogels obtained in well-identified processing conditions display excellent mechanical and rheological properties that are even superior to those of thermal gels, making them very promising materials [12,13].

Nevertheless, to the best of our knowledge, no studies on the physical stability of these novel structures have been carried out, the determination of which is of the utmost importance in view of improving the design of these biomaterials, optimizing the processing conditions, and proposing their application in different industrial sectors.

This work aimed at evaluating the stability and physicomechanical characteristics of starch-based HPP hydrogels with time by applying different stress conditions, using either specific conventional or accelerated detection methods, and determining the shelf life of hydrogels over short storage periods. Bearing in mind that the guidelines for stability testing (International Conference on Harmonization, ICH, [26]) commonly used to predict the physical stability of cosmeceutical and pharmaceutical products under different ambient conditions (temperature, relative humidity) utilize long sampling periods even in accelerated conditions (from 3 to 6 months), are expensive and time-consuming, and require considerable scientific expertise and knowledge of the materials and methodologies involved [26], we attempted to develop specific testing methods to determine the stability of the new materials. These methods, which are less expensive and time-consuming, allow reproducible and high-quality data correlated with extreme environmental conditions to be obtained that are useful for innovative product development.

2. Results and Discussion

2.1. Conventional Evaluations

2.1.1. Microbiological Analysis

Growth curves of aerobic mesophilic microorganisms, yeasts, and molds on starch-based HPP hydrogels stored at 20 °C are shown in Figure 1.

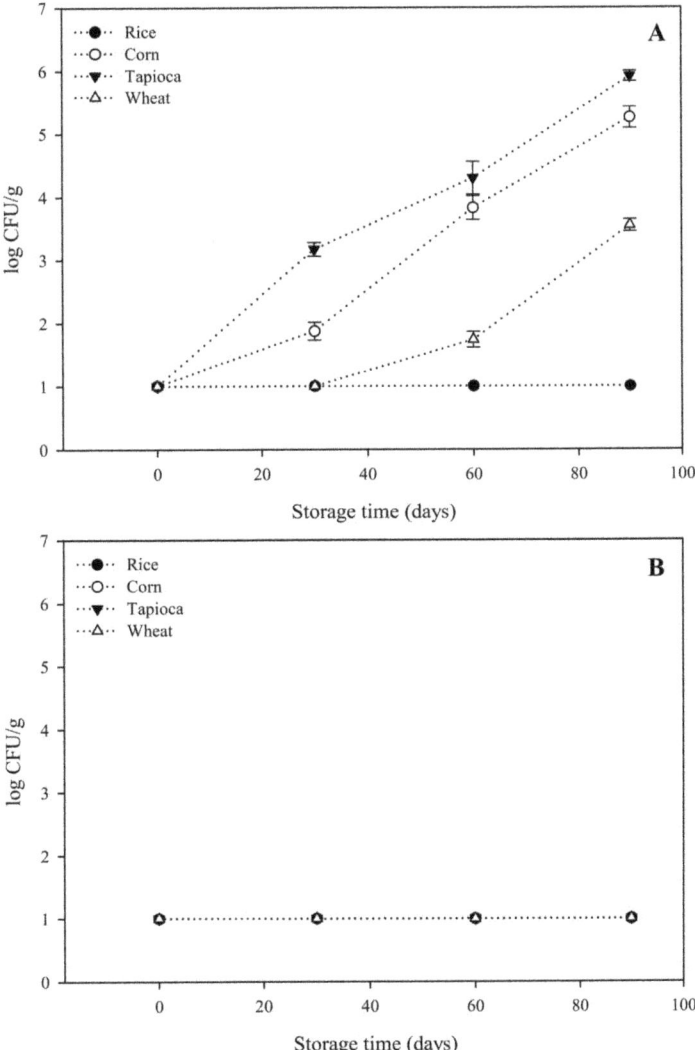

Figure 1. Growth curves of aerobic mesophilic total count (**A**) and yeasts and molds (**B**) of starch-based HPP hydrogels during the storage period. Symbols are means of three measurements ± standard deviations (S.D).

The initial microbial counts in all starch-based hydrogels tested were within the undetectable range, namely < 1 log CFU/g (Figure 1). Moreover, no yeasts or molds were detected, and the number of survivors was always below the detection limit (<1.0 log CFU/g) during the entire storage period (Figure 1B), confirming the effectiveness of HPP to inactivate this group of microorganisms. It has been extensively demonstrated that yeasts and molds are very sensitive to HPP due to the instantaneous impact of pressure over the nuclear membrane of yeast and mold populations, which causes a lethal injury to microbial cells, leading to their complete inactivation [27,28]. In contrast, although immediately after HPP treatments the microbial load was below the detection limit, significant growth of aerobic mesophilic microorganisms in almost all starch-based hydrogels was detected during storage (Figure 1A). While the microbial load of rice HPP hydrogels remained below detection limits (<1.0 log CFU/g) during the entire storage period, in tapioca, corn, and wheat hydrogels we detected exponential growth of aerobic mesophilic bacteria from the first month of storage onwards. After 90 days of storage, the microbial counts of the three hydrogels were above 6, 5, and 3 log CFU/g, respectively. These results confirmed the well-known findings that HPP treatments cause sublethal injuries on certain resistant microorganisms that are able to repair cellular damages due to pressure and recover their viability during storage, as reported elsewhere [28,29]. In this regard, it can be hypothesized that the resistant survivors detected in corn, wheat, and tapioca-starch-based HPP hydrogels could belong to pressure-resistant microbial subpopulations such as bacterial spores [30], which seemingly differ based on the type of starch source and should be addressed in further experiments. Bacterial spores are extremely resistant life forms triggered by stress scenarios such as high-pressure conditions, returning to active growth during storage at optimal temperature conditions such as 20 °C [30,31]. However, further experiments on the identification and individual aspects of the pathogenicity of the microbial populations in starch-based hydrogels should be performed.

Furthermore, the results obtained gave interesting insights into the microbiological fingerprint of starch-based hydrogels produced by HPP, indicating that microbiological stabilization of these materials should be a matter of concern during processing.

2.1.2. Swelling Stability

Water is the main solvent that influences the processing, quality, texture properties, and stability of starch-based hydrogels [23,32–34]. The ability of starches to bind water during gel formation and retain it during storage can be assessed by measuring parameters such as the swelling power (SP) and water activity (A_w). Figure 2 depicts the evolution (delta of current and initial values) of SP and A_w values of corn, rice, wheat, and tapioca HPP hydrogels during storage at 20 °C.

At the beginning of the storage period (0 days), starch-based HPP hydrogels presented similar A_w values (in the range of 0.974–0.969). Tapioca starch hydrogels showed the highest initial SP values (7.4 g H_2O/g $_{dry\ starch}$), followed by rice (6.5 g H_2O/g $_{dry\ starch}$), corn (5.4 g H_2O/g $_{dry\ starch}$), and wheat starch hydrogels (5.16 g H_2O/g $_{dry\ starch}$). These results can be explained by the higher capacity of tapioca starch granules to swell and solubilize with respect to cereal starch granules and the absence of amylose–lipid complex formation, as reported elsewhere [14,22,35].

Nonetheless, an evident syneresis phenomenon was observed, as shown in Figure 2A, where for all samples the SP values decreased with increasing the storage time. Wheat and corn starch hydrogels displayed better swelling stability than rice and tapioca starch hydrogels. For the latter structures, an abrupt reduction of water binding capacity from the beginning of the storage period was detected and losses at the end of the storage period (90 days) of 1.3 g H_2O/g $_{dry\ starch}$ for rice and −2.2 g H_2O/g $_{dry\ starch}$ for tapioca starch hydrogel were measured. These results are in slight agreement with the ones reported by Torres et al. [35] for potato starch gels obtained by thermal treatments utilizing different types of water. The authors observed a syneresis phenomenon after 60 days of cold storage (4 °C) in samples prepared with distilled water, at potato starch concentrations of 10% and 20% *w/w*. However, it should be emphasized that either the gel formation conditions (60 °C for 30 min) or the storage temperature (4 °C) was more severe, and could have delayed the physical and microbiological deterioration of the hydrogels. In our case, it could be assumed that the differences observed among samples can be attributed to the different starch retrogradation extent in hydrogel samples. Retrogradation has been defined as a structural reorganization of the starch chains due to the crystallization of amylose and amylopectin molecules during storage, causing gel shrinking and phase separation (syneresis) [36]. The rate and extent of retrogradation are sensitive to the water contents of starch gels and baked products, because water acts as a plasticizer of the amorphous region [37,38]. Therefore, it can be assumed that the starch retrogradation extent was higher in hydrogels with higher initial SP values, such as tapioca and rice starch hydrogels, than in hydrogels with lower initial SP values, such as corn and wheat starch hydrogels. Microbial spoilage may accelerate retrogradation in tapioca starch hydrogels, promoting higher syneresis, confirming the results of Figures 1A and 2A.

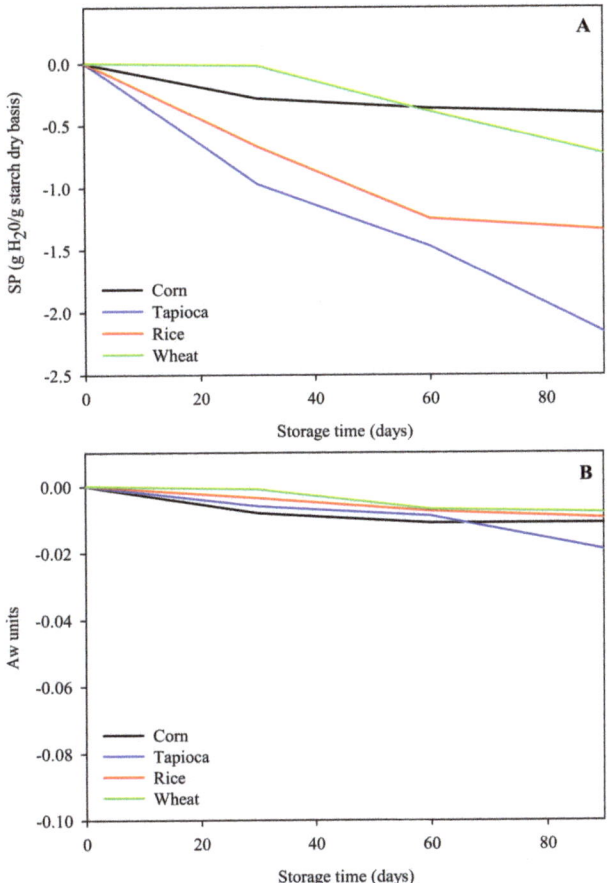

Figure 2. Influence of storage time on the swelling power (**A**) and A_w (**B**) values of starch-based HPP hydrogels.

Measurements of A_w, which accounts for the status and dynamics of water in association or interaction with other molecules in complex matrices [39], were carried out. As shown in Figure 2B, in all samples a slight and constant decrease in A_w was observed during storage. Through A_w measurements, it was possible to evaluate the free water present on the surface of the sample, which changed only marginally with storage time. On the contrary, water loss, also quantified as free water in SP measurements, during storage time was much higher than that detected through A_w due to the centrifugal forces applied in this method. These discrepancies suggest that the water entrapped in the polymeric network experienced different changes upon storage, moving outwards layer by layer, reaching the hydrogel surface from where it was easily removed during SP measurements.

2.1.3. Texture Profile Analysis (TPA)

The physical stability of polymeric materials is defined as the extent to which a product retains, within the specified limits, the same properties and physical characteristics possessed at the time of its packaging [26]. Texture measurements represent valuable tools to quantify the physical properties of polymeric materials such as starch-based hydrogels. Moreover, based on our previous findings [12], the firmness and adhesiveness are the parameters characterizing the texture of starch-based HPP hydrogels. Consequently, the trends of these two parameters were determined for rice, wheat, corn, and tapioca starch hydrogels during the storage period and the data are reported in Figure 3 as the delta of current and initial values.

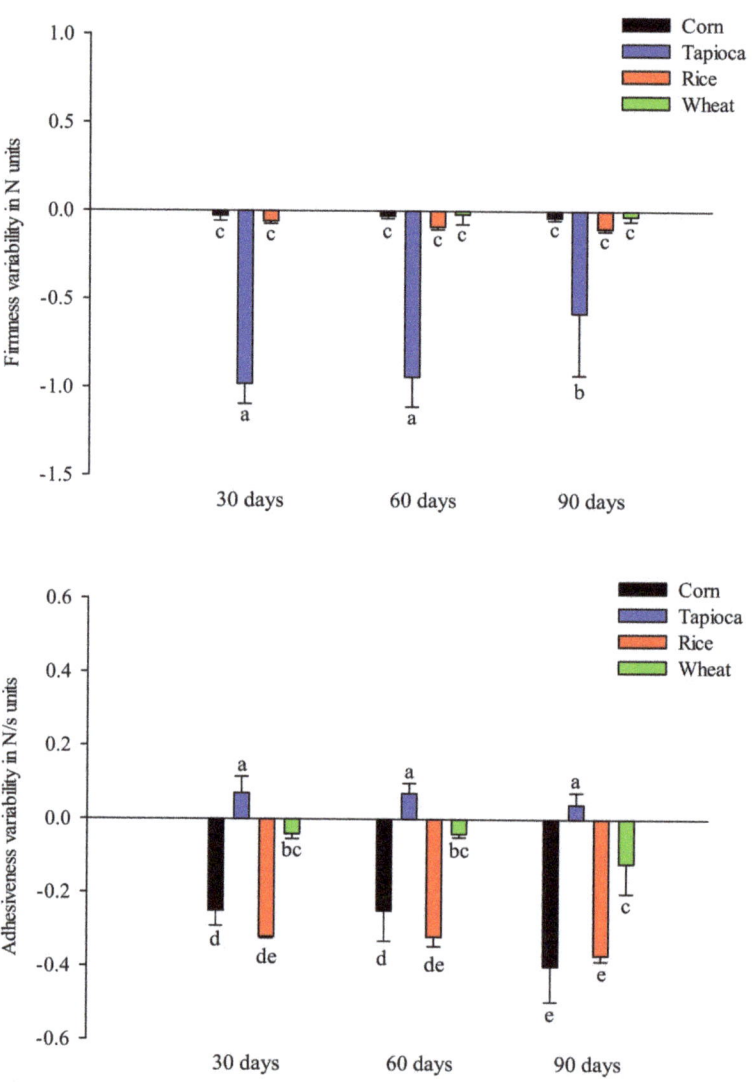

Figure 3. Influence of storage time on firmness and adhesiveness values of starch-based HPP hydrogels. Different letters above the bars indicate significant differences among the mean values (LSD, $p \leq 0.05$).

The firmness gives information on the strength of a gel structure, whereas the adhesiveness accounts for the adhesive forces present in starch-based hydrogels [14]. As can be seen in Figure 3, the texture of starch-based HPP hydrogels was strongly affected by storage time. In particular, the firmness values of all hydrogel samples decreased throughout the storage period, with tapioca starch hydrogels showing the highest reductions in this parameter (>70%) after 30 days of storage at 20 °C. Rice, corn, and wheat starch hydrogels showed slight reductions in firmness during the entire storage period (90 days). These results agree with those previously discussed (Figures 1 and 2), confirming that the physical stability of these structures is strongly influenced by factors triggered during the storage period, such as microbiological spoilage and starch retrogradation, which cause strong structural changes, such as network weakening, phase separation, and syneresis.

Similarly, as the storage time increased, the adhesiveness of rice, corn, and wheat starch hydrogels decreased due to the viscoelasticity losses of the gel network. Interestingly, the increased adhesiveness of tapioca starch hydrogels after 30 days of storage can be attributed to the leaching of water, amylose, and amylopectin, which determine adhesive forces in starch gels [40].

2.1.4. Organoleptic Property Evaluation

Data for the organoleptic properties of rice, corn, tapioca, and wheat starch hydrogels measured during storage are reported in Table 1 and Figure 4.

Table 1. Influence of storage time on the organoleptic properties of starch-based HPP hydrogels.

HHP Hydrogel	Storage Time (Days)			
	0	30	60	90
Rice	Homogeneous Brilliant	N.M.	N.M.	More liquid
Corn	Homogeneous Opaque	N.M.	Yellowish	More liquid and yellowish
Tapioca	Homogeneous Compact	Evident Syneresis and Phase separation		Broken structure
Wheat	Homogeneous opaque	N.M.	N.M.	Some lumps

N.M.: No modifications.

In agreement with the results of the previous sections, tapioca starch hydrogels showed more significant organoleptic changes than rice, corn, and wheat hydrogels. Syneresis and phase separation after the first month of storage, and consequently, the rupture of the structure after 60 days, were detected via the organoleptic evaluation as the main physical changes of tapioca starch hydrogels during storage. Interestingly, only slight organoleptic modifications of rice starch hydrogels were observed during the storage period (Table 1, Figure 4). These findings suggest that the organoleptic changes in starch-based hydrogels are strongly influenced by the actions of microorganisms, which produce network deteriorations and facilitate starch retrogradation. Moreover, the level of the organoleptic changes depended on the structural characteristics of each hydrogel (Figure 4). For instance, in cream-like structures such as rice, corn, and wheat hydrogels, the organoleptic changes are less visible than in rubber-like structures such as tapioca starch hydrogels, highlighting the importance of instrumental determination in the evaluation of the physical stability of such gel structures.

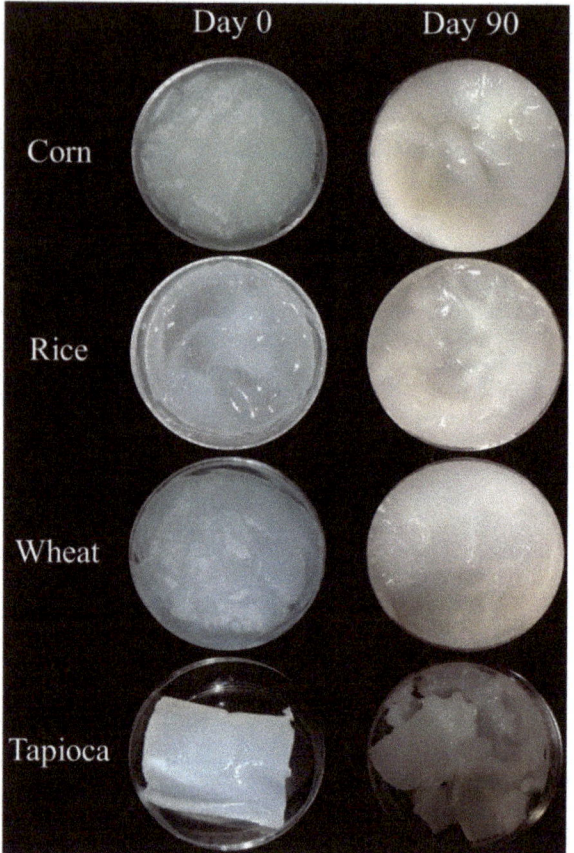

Figure 4. Influence of storage time on the physical appearance of starch-based HPP hydrogels.

Furthermore, from conventional stability results, it can be hypothesized that the stiffer the starch-based HPP hydrogel is, the more marked the structural deterioration, meaning weaker physical stability can be expected when microbial spoilage promotes starch retrogradation. Starch, on which the hydrogels under investigation in this work are based, is a polymer chain of glucose molecules linked to each other through glycosidic bonds that can undergo important modifications during storage. The structural reorganization of its macromolecules and the attack of spore fungiform microorganisms that normally possess amylases [41] can synergically produce the weakening of the internal structure of the starch hydrogels due to the rupture of the physical bonds or interactions, causing organoleptic changes, firmness reductions, syneresis, and phase separation.

2.2. Accelerated Stability Tests

In accelerated stability tests, a product is subjected to high-stress conditions. For the sake of completeness, the applied stress required to cause starch-based hydrogels structure failure was determined in this work. These data can be utilized to predict the capability of products to keep their initial characteristics, as well as for processing improvements [26]. In our experimental trials, parameters such as temperature and mechanical stress were set as stress conditions during accelerated stability testing, which are widely used to assess the stability of gels and pharmaceutical products [42,43].

2.2.1. Temperature Cycling Test

Temperature cycling tests are accelerated physical methods that are commonly utilized in pharmaceutical sciences to provide information on a product's instabilities that is not provided by isothermal tests [44]. Cyclic temperature tests are designed based on the characteristics of the products. In this study, starch-based HPP hydrogels were packaged and stored at 4 °C and 40 °C, and the temperature was changed every 24 h for 7 days, simulating extreme storage conditions [43]. The results of temperature cycling tests on the texture parameters of starch-based HPP hydrogels are reported in Table 2.

Table 2. Influence of temperature cycling tests on the firmness and adhesiveness of starch-based hydrogels immediately after HPP treatments.

HPP Hydrogel	Time (Day)	Firmness (N)	Adhesiveness (−N *s)
Corn	0	0.11 ± 0.01 [c]	0.82 ± 0.11 [a]
	7	0.06 ± 0.01 [d]	0.36 ± 0.05 [c]
Tapioca	0	1.36 ± 0.09 [a]	-
	7	0.24 ± 0.07 [b]	0.06 ± 0.00 [e]
Rice	0	0.17 ± 0.08 [b]	0.85 ± 0.01 [a]
	7	0.06 ± 0.01 [d]	0.50 ± 0.02 [b]
Wheat	0	0.11 ± 0.01 [c]	0.44 ± 0.05 [bc]
	7	0.06 ± 0.00 [d]	0.25 ± 0.02 [d]

[a–e] Different letters in the same column indicate significant differences among samples (LSD, $p \leq 0.05$).

The data reported in Table 2 show that the textures of all hydrogels were significantly affected by the temperature cycling test conditions. Rice, corn, and wheat starch hydrogels showed significant reductions in firmness and adhesiveness values ($p \leq 0.05$), whereas a sharp decrease in firmness was observed in tapioca starch hydrogels (−88%) ($p < 0.05$), as well as an increased adhesiveness ($p \leq 0.05$) after the cycling tests. These results could be attributed to the leakage of water and solubilized amylose, or amylopectin molecules promoted by temperature changes (cycles from 4 °C to 40 °C), causing changes in the interactions occurring in the gel network and the adhesive forces. This is in agreement with the findings reported by Schirmer et al. [45]. The authors stated that viscosity changes can be detected in starch–water systems in the temperature range between 30 °C and 50 °C as a result of the solubilization (leaching) of macromolecules.

Furthermore, the results of the temperature cycling tests showed that physical changes in cream-like hydrogels are more related to the network weakening than to rupture, differently from what was observed in tapioca starch hydrogels. Remarkably, these physical changes are similar to those observed after the conventional storage period, highlighting the effectiveness and accuracy of this accelerated method to predict the physical instabilities of starch-based HPP hydrogels in shorter time periods.

2.2.2. Rheology

Rheological tests have been proposed to predict the stability of products and provide useful information to improve formulations or overcome the occurrence of certain instabilities in gels structures [43]. For this purpose, temperature and stress sweep tests were carried out, which are valuable tools for predicting the physical stability of emulsions, gels, and pharmaceutical products [43,46,47].

Temperature Sweep Tests

To predict the thermal stability of the starch-based HPP hydrogels studied in this investigation, temperature sweep tests from 25 °C to 60 °C at a heating rate of 1 °C/min were performed. For the sake of comparison, thermal gels were also tested. Figure 5 depicts the elastic responses (G′) of corn, rice, tapioca, and wheat hydrogels undergoing temperature stress.

From Figure 5, it can be observed that the temperature increments influenced the rheological properties of starch-based HPP hydrogels to different extents. Elastic instabilities occurred in corn and wheat HPP hydrogels over 47 °C, whereas the same instabilities were detected in rice and tapioca HPP hydrogels at 39 °C and 35 °C, respectively. Moreover, the inflection points of G′ curves, corresponding to the temperatures affecting the gel-like profiles of hydrogels, were always observed at higher temperature values (±6.5 °C average) in hydrogels produced using high-pressure processing than in those obtained with thermal processes, suggesting a superior thermal stability of the former ones.

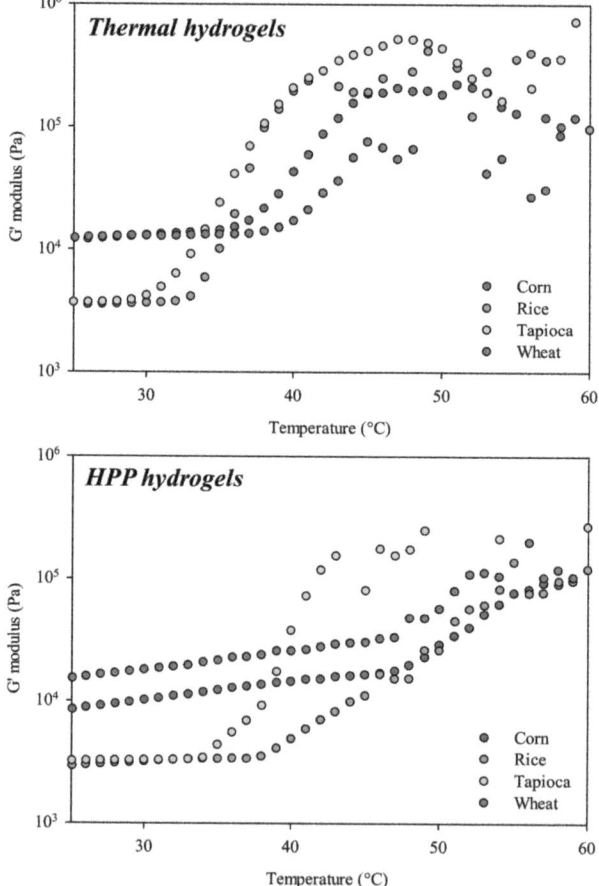

Figure 5. The influence of temperature stress on the elastic response (G′ modulus) of starch-based hydrogels immediately after HPP treatments and thermal treatments.

It is well known that gels obtained with physical methods exhibit a strong dependence of the G′ modulus on temperature, with significant decreases in elastic domains observed with increasing temperature, mainly due to the loss of interconnectivity of the network constituents [43]. Interestingly, an opposite trend can be observed in Figure 5, with the elastic domains (G′) of starch-based hydrogels being wider and the inflection points moving towards higher temperature values. This can be attributed to the retrogradation phenomenon typically observed in starch gels produced with thermal treatments, and to a less extent in HPP-treated starch [48–52].

Moreover, some studies have demonstrated that differing from thermal gelatinization, in HPP starch gels a characteristic retrogradation phenomenon is likely to occur due to the different dynamics of water and the presence of almost intact starch granules (absence of stirring). This prevents lower amylose leaching, which in turn is less prone to retrograde [48–50]. This may also partially explain the higher thermostability of starch-based HPP hydrogels.

Stress Sweep Tests

The network strength of starch-based HPP hydrogels, and for the sake of comparison of a commercial hydrogel, namely Carbopol, was evaluated through stress sweep tests. In Figure 6, the data of the viscoelastic responses (G′ elastic and G″ viscous moduli) of corn, rice, tapioca, wheat hydrogels, and Carbopol subjected to a deformation (oscillation) stress ramp are reported.

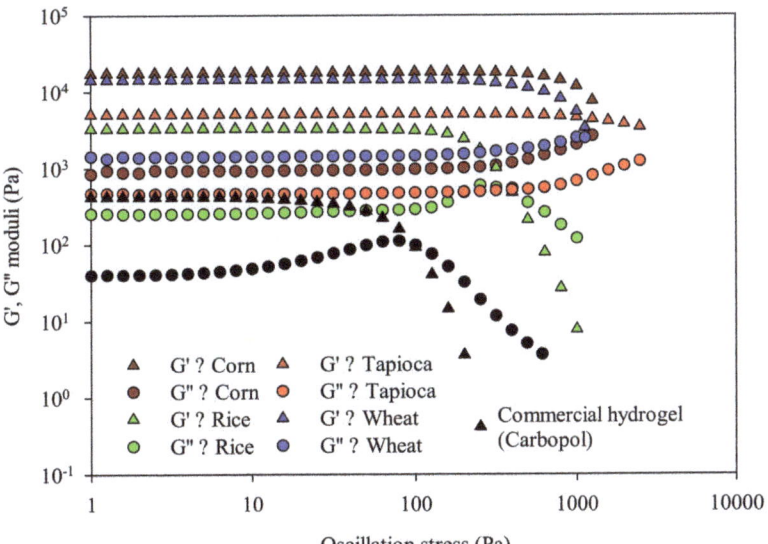

Figure 6. Influence of deformation stress on the viscoelastic responses (G′, G″ moduli) of starch-based hydrogels immediately after HPP treatments.

All samples showed the typical behavior of gels structures, with elastic properties predominating over the viscous ones (G′ >> G″) [13,14,53]. Interestingly, G′ and G″ values of starch-based HPP hydrogels were significantly higher than those of the commercial gel, indicative of a highly structured profile [54]. Moreover, considering the gel strength, the linear viscoelasticity range (LVR) reveals the range of deformation stress that a polymeric network withstands; thus, the wider the extension of the LVR, the higher the network strength. [47]. Remarkably, all starch-based HPP hydrogels investigated showed a wider LVR compared to Carbopol. Carbopol had already lost its gel-like behavior (G′ = G″) at 100 Pa, while in tapioca, corn, wheat, and rice starch HPP hydrogels the rheological instabilities occurred above 5000, 2000, 1000, and 300 Pa, respectively. These findings indicate the greater network strength of the latter natural structures, resulting in greater physical stability than that of commercial hydrogels. Considering the governing principles of HPP, in gels obtained utilizing this technology the physical crosslinking networks formed by the inter- and intramolecular interactions between starch and water molecules are significantly favored, and the resulting gels are characterized by strong and dense networks.

3. Conclusions

Starch-based HPP hydrogels are structures with high application potential in pharma, cosmetic, and food sectors. These novel structures are required to be physically stable at room temperature, and microbiological spoilage should be avoided. Thus, the determination of their stability is crucial to assess their performance with storage time. In this study, conventional and accelerated methods for the evaluation of starch-based HPP hydrogels were used. Conventional techniques demonstrated that the physical stability of starch-based HPP hydrogels depended on the occurrence of microbial spoilage and starch retrogradation, which synergically triggered the physical deterioration of the gel structures, such as network weakening and phase separation upon storage at room temperature.

Accelerated methods were proven adequate to evaluate the stability of hydrogels, requiring a shorter testing period. They allowed us to predict good physical and thermal stability and good network strength for starch-based HPP hydrogels, superior to the results for commercial hydrogels. Rheological alterations of rice, wheat, and corn HPP hydrogels were demonstrated to take place only at temperatures higher than 39 °C.

In conclusion, data obtained in this study enabled an understanding of the limits of these novel biomaterials and predictions of their physical stability, as well as improvements in their design and forecasting of their novel applications.

4. Materials and Methods

4.1. Starch-Based Hydrogel Preparation via High-Pressure Processing (HPP)

4.1.1. Materials

Rice (S7260) (17.7% amylose content, 7.2% water content), wheat (S5127) (26.96% amylose content, 7.8% water content), and corn (S4126) (21.17% amylose content, 8.3% water content) starch powders were purchased from Sigma Aldrich (Steinheim, Germany). Tapioca starch (20.2% amylose content, 9.6% water content) was obtained from Rudolf Sizing Amidos do Brazil (Ibirarema—Sao Paulo, Brazil). Starch moisture content was determined according to AOAC guidelines (925.10). Amylose content was determined by an enzymatic rapid assay kit (Megazyme International Ireland Ltd., Wicklow, Ireland).

4.1.2. Samples Preparation

Based on previous experimental findings [12,13], different starch–water suspensions were formed at a concentration of 20% (w/w). To ensure sample homogeneity and avoid particles sedimentation, the starch suspensions were prepared immediately before HPP treatments.

4.1.3. High-Pressure Processing Treatments

HPP treatments were performed in a U22 laboratory-scale high pressure unit (Institute of High Pressure Physics, Polish Academy of Sciences, Unipress Equipment Division, Poland) as described elsewhere [55].

In each test, 10 g of the starch suspension was de-aerated, thoroughly mixed, and packed in flexible pouches, which were then sealed, loaded in the pressure vessel, and HPP-treated at 600 MPa for 15 min at 25 °C. This processing condition was sufficient for complete gelatinization and proper gel formation [14].

HPP hydrogels were stored at room temperature (20 °C) before analyses.

4.2. Stability Measurements

4.2.1. Conventional Evaluations

Rice, wheat, corn, and tapioca HPP hydrogels were stored at 20 ± 0.2 °C in an incubator (Thermocycle, Pbi international, Milano, Italy). Microbial counts, organoleptic and texture properties, as well as swelling power and water activity levels were evaluated every four weeks. For all experimental determinations, three different samples of each starch-based hydrogel were used. Unless otherwise stated, all measurements were performed in triplicate.

Microbial Count

Hydrogels samples were analyzed to enumerate the numbers of mesophilic aerobic microorganisms, yeasts, and molds. Each bag containing 10 g of starch-based HPP hydrogels was aseptically opened, and together with 90 mL of buffered peptone water (VWR International, Leuven, Belgium), was introduced in an aseptic stomacher bag. A Stomacher 400 unit (Seward Laboratory, London, UK) at 260 rpm was used for the complete homogenization of the samples. Further decimal dilutions were prepared with the same diluent and plated on appropriate growth media.

To enumerate aerobic mesophilic microorganisms, 1.0 mL of each dilution was pour-plated in Plate Count Agar (PCA, Merck, Darmstadt, Germany) and incubated at 30 °C for 72 h. To count yeasts and molds, 1.0 mL of each dilution was spread-plated on Dichloran Rose Bengal Chloramphenicol medium agar (DRBC, Oxoid, Basingstoke, Hampshire, England) and then incubated for 3–5 days at 25 °C.

The microbial colonies were identified and quantified by standard methods. Microbiological data were transformed into logarithms of the number of colony-forming units (log CFU/g). The detection limit was 10 CFU/g (1.0 log CFU/g).

Evaluation of Organoleptic Properties

The organoleptic properties—namely the appearance, homogeneity, integrity, color, and odor—of the samples during storage were monitored using visual observations. Additionally, the physical appearances of samples were recorded using a digital camera (Sony Corp, Japan) in angular mode. Original pictures without editing and filtering were stored.

Swelling Properties

Swelling Power (SP)

The swelling power levels of hydrogels were determined according to the method reported by Kusumayanti et al. [56] and modified by Larrea-Wachtendorff et al. [14].

Each hydrogel sample was centrifuged (PK130R, ALC, Winchester, VA) at $1351\times g$ for 10 min, the supernatant was removed, and the pellet was weighed before and after drying for 6 h at 105 °C. The swelling power was evaluated as follows:

$$\text{Swelling power (g/g)} = \frac{\text{Weight of the wet pellet (g)}}{\text{Dry weight of hydrogel sample (dry basis) (g)}} \quad (1)$$

Determination of Water Activity (A_w)

The water activity (A_w) levels of samples were measured using a Novasina water activity instrument (TH-500, Pfäffikon, Lachen, Switzerland). Before A_w measurements, the instrument was calibrated at 25 °C using standard patterns. For the measurements, 1 g of hydrogel was poured into the center of the sampling chamber and consecutive readings were carried out until constant A_w values were attained.

Texture Profile Analysis (TPA)

TPA of all HPP hydrogels was performed in a TA.XT2 texture analyzer (Stable Micro Systems, Surrey, UK) equipped with a 5 kg load cell, connected to a microcomputer. The textures of hydrogel samples were determined according to the procedure reported by Larrea-Wachtendorff et al. [13]. Briefly, 6 g of hydrogel samples were poured into a cylindrical cell (24 mm height and 25 mm of internal diameter), and tests were carried out at room temperature using a P10 probe (10 mm diameter) at a rate of 1 mm/s until 50% sample deformation was attained. The compression runs were repeated twice, at a decompression rate of 1 mm/s and a delay of 5 s between two bites, to generate force–time curves. Hardness, adhesiveness, springiness, chewiness, cohesiveness, and gumminess values of hydrogels were calculated from the recorded penetration data.

4.2.2. Accelerated Evaluations

Cycling Test

The cycling test was performed according to the method reported by Almeida and Bahia [43] with slight modifications. Hydrogels samples, obtained immediately after HPP treatments (day 0), were stored in a climatic chamber (TCN-50 plus, ARGO Lab, China) and the temperature was changed between 4 and 40 °C every 24 h for 7 days. Before undergoing TPA, all samples were stored at 20 °C for 24 h.

Rheological Analysis

The rheological analysis of samples was carried out in a controlled stress rheometer (AR2000-TA instruments, New Castle, DE, USA) in a plate–cone geometry configuration (40 mm diameter, 2°) with a gap of 52 μm. The viscoelastic responses of hydrogel samples were recorded (immediately obtained after HPP treatments), namely the elastic (G′) and viscous (G″) moduli under different stress (0.01 to 1000 Pa at 25 °C) and temperature (25 °C to 80 °C at 1°C/min) conditions. Additionally, for the sake of comparison, the same measurements were also carried out on commercial hydrogel samples (Diclac®, Eurofarma, Chile) and thermal gels. The latter were produced according to the method and processing conditions described by Larrea-Wachtendorff et al. [13].

4.3. Statistical Analysis

The results were analyzed using statistical descriptive analysis (mean ± SD), one-way ANOVA, and post-hoc comparison using the Fisher least significant difference (LSD) test to determine significant differences among experiments (p-value was <0.05). All analyses were performed using Statgraphic Centurion XVI Statistical Software (Statistical Graphics Corp., Herdon, VA, USA).

Author Contributions: Conceptualization, D.L.-W. and G.F.; methodology, D.L.-W.; software, D.L.-W.; validation, D.L.-W., V.D.G. and G.F.; formal analysis, D.L.-W. and G.F.; investigation, D.L.-W. and V.D.G.; resources, G.F.; data curation, D.L.-W.; writing—original draft preparation, D.L-W.; writing—review and editing, D.L.-W., V.D.G. and G.F.; visualization, D.L.-W. and G.F.; supervision, G.F.; project administration, G.F.; funding acquisition, G.F. All authors have read and agreed to the published version of the manuscript.

Funding: This work was financially supported by ProdAl Scarl and the University of Salerno, Italy.

Institutional Review Board Statement: Not applicable.

Informed Consent Statement: Not applicable.

Data Availability Statement: The data presented in this study are available in the article.

Conflicts of Interest: The authors declare no conflict of interest.

References

1. Biduski, B.; Max, W.; Colussi, R.; Lisie, S.; El, D.M.; Lim, L.; Renato, Á.; Dias, G.; Zavareze, R. International Journal of Biological Macromolecules Starch hydrogels: The influence of the amylose content and gelatinization method. *Int. J. Biol. Macromol.* **2018**, *113*, 443–449. [CrossRef] [PubMed]
2. Mahinroosta, M.; Jomeh, Z.; Allahverdi, A.; Shakoori, Z. Hydrogels as intelligent materials: A brief review of synthesis, properties and applications. *Mater. Today Chem.* **2018**, *8*, 42–55. [CrossRef]
3. McClements, D.J. Recent progress in hydrogel delivery systems for improving nutraceutical bioavailability. *Food Hydrocoll.* **2017**, *68*, 238–245. [CrossRef]
4. Mohammadinejad, R.; Maleki, H.; Larrañeta, E.; Fajardo, A.R.; Nik, A.B.; Shavandi, A.; Sheikhi, A.; Ghorbanpour, M.; Farokhi, M.; Govindh, P.; et al. Status and future scope of plant-based green hydrogels in biomedical engineering. *Appl. Mater. Today* **2019**, *16*, 213–246. [CrossRef]
5. Ismail, H.; Irani, M.; Ahmad, Z. Starch-based hydrogels: Present status and applications. *Int. J. Polym. Mater. Polym. Biomater.* **2013**, *62*, 411–420. [CrossRef]

6. Dong, G.; Mu, Z.; Liu, D.; Shang, L.; Zhang, W.; Gao, Y.; Zhao, M.; Zhang, X.; Chen, S.; Wei, M. Starch phosphate carbamate hydrogel based slow-release urea formulation with good water retentivity. *Int. J. Biol. Macromol.* **2021**, *190*, 189–197. [CrossRef]
7. García-astrain, C.; Avérous, L. Synthesis and evaluation of functional alginate hydrogels based on click chemistry for drug delivery applications. *Carbohydr. Polym.* **2018**, *190*, 271–280. [CrossRef] [PubMed]
8. Mun, S.; Kim, Y.R.; McClements, D.J. Control of β-carotene bioaccessibility using starch-based filled hydrogels. *Food Chem.* **2015**, *173*, 454–461. [CrossRef] [PubMed]
9. Van Nieuwenhove, I.; Salamon, A.; Adam, S.; Dubruel, P.; Van Vlierberghe, S.; Peters, K. Gelatin- and starch-based hydrogels. Part B: In vitro mesenchymal stem cell behavior on the hydrogels. *Carbohydr. Polym.* **2017**, *161*, 295–305. [CrossRef]
10. Qi, X.; Wei, W.; Li, J.; Su, T.; Pan, X.; Zuo, G.; Zhang, J.; Dong, W. Design of Salecan-containing semi-IPN hydrogel for amoxicillin delivery. *Mater. Sci. Eng. C* **2017**, *75*, 487–494. [CrossRef] [PubMed]
11. Xiao, X.; Yu, L.; Xie, F.; Bao, X.; Liu, H.; Ji, Z.; Chen, L. One-step method to prepare starch-based superabsorbent polymer for slow release of fertilizer. *Chem. Eng. J.* **2017**, *309*, 607–616. [CrossRef]
12. Larrea-Wachtendorff, D.; Di Nobile, G.; Ferrari, G. Effects of processing conditions and glycerol concentration on rheological and texture properties of starch-based hydrogels produced by high pressure processing (HPP). *Int. J. Biol. Macromol.* **2020**, *159*, 590–597. [CrossRef] [PubMed]
13. Larrea-Wachtendorff, D.; Tabilo-Munizaga, G.; Ferrari, G. Potato starch hydrogels produced by high hydrostatic pressure (HHP): A first approach. *Polymers* **2019**, *11*, 1673. [CrossRef] [PubMed]
14. Larrea-Wachtendorff, D.; Sousa, I.; Ferrari, G. Starch-Based Hydrogels Produced by High-Pressure Processing (HPP): Effect of the Starch Source and Processing Time. *Food Eng. Rev.* **2020**, *13*, 622–633. [CrossRef]
15. Barba, F.J.; Terefe, N.S.; Buckow, R.; Knorr, D.; Orlien, V. New opportunities and perspectives of high pressure treatment to improve health and safety attributes of foods. A review. *Food Res. Int.* **2015**, *77*, 725–742. [CrossRef]
16. Knorr, D.; Heinz, V.; Buckow, R. High pressure application for food biopolymers. *Biochim. Biophys. Acta-Proteins Proteom.* **2006**, *1764*, 619–631. [CrossRef]
17. Stute, R.; Klingler, R.W.; Boguslawski, S.; Eshtiaghi, M.N.; Knorr, D. Effects of high pressures treatment on starches. *Starch/Starke* **1996**, *48*, 399–408. [CrossRef]
18. Błaszczak, W.; Fornal, J.; Kiseleva, V.I.; Yuryev, V.P.; Sergeev, A.I.; Sadowska, J. Effect of high pressure on thermal, structural and osmotic properties of waxy maize and Hylon VII starch blends. *Carbohydr. Polym.* **2007**, *68*, 387–396. [CrossRef]
19. Błaszczak, W.; Valverde, S.; Fornal, J. Effect of high pressure on the structure of potato starch. *Carbohydr. Polym.* **2005**, *59*, 377–383. [CrossRef]
20. Błaszczak, W.; Buciński, A.; Górecki, A.R. In vitro release of theophylline from starch-based matrices prepared via high hydrostatic pressure treatment and autoclaving. *Carbohydr. Polym.* **2015**, *117*, 25–33. [CrossRef]
21. Buckow, R.; Heinz, V.; Knorr, D. High pressure phase transition kinetics of maize starch. *J. Food Eng.* **2007**, *81*, 469–475. [CrossRef]
22. Katopo, H.; Song, Y.; Jane, J.L. Effect and mechanism of ultrahigh hydrostatic pressure on the structure and properties of starches. *Carbohydr. Polym.* **2002**, *47*, 233–244. [CrossRef]
23. Kawai, K.; Fukami, K.; Yamamoto, K. Effect of temperature on gelatinization and retrogradation in high hydrostatic pressure treatment of potato starch—water mixtures. *Carbohydr. Polym.* **2012**, *87*, 314–321. [CrossRef] [PubMed]
24. Li, W.; Tian, X.; Liu, L.; Wang, P.; Wu, G.; Zheng, J.; Ouyang, S.; Luo, Q.; Zhang, G. High pressure induced gelatinization of red adzuki bean starch and its effects on starch physicochemical and structural properties. *Food Hydrocoll.* **2015**, *45*, 132–139. [CrossRef]
25. Li, W.; Bai, Y.; Mousaa, S.A.S.; Zhang, Q.; Shen, Q. Effect of High Hydrostatic Pressure on Physicochemical and Structural Properties of Rice Starch. *Food Bioprocess Technol.* **2012**, *5*, 2233–2241. [CrossRef]
26. Bajaj, S.; Singla, D.; Sakhuja, N. Stability testing of pharmaceutical products. *J. Appl. Pharm. Sci.* **2012**, *2*, 129–138. [CrossRef]
27. Reyes, J.E.; Guanoquiza, M.I.; Tabilo-Munizaga, G.; Vega-Galvez, A.; Miranda, M.; Pérez-Won, M. Microbiological stabilization of Aloe vera (*Aloe barbadensis* Miller) gel by high hydrostatic pressure treatment. *Int. J. Food Microbiol.* **2012**, *158*, 218–224. [CrossRef]
28. Daryaei, H.; Yousef, A.E.; Balasubramaniam, V.M. Microbiological aspects of high pressure food processing: Inactivation of vegetative microorganisms and spores. In *High Pressure Processing of Food*; Springer: New York, NY, USA, 2016; pp. 271–294. [CrossRef]
29. Reyes, J.E.; Tabilo-Munizaga, G.; Pérez-Won, M.; Maluenda, D.; Roco, T. Effect of high hydrostatic pressure (HHP) treatments on microbiological shelf-life of chilled Chilean jack mackerel (*Trachurus murphyi*). *Innov. Food Sci. Emerg. Technol.* **2015**, *29*, 107–112. [CrossRef]
30. Modugno, C.; Peltier, C.; Simonin, H.; Dujourdy, L.; Capitani, F.; Sandt, C.; Perrier-Cornet, J.M. Understanding the Effects of High Pressure on Bacterial Spores Using Synchrotron Infrared Spectroscopy. *Front. Microbiol.* **2020**, *10*, 1–10. [CrossRef]
31. Nguyen Thi Minh, H.; Dantigny, P.; Perrier-Cornet, J.M.; Gervais, P. Germination and inactivation of Bacillus subtilis spores induced by moderate hydrostatic pressure. *Biotechnol. Bioeng.* **2010**, *107*, 876–883. [CrossRef]
32. Kawai, K.; Fukami, K.; Yamamoto, K. State diagram of potato starch-water mixtures treated with high hydrostatic pressure. *Carbohydr. Polym.* **2007**, *67*, 530–535. [CrossRef]
33. Kawai, K.; Fukami, K.; Yamamoto, K. Effects of treatment pressure, holding time, and starch content on gelatinization and retrogradation properties of potato starch-water mixtures treated with high hydrostatic pressure. *Carbohydr. Polym.* **2007**, *69*, 590–596. [CrossRef]

34. Chung, Y.L.; Lai, H.M. Water status of two gelatin gels during storage as determined by magnetic resonance imaging. *J. Food Drug Anal.* **2004**, *12*, 221–227. [CrossRef]
35. Torres, M.D.; Fradinho, P.; Raymundo, A.; Sousa, I.; Falqué, E.; Domínguez, H. The key role of thermal waters in the development of innovative gelled starch-based matrices. *Food Hydrocoll.* **2021**, *117*, 106697. [CrossRef]
36. BeMiller, J.N.; Whistler, R. (Eds.) *Starch: Chemistry and Technology*, 3rd ed.; Academic Press: Cambridge, MA, USA, 2009; ISBN 9780127462752.
37. Wang, S.; Li, C.; Copeland, L.; Niu, Q.; Wang, S. Starch Retrogradation: A Comprehensive Review. *Compr. Rev. Food Sci. Food Saf.* **2015**, *14*, 568–585. [CrossRef]
38. BeMiller, J.N. *Carbohydrate Chemistry for Food Scientists*; Woodhead Publishing and AACC International Press: Cambridge, UK, 2018; ISBN 9780128120699.
39. Scott, W.J. Water Relations of Food Spoilage Microorganisms. *Adv. Food Res.* **1957**, *7*, 83–127. [CrossRef]
40. Pycia, K.; Gałkowska, D.; Juszczak, L.; Fortuna, T.; Witczak, T. Physicochemical, thermal and rheological properties of starches isolated from malting barley varieties. *J. Food Sci. Technol.* **2015**, *52*, 4797–4807. [CrossRef] [PubMed]
41. Wang, P.; Wang, P.; Tian, J.; Yu, X.; Chang, M.; Chu, X.; Wu, N. A new strategy to express the extracellular α-amylase from Pyrococcus furiosus in Bacillus amyloliquefaciens. *Sci. Rep.* **2016**, *6*, 22229. [CrossRef] [PubMed]
42. Kommanaboyina, B.; Rhodes, C.T. Trends in stability testing, with emphasis on stability during distribution and storage. *Drug Dev. Ind. Pharm.* **1999**, *25*, 857–868. [CrossRef] [PubMed]
43. Almeida, I.F.; Bahia, M.F. Evaluation of the physical stability of two oleogels. *Int. J. Pharm.* **2006**, *327*, 73–77. [CrossRef]
44. Carstensen, J.T.; Rhodes, C.T. Cyclic testing in stability programs. *Drug Dev. Ind. Pharm.* **1986**, *12*, 1219–1225. [CrossRef]
45. Schirmer, M.; Jekle, M.; Becker, T. Starch gelatinization and its complexity for analysis. *Starch/Starke* **2015**, *67*, 30–41. [CrossRef]
46. Taherian, A.R.; Fustier, P.; Ramaswamy, H.S. Steady and dynamic shear rheological properties, and stability of non-flocculated and flocculated beverage cloud emulsions. *Int. J. Food Prop.* **2008**, *11*, 24–43. [CrossRef]
47. Abdul Rahman, M.N.; Qader, O.A.J.A.; Sukmasari, S.; Ismail, A.F.; Doolaanea, A.A. Rheological characterization of different gelling polymers for dental gel formulation. *J. Pharm. Sci. Res.* **2017**, *9*, 2633–2640.
48. Doona, C.J.; Feeherry, F.E.; Baik, M.Y. Water dynamics and retrogradation of ultrahigh pressurized wheat starch. *J. Agric. Food Chem.* **2006**, *54*, 6719–6724. [CrossRef] [PubMed]
49. Hu, X.; Xu, X.; Jin, Z.; Tian, Y.; Bai, Y.; Xie, Z. Retrogradation properties of rice starch gelatinized by heat and high hydrostatic pressure (HHP). *J. Food Eng.* **2011**, *106*, 262–266. [CrossRef]
50. Yang, Z.; Chaib, S.; Gu, Q.; Hemar, Y. Food Hydrocolloids Impact of pressure on physicochemical properties of starch dispersions. *Food Hydrocoll.* **2017**, *68*, 164–177. [CrossRef]
51. Biliaderis, C.G. Structural Transitions and Related Physical Properties of Starch. In *Starch*; Academic Press: Cambridge, MA, USA, 2009; pp. 293–372, ISBN 9780127462752.
52. Fang, F.; Tuncil, Y.E.; Luo, X.; Tong, X.; Hamaker, B.R.; Campanella, O.H. Shear-thickening behavior of gelatinized waxy starch dispersions promoted by the starch molecular characteristics. *Int. J. Biol. Macromol.* **2019**, *121*, 120–126. [CrossRef] [PubMed]
53. Torres, M.D.; Chenlo, F.; Moreira, R. Rheological Effect of Gelatinisation Using Different Temperature-Time Conditions on Potato Starch Dispersions: Mechanical Characterisation of the Obtained Gels. *Food Bioprocess Technol.* **2018**, *11*, 132–140. [CrossRef]
54. Lapasin, R. Rheological Characterization of Hydrogels. In *Polysaccharide Hydrogels*; Matricardi, P., Alhaique, F., Coviello, T., Eds.; Jenny Stanford Publishing: Dubai, United Arab Emirates, 2016; pp. 99–154, ISBN 9780429069338.
55. De Maria, S.; Ferrari, G.; Maresca, P. Effect of high hydrostatic pressure on the enzymatic hydrolysis of bovine serum albumin. *J. Sci. Food Agric.* **2017**, *97*, 3151–3158. [CrossRef] [PubMed]
56. Kusumayanti, H.; Handayani, N.A.; Santosa, H. Swelling Power and Water Solubility of Cassava and Sweet Potatoes Flour. *Procedia Environ. Sci.* **2015**, *23*, 164–167. [CrossRef]

Article

Multivalent Allylammonium-Based Cross-Linkers for the Synthesis of Homogeneous, Highly Swelling Diallyldimethylammonium Chloride Hydrogels

Tim B. Mrohs and Oliver Weichold *

Institute of Building Materials Research, RWTH Aachen University, Schinkelstraße 3, 52062 Aachen, Germany; mrohs@ibac.rwth-aachen.de
* Correspondence: weichold@ibac.rwth-aachen.de

Abstract: N,N'-methylenebisacrylamide (BIS) is a very popular cross-linker for the radical polymerisation in water. It is highly reactive but prone to alkaline hydrolysis and suffers from a low solubility. This study shows that with slow polymerising systems such as N,N-diallyldimethylammonium chloride, only inhomogeneous networks are formed. As a consequence, gels with very low cross-linking densities, i.e., high swelling capacities, disintegrate during the swelling test and firm, coherent gels are not accessible due to the solubility limit. A promising alternative are multivalent tetraallyl-based compounds, of which tetraallylammonium bromide (TAAB), N,N,N',N'-tetraallylpiperazinium dibromide (TAPB) and N,N,N',N'-tetraallyltrimethylene dipiperidine dibromide (TAMPB) are the subject of this study. With these, the cross-linking polymerisation appears to be statistical, as gels formed at low monomer conversion have essentially the same swelling properties as those formed at high conversions. This is not observed with BIS. However, gelation with the tetraallyl cross-linkers is much slower than with BIS and follows the order TAPB < TAMPB < TAAB, but the differences become significantly smaller with increasing content. At low contents, all three allow the preparation of gels with high swelling capacities of up to 360 g/g.

Keywords: hydrogel; copolymer; reactivity ratios; DADMAC; gelation; synthesis

1. Introduction

Hydrogels are crosslinked polymer networks which can absorb water from the environment, while maintaining their three-dimensional structure [1]. Hydrogels are divided into two classes, namely ionic and non-ionic hydrogels [2,3]. These two classes differ not only in their molecular structure, but also in their swelling behaviour and the nature of the intermolecular interactions [4]. The swelling characteristics of non-ionic hydrogels are highly dependent on the cross-linker concentration, but are only minimally affected by the salt concentration and the pH value of the surrounding medium [5]. These hydrogels are, therefore, often used in biomedical applications [6] or protein analysis [7]. For ionic hydrogels, the driving forces for swelling in water are the repulsive interactions between the individual charges of the polyelectrolyte chains. Hence, ionic hydrogels usually show a significantly higher swelling capacity compared to non-ionic hydrogels [8]. The synthesis of such hydrogels is usually very simple and inexpensive, so that this class of hydrogels is increasingly used in industrial applications. Some examples are poly (sodium acrylate) as superabsorbers for diapers [9], ion exchange resins for the extraction of toxins such as arsenate [10], or for the reduction of the autogeneous shrinking during the hardening of concrete [11]. A highly alkaline hydrogel for the rehabilitation of reinforced concrete was first introduced in 2018 [12]. This particular hydrogel was synthesised via free radical polymerisation of diallyldimethylammonium hydroxide (DADMAOH), a comonomer and N,N'-methylenebisacrylamide (BIS) as cross-linker. The gel itself forms a stationary cationic polyelectrolyte backbone with mobile hydroxide counterions. This mobility allows the hydroxide ions in the gel to be exchanged with the carbonate in old concrete, which is a common cause for corrosion of the steel reinforcement. Overall, the cementitious matrix is "realkalised". This protects the reinforcement from further corrosion. A similar hydrogel was recently used as coupling material for electrochemical chloride extraction [13], an excellent method to counteract chloride-induced steel corrosion. However, the use of BIS as cross-linker in these allyl-based systems can be precarious. As an amide, it is potentially liable to alkaline hydrolysis at prolonged application times, which then produces toxic compounds such as formaldehyde [14]. As an acrylate derivative, it polymerises faster than the notoriously slow allyl compounds. In addition, BIS has a comparatively low solubility of only 20 g/L in water at 20 °C, which makes it less versatile when aiming for firmer gels. On the other hand, the diallylammonium subunit offers a number of synthetic possibilities to prepare cationic and, hence, water-soluble cross-linkers. For example, tetraallylammonium bromide was successfully polymerised with vinylpyrrolidone to give a hydrolytically stable hydrogel which was used to remove dyes from aqueous solutions [15]. N,N,N',N'-tetraallyl piperazinium dichloride, which is resistant to acidic hydrolysis, was used to cross-link acrylamide and acrylic acid [16,17]. Furthermore, a copolymer of N,N,N',N'-tetraallyl trimethylene dipiperidine dichloride with diallyldimethylammonium chloride (DADMAC) was used as a catalytically active ion exchange resin to convert phenols into aromatic ethers [18].

Although some allyl-based cross-linkers are known, the gel properties such as the rheological or swelling properties are largely unexplored, and an evaluation of the cross-linker performance is missing. For a deeper understanding of the important class of cationic hydrogels, these are important parameters. The goal of this study is, therefore, to assess the performance of three quaternary, tetraallylammonium-based cross-linkers, namely tetraallylammonium bromide (TAAB), N,N,N',N'-tetraallylpiperazinium dibromide (TAPB) and N,N,N',N'-tetraallyltrimethylene dipiperidine dibromide (TAMPB) and gels made with the neutral monomer diallyldimethylammonium chloride (DADMAC). These gels are then compared with those containing N,N'-methylenebisacrylamide (BIS).

2. Results and Discussion

The tetraallyl-based cross-linkers tetraallyl ammonium bromide (TAAB, 1a), N,N,N',N'-tetraallyl piperazinium dibromide (TAPB, 1b) and N,N,N',N'-tetraallyl trimethylene dipiperidine dibromide (TAMPB, 1c) were synthesised by allylation of the appropriate amines. However, exhaustive allylation of the secondary amines in the presence of KOH or K_2CO_3 was not found to be a viable method. The products from these reactions contain double-digit percentages of KBr, which can only be removed at great loss of product. It is far better to start from the tertiary amines. In the case of TAAB, this is commercially available, and for TAMP and TAPB they can be made from the commercial secondary amines using only a slight excess of allyl bromide. The tertiary amines separate readily from the aqueous reaction mixture and are obtained in sufficient purity to be able to be converted into the quaternary ammonium salts in a subsequent step. The procedure is exemplified for TAPB (1b) in Figure 1.

Following the previously published procedure for cross-linking the polymerisation of DADMAOH with BIS [12], the combination potassium peroxodisulphate ($K_2S_2O_8$, KPS) and sodium disulphite ($Na_2S_2O_5$) was chosen as a redox initiation system [19,20], since it was found to be very reliable in this system. In the initial experiments using 2 mol% cross-linker based on DADMAC, gels could be obtained with all four cross-linkers shown in Figure 1. To get more insight into the copolymerisation, the reactivity ratios were calculated using the method of Alfrey and Price [21], Unlike the reactivity ratios, which are only defined for a pair of monomers, Alfrey and Price introduced semiempirical parameters for the reactivity (Q) and polarity (e) of individual monomers. The reactivity ratios $r_{1,2}$ of an unknown pair of monomers can then be calculated from $Q_{1,2}$ and $e_{1,2}$ according to e. g.,

$$r_1 = \frac{Q_1}{Q_2} e^{-e_1(e_1 - e_2)}$$

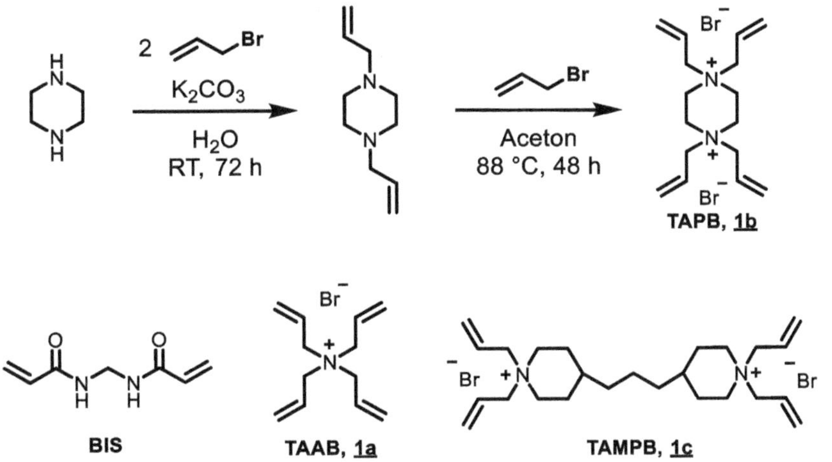

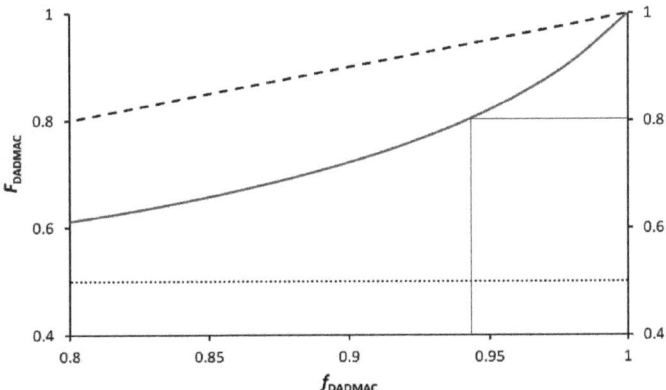

Figure 1. Top: Outline of the two-step synthesis of 1b, c using 1b as example; bottom: structures of the other three cross-linkers used in this study.

For DADMAC, the parameters $Q_1 = 0.32$, $e_1 = -0.22$ are known in the literature [12]. For BIS, these parameters are not known and, thus, acrylamide with $Q_2 = 1.18$ and $e_2 = 1.30$ was used as a substitute, assuming a similar polarity Q and reactivity e [22]. From these, the reactivity ratios $r_1 = 0.194$ and $r_2 = 0.511$ can be calculated. That is, both monomers, when added to the chain end, react favorably with the other monomer although this is more pronounced for DADMAC than for BIS. The product $r_1\,r_2 = 0.099$ indicates a certain tendency towards alternating copolymerisation. In the copolymerisation diagram (Figure 2) this can be seen as deviation from the statistical (dashed line) towards the alternating (dotted line) copolymerisation. In total, BIS is incorporated faster into the polymer than DADMAC.

Figure 2. Extract of the calculated copolymerisation diagram of DADMAC with acrylamide, which is used as a substitute for BIS (the full diagram in shown in Figure S16 in the Supporting Information). The solid line indicates the calculated mole fraction of DADMAC in the polymer F_{DADMAC} as function of the mole fraction in solution f_{DADMAC}. The dashed line shows the polymer composition for the statistical copolymerisation and the dotted line the alternating type. The latter are for reference only. The marks indicate the example containing 3 mol% BIS mentioned in the text.

In an assumed monomer solution containing 3 mol% BIS, the effective mole fraction of acrylamide groups in solution is $f_{BIS} = \frac{2 \cdot [BIS]}{[BIS]+[DADMAC]} = 0.057$, and as a consequence

$f_{DADMAC} = 1 - f_{BIS} = 0.943$. As can be seen from Figure 2, chains formed at low conversion (i.e., with little change in the solution composition) contain a mole fraction of DADMAC in the polymer $F_{DADMAC} = 0.803$, and therefore approx. 20 mol% BIS (cf. marking lines in Figure 2). This quickly depletes the reaction mixture of BIS so that the network formed at higher conversion exhibits a significantly lower cross-linking density. In contrast, assuming that the allyl-based cross-linkers 1a–c have reactivity ratios similar to DADMAC, these would be incorporated statistically, i. e., $f = F$ at any composition (dashed line in Figure 2), thus forming a homogeneous network.

When qualitatively assessing the gels, the ones containing TAAB (1a) were clearly softer than those containing 1b, c or BIS. To get more insight into the gelation process, the reactions were followed using dynamic rheological measurements that provided the storage and loss modulus as a function of time. In addition, the cross-linker content was varied in the range of 1 to 5 mol%. Mixtures containing 4 mol% of BIS could not be analysed due to the fast gelation, and also turn out to be extremely crumbly; higher BIS contents are inaccessible due to the solubility limit. Gels with high contents of the tetraallyl cross-linkers range from pastelike (TAAB) to firm (TAPB), but coherent. The gelation time, defined as the time of intersection of the storage and loss-modulus curves, decreases for all cross-linkers with increasing concentration, because more cross-linking points are formed in a shorter time at the same polymerisation rate (Figure 3). This way, the elastic properties of the systems develop faster. For all cross-linker contents, the gelation times follow the order BIS < TAPB (1b) < TAMPB (1c) < TAAB (1a), although the differences decrease with increasing content.

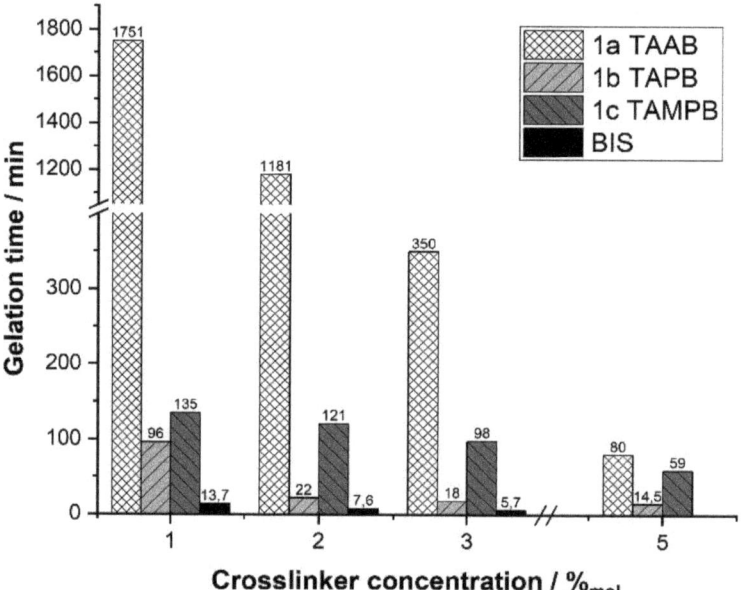

Figure 3. Gelation time of the cross-linking copolymerisation of DADMAC at 20 °C as a function of the cross-linker content. Reproducibility of the gel points is approx. ±10%.

It is remarkable that the gelation times of samples containing TAAB (1a) are considerably longer than those containing TAPB or TAMPB. Similar to DADMAC, for the cross-linkers 1a–c the addition of a radical to one of the allylic double bonds leads to the formation of a 5-membered ring by a 5-exo-dig attack. However, in the case of TAAB (1a) the addition of a second radical, which causes cross-linking of the polymer chains, forms a strained spiro[4,4] structure (Figure 4). It is therefore assumed that this is the rate-determining step for gelation with 1a and it appears to be rather slow. In contrast, cross-linking using TAMPB and TAPB gives rise to spiro[4,5] compounds. This reduces the strain at the spiro centre, because the six-membered ring allows a larger bond angle and the gelation is, therefore, assumed to be faster. This assumption is corroborated by observations made by Blicke and Hotelling during the synthesis of spiro[4,4] and spiro[4,5] ammonium compounds. Under similar conditions, the former is obtained in a 45% yield, while the latter is obtained in a 98% yield [23]. Generally, all three tetraallyl cross-linkers are slower than BIS. Besides the less reactive allylic double bond, the formation of spiro compounds might also contribute to this observation.

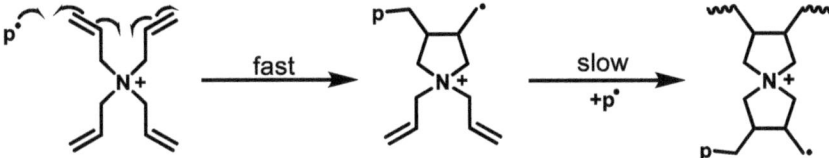

Figure 4. Ring closing reaction of the polymerisation of TAA$^+$.

Further evidence for the slow cross-linking reaction when using TAAB 1a is gained by comparing the evolution of the storage G' and loss moduli G'' of all four polymerisations (Figure 5). The storage modulus G' indicates the elastic properties of the mixture, which is mainly influenced by the cross-linking, while the loss modulus G'' represents viscous properties associated with the irreversible displacement of the free chains.

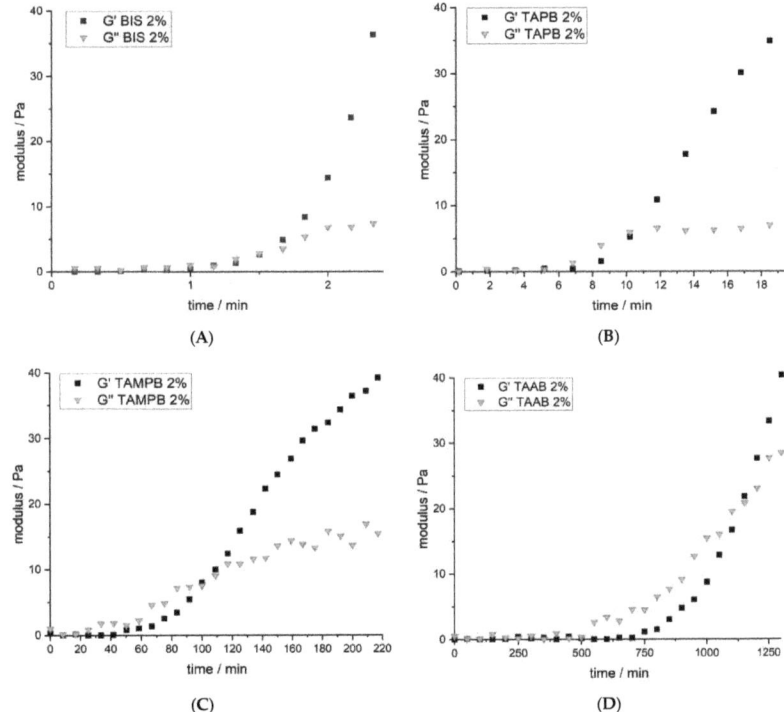

Figure 5. Storage G' and loss modulus G'' as a function of time for mixtures containing DADMAC and 2 mol% of the four cross-linkers (**A**) BIS, (**B**) TAPB, (**C**) TAMPB and (**D**) TAAB shown in Figure 1 at 20 °C, recorded at an amplitude of 1 Hz and 1% deformation. Note the different scale of the x-axes.

As can be seen from the copolymerisation diagram, gels made with BIS incorporate a disproportionate amount of cross-linker early in the polymerisation, which leads to a rapid build-up of G' (high cross-linking density), while G'' increases only marginally before the gel point (Figure 5A). After that, G'' remains almost constant, while G' continues to increase rapidly. Similar trends can be observed for TAPB (1b, Figure 5B) and TAMPB (1c, Figure 5C), albeit at extended times, and in the case of 1c at notably higher moduli. This is due to the larger and more flexible structure of TAMPB (1c), which introduces more degrees of freedom and a higher flexibility to the network compared to BIS and 1b. Of the three cationic cross-linkers, TAAB (1a) is expected to have the most rigid structure. However, at the gel point, G'' has already climbed to more than 20 N m^{-2} (Figure 5D), almost twice the value for 1c and almost 4 times the value of 1b. This is indicative of a much larger number of free polymer chains or longer chain segments connecting the cross-links in mixtures containing 1a as opposed to 1b, c or BIS. In addition, G'' is not constant after the gel point but still increases, which suggests that even more free chains are formed. Compared to the other three cross-linkers, G' also increases more slowly. Both features account for the previously mentioned softer texture of gels containing 1a and support the mechanism proposed in Figure 4.

^{1}H- and ^{13}C-NMR spectra of copolymer gels containing TAAB (1a) swollen in D_2O were not decisive, as these vinyl signals partially overlap with the broad signals of the DADMA$^+$ backbone, and the cross-linker concentrations are generally rather small (Figures S17 and S18 in the Supporting Information). Double-bond signals found in the ^{13}C-NMR spectra could be addressed to the unreacted DADMAC monomer. However, the TAAB homopolymer clearly showed double-bond signals in the ^{1}H-NMR spectrum (Figure S19 in the Supporting Information), which are not identical with the TAAB monomer. The only other explanation is unilaterally incorporated TAAB units. It can be deduced from the integrals that every sixth to seventh unit was only half-incorporated. For the homopolymers of TAPB (1b) and TAMPB (1c), the double-bond signals are significantly lower. This indicates a more homogeneous reactivity of both crosslinker sides. Clearly, this cannot be directly transferred to the cross-linking polymerisation of DADMAC with TAAB, but further supports the above indications for a slow second cyclisation of TAAB, as outlined in Figure 4.

Another important point in time is the end of the reaction. The polymerisation for up to 48 h on the rheometer showed no significant slowdown of the reaction rate based on the increase of G' and the complex viscosity. Monitoring over longer periods of time is not possible on the rheometer, as this leads to massive dehydration, even with the use of a solvent trap. This made the acquisition of useful data impossible. Therefore, Electrochemical Impedance Spectroscopy (EIS) was used to follow the polymerisation over the course of 14 d. The use of EIS is possible, as the gel formation depletes the solution of positively charged monomers and immobilises them in the polymer network. Consequently, the impedance Z increases as the polymerisation progresses. From that, the real part Z_{real} can be calculated (Figure 6).

Immediately after initiation, all four solutions show a similar resistance of $Z_{real} \approx 6\ \Omega$, as this depends only on the starting concentration. The slight differences are due to the different sizes of the cross-linkers, and the solutions containing TAAB show the lowest Z_{real}, as this is presumed to have the highest mobility. Initially, Z_{real} increases for all four samples in accordance with the theory. Following that, the reaction mixture containing BIS continues to show the expected behaviour in the form of a saturation curve, which levels off after approx. 7 d. This indicates continued monomer depletion of the solution. However, the curves of pure DADMAC and the samples containing the allyl-based cross-linkers 1a–c exhibit a sharp bend after significantly less than a day. In terms of the measured quantity Z_{real} this means that the number of ions stays rather constant. Transferred to the polymerisation this means that the conversion of monomer has subsided or proceeds only at a very low rate. This seems to contradict the rheological measurements in as far as e. g. for TAAB (1a), the build-up of G' is still accelerating after approx. 1 d (cf. Figure 5 and description above). Apart from BIS and the other cross-linkers 1a–c, one reason for the difference between rheological measurements and EIS could be that the mechanism postulated in Figure 4 holds true, to a certain extent, for all allyl-based cross-linkers. In that case, the DADMAC monomer and the cross-linkers are first incorporated into largely linear chains with only little cross-linking. This is possible if the rate of polymerisation of DADMAC and the first two allyl units of the cross-linkers 1a–c is similar as postulated above. The ratio between chain growth and cross-linking appears to vary between 1a, b, and c as the build-up of G' occurs at different rates (Figure 5) and the Z_{real} curves do not run on top of each other. After larger amounts of the low-molecular weight monomers have been immobilised, polymerisation shifts increasingly to the cross-linking reaction. This can only be detected with difficulty by the EIS, as the number of mobile charges does not change much. Cationic hydrogels are still highly electrically conductive, even in deionised water [24]. The increasing viscosity, as recorded by the rheometer, additionally hampers the movement of remaining charges and together this accounts for the slight and steady increase of Z_{real} in Figure 6. However, the final gels appear to contain approx. 7–9 % residual monomer. In contrast, samples containing BIS incorporate much more cross-linker—and less DADMAC—in early stages of the reaction, which leads to much higher viscosities after shorter times. This limits the mobility of the leftover ions, which accounts for the increase in Z_{real} and decreasing rates of polymerisation. As a consequence, DADMAC consumption continues over longer periods of time.

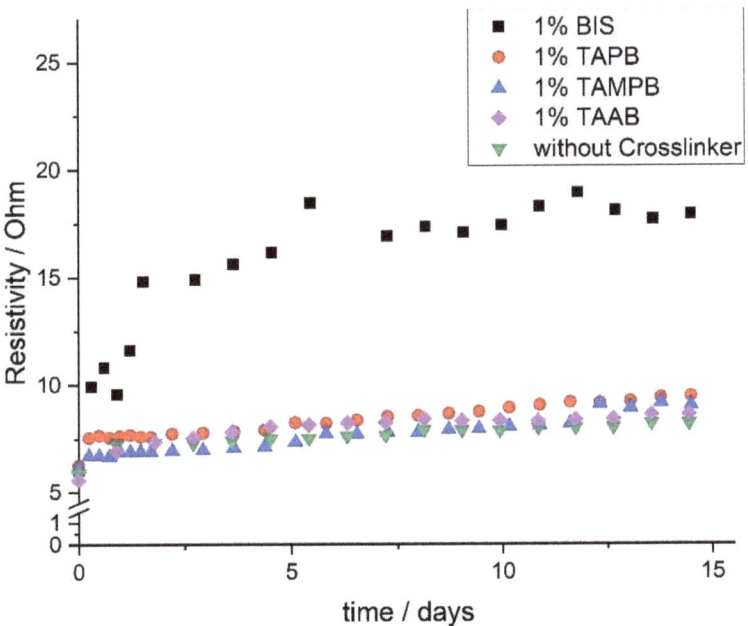

Figure 6. Evolution of the solution resistance Z_{real} of pure poly(DADMAC) (▼) and gels cross-linked with 1 mol% BIS, TAAB (1a), TAPB (1b), or TAMPB (1c) over the course of 14 days.

Based on these findings, gels used for the swelling tests were allowed to react for 14 days in order to ensure extensive polymerisation. The cross-linkers were used in ratios of 0.25 mol% to 7 mol% (Figure 7). However, gels with low ratios of TAAB (<0.4 %) or BIS (<0.75 %) were not considered. Although these appear to be coherent during synthesis, they dissolve in double-distilled water. For TAAB, this is another indication of the low crosslinking efficiency, which was already assumed above. As the copolymerisation of DADMAC with BIS is considered to form inhomogeneous networks, very low amounts of BIS in the sample will be quickly consumed at the beginning of the reaction so that the "network" formed later contains little to no cross-links. The highly cross-linked parts appear to be smaller than the mesh size of tea-bags (90 μm), and are lost during the swelling experiments. As mentioned before, gels with > 4 mol% BIS could not be prepared due to the solubility limit.

In general, the swelling capacity decreased with increasing the cross-linker content up to approx. 5 mol%, after which it remained constant at approx. 10 to 18 g/g depending on the actual cross-linker. Gels based on the allyl cross-linkers, particularly 1b, c, could be synthesized with very low amounts of cross-linker, giving rise to very high swelling capacities of up to 360 g/g. Such a significant increase in the degree of swelling at low crosslinking densities has already been observed for other crosslinked hydrogel systems [25]. These values are comparable to common acrylate superabsorbent polymers [26,27], and are another indication of the homogeneous nature of these gels. In contrast, the coherent gel with the lowest concentration of BIS could only absorb 116 g/g.

Finally, to test whether the assumption that BIS leads to inhomogeneous networks is true, gels containing 1 mol% of the cross-linkers 1a–c and BIS were polymerised for only 1 d and their swelling behaviour was compared to that in Figure 7 (polymerisation time 14 d). Gels containing the allyl-based cross-linkers 1a–c showed virtually no difference in the swelling capacity, whereas for those containing BIS, the value doubled (Figure 8). This strongly supports the postulated inhomogeneous network. As outlined above, the reactivity ratios indicate the preferred incorporation of BIS into the network. At low conversions, this leads to more strongly cross-linked structures which exhibit lower swelling capacities. This also depletes the solution of BIS so that the network formed at later stages will have increasingly lower cross-linking densities, but at the same time higher swelling capacities. In total, the BIS gels are highly inhomogeneous with increasing swelling capacity over time.

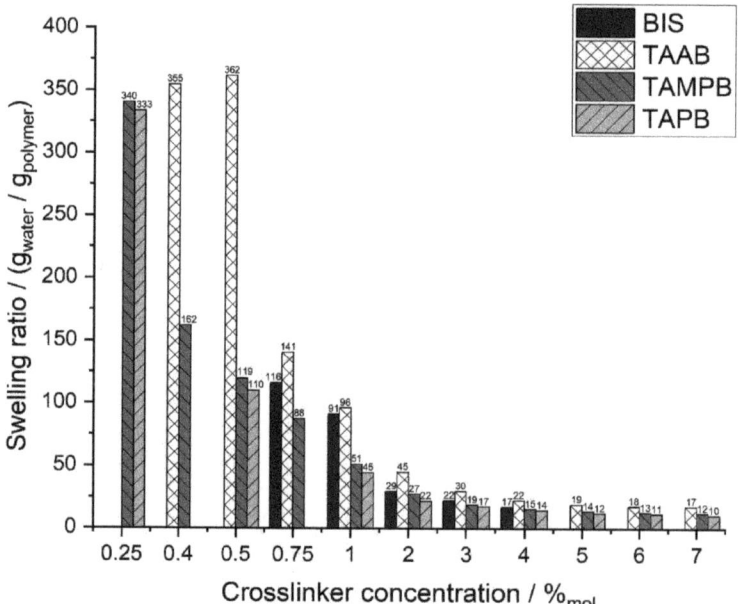

Figure 7. Swelling ratios of cationic poly(DADMAC) gels with different cross-linkers as a function of the cross-linker concentration. Reproducibility is approx. ±5 % for values <200 g/g and approx. ±14% above that.

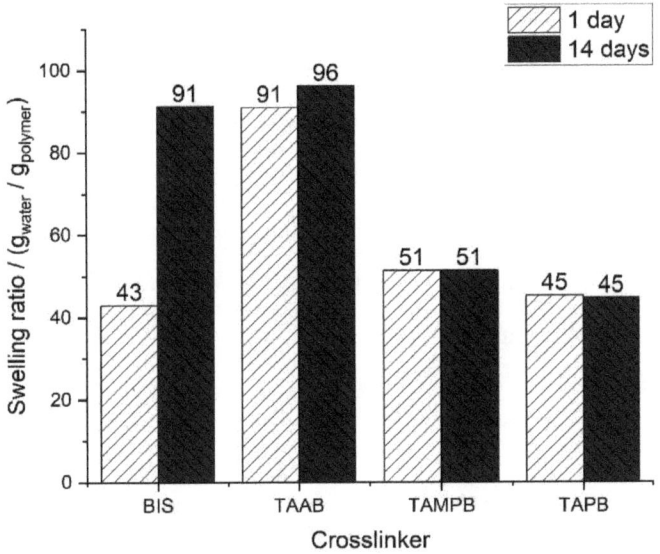

Figure 8. Swelling ratios of cationic poly(DADMAC) gels containing 1 mol% of the cross-linkers 1a-c and BIS after 1 day and 14 days polymerisation time.

3. Conclusions

The three tetraallylammonium-based cross-linkers tetraallylammonium bromide, N,N,N',N'-tetraallylpiperazinium dibromide and N,N,N',N'-tetraallyltrimethylene dipiperidine dibromide are excellent alternatives to the commonly used N,N'-methylenebisacrylamide for the cross-linking polymerisation of slow monomers such as N,N-diallyldimethylammonium chloride and similar monomers in water. In these cases, the incorporation of monomers into the network proceeds statistically. The combination of rheological and EIS data indicates that, initially, largely linear chains with little cross-linking form, and that cross-linking is a slower process due to the formation of spiro structures. Both factors favour the formation of homogeneous networks. Homogeneity is a prerequisite for the formation of coherent gels with low cross-linking densities, i.e., high swelling capacities, as well as firm, resilient gels. The former have potential applications in biomedical and hygiene products, while the latter are more suitable as ion-exchange resins and in soil applications.

4. Materials and Methods

4.1. Materials

The anion exchange-resin Lewatit Monoplus MP 800 was provided by Lanxess (Leverkusen, Germany). Double-distilled water, potassium hydroxide, chloroform ($\geq$99%), acetone ($\geq$99%), dichloromethane ($\geq$99%), potassium persulfate, sodium metabisulphite, sodium hydroxide (97%) and N,N'-methylenebisacrylamide were obtained from VWR International GmbH (Darmstadt, Germany). Potassium carbonate (99%), allyl bromide (99%), methanol (99%) and 1,3-bis(4-piperidyl)propane (97+%) were obtained from Alfa Aesar (Kandel, Germany), diallyldimethylammonium chloride (65 wt% in H_2O) and triallylamine (99%) were obtained from Sigma Aldrich. Piperazine, diallylamine (97%) and D_2O (99.9%) were purchased from Merck KGaA (Darmstadt, Germany). The polyester filter bags with a maximum mesh size of 90 µm were bought from Rosin Tech Products (Bethpage, NY, USA).

4.2. Syntheses

4.2.1. Preparation of Tetraallylammonium Bromide 1a (TAAB)

A solution of 7.41 mL (6.00 g, 43.72 mmol) triallylamine in 50 mL acetone was placed in a 250 mL round-bottom flask and cooled in an ice-water bath. To this, 4.53 mL (6.35 g,

52.47 mmol, 1.2 equiv.) allyl bromide were added over the period of one hour. Afterwards, the solution was heated to 88 °C for 48 h. The solution was filtered and the resulting solid was washed several times with acetone. Tetraallyl ammonium bromide was obtained as a white crystalline solid. Both thermogravimetry and differential scanning calorimetry show unanimously that 1a decomposes at 185 °C. Previous reports listed this as melting point [28].

Yield: 6.90 g (61%); Decomposition at 185 °C.
^{1}H-NMR (400 MHz, D$_2$O) δ 6.09 (m, 4H), 5.81–5.69 (m, 8H), 3.93 (d, J = 7.3 Hz, 8H).
(Figure S2 in the Supporting Information).
^{13}C-NMR (300 MHz, D$_2$O) δ 128.52, 123.98, 60.32.
(Figures S3 and S4 in the Supporting Information).
ATR-IR: 2937 cm^{-1} ν(C-H), 1447 cm^{-1} γ (CH2), 1020 cm^{-1} ν(C-N).
(Figure S5 in the Supporting Information).

4.2.2. Preparation of Diallylpiperazine

10 g (116.1 mmol) of piperazine was dissolved in 31 mL saturated potassium carbonate solution in a 250 mL round-bottom flask. The solution was cooled in an ice-water bath and 22.0 mL (30.8 g, 255 mmol, 2.2 equiv. with regard to piperazine) allyl bromide was slowly added over the course of 1 h. When approx. half of the allyl bromide had been added, the solution separated into a lower white suspension and an upper oily yellow phase formed. After complete addition, the reaction mixture was stirred at room temperature for 72 h before adding 50 mL Dichloromethane. The phases were separated and the aqueous phase was extracted three times with Dichloromethane. The combined organic phases were dried with magnesium sulphate, filtered, and the solvent was removed under reduced pressure. 11.43 g (54%) of the crude diallylpiperazine was obtained as pale, yellow, oily liquid.

^{1}H-NMR (400 MHz, Chloroform-d) δ 5.85 (m, 2H), 5.13 (m, 4H), 2.98 (d, J = 6.6 Hz, 4H), 2.47 (s, 8H).
(Figure S6 in the Supporting Information).

4.2.3. Preparation of N,N,N',N'-Tetraallyl Piperazinium Dibromide 1b (TAPB)

11.43 g (68.74 mmol,) diallylpiperazine was dissolved in 95 mL of acetone. To this, 17.82 mL (24.95 g, 206.23 mmol, 3 equivalents) allyl bromide was slowly added. The reaction mixture was refluxed for 72 h at 88 °C. During this, the crude product precipitated as a beige solid, which was filtered, washed with acetone, and recrystallized from methanol to yield 28.06 g (65%) tetraallylpiperazinedibromide as white crystalline solid. Similar to 1a, the dibromide 1b was found to decompose at 210 °C, previously reported as m.p. 207 °C [28])

^{1}H-NMR (400 MHz, D$_2$O) δ 6.10 (m, 4H), 5.90 (m, 8H), 4.26 (d, J = 7.2 Hz, 8H), 3.99 (s, 8H).
(Figure S7 in the Supporting Information).
^{13}C-NMR (300 MHz, D$_2$O) δ 130.78, 122.22, 50.66.
(Figures S8 and S9 in the Supporting Information).
ATR-IR: 2988 cm^{-1} ν(C-H), 1473 cm^{-1} γ (CH2), 1150 cm^{-1} ν(C-N).
(Figure S10 in the Supporting Information).

4.2.4. Preparation of N,N,N',N'-Tetraallyl Trimethylene Dipiperidine Dibromide 1c (TAMPB)

The preparation follows the two-step procedure outlined for TAPB. 16 g (81.49 mmol) trimethylenedipiperidine afforded 8.51 g (38%) crude diallyltrimethylendipiperidine (DAMP) as yellow, oily liquid.

^{1}H-NMR (600 MHz, Chloroform-d) δ 5.82 (m, 2H), 5.07 (m, 4H), 2.91 (d, J = 6.7 Hz, 4H), 2.85 (d, J = 11.5 Hz, 4H), 1.81 (t, J = 11.1 Hz, 4H), 1.60 (d, J = 9.5 Hz, 4H), 1.19 (m, 12H).
(Figure S11 in the Supporting Information).

In the second step, 8.51g DAMP yielded after recrystallization in methanol 11.53 g (72%) TAMPB as beige crystalline solid. Similar to 1a, the dibromide 1c was found to decompose at 220 °C.

^{1}H-NMR (400 MHz, D$_2$O) δ 6.07 (m, 4H), 5.73 (m, 8H), 4.00 (d, J = 7.3 Hz, 4H), 3.89 (d, J = 7.4 Hz, 4H), 3.52 (d, J = 12.2 Hz, 4H), 3.28 (t, J = 11.2 Hz, 4H), 1.93 (d, J = 9.5 Hz, 4H), 1.68 (m, 6H), 1.40 (m, 6H).

(Figure S12 in the Supporting Information).

^{13}C-NMR (300 MHz, D$_2$O) δ 128.61, 128.22, 123.77, 65.22, 56.14, 34.31, 32.15, 25.25, 22.51.

(Figures S13 and S14 in the Supporting Information).

ATR-IR: 3013 cm^{-1} ν(C-H), 1497 cm^{-1} γ (CH2), 1037 cm^{-1} ν(C-N).

(Figure S15 in the Supporting Information).

4.2.5. Copolymerisation of Crosslinked DADMAC-Hydrogels

The method is described using 2 mol% crosslinker as an example. Details on the compositions and further experimental details can be found in Table S1 in the Supporting Information. A mixture of 5 g of a DADMAC solution (65 wt% in water, 20 mmol), 30 mg sodium disulfite (0.16 mmol), and 62 mg N,N'-methylenebisacrylamide (0.4 mmol, 2 mol% related to the DADMAC content) was stirred until the crosslinker had completely dissolved. Meanwhile, 60 mg KPS were dissolved separately in 1.5 mL H$_2$O and then added to the monomer solution. The mixtures containing **1a–c** were stirred for 10 min and those with BIS for shorter times, as gelation with BIS begins earlier. An example image of a swollen polyDADMAC gel is given in the Supporting Information (Figure S21).

4.3. Infrared Spectroscopy (FTIR-ATR)

Infrared spectra were recorded using a Perkin Elmer Spectrum Two UATR FTIR spectrometer equipped with a diamond ATR (attenuated total reflection) window. All spectra were recorded in the spectral range of 4000–400 cm^{-1} with 12 scans at a spectral resolution of 4 cm^{-1}. Before each measurement, the diamond ATR crystal was cleaned with isopropanol.

4.4. Rheology

The rheological data were recorded on an Anton Paar Modular Compact Rheometer 102. A plate-plate geometry with a diameter of 25 mm made of stainless steel was used. All samples were measured with a plate gap of 1 mm at 20 °C. The isothermal oscillating time-dependent measurements were performed at 1% amplitude and an angular frequency of 1 Hz over a period of 16 h. The set-up was covered with a solvent trap over this period.

Sample preparation follows the above procedure for the preparation of DADMAC hydrogels. After adding the KPS initiator, the mixture was stirred intensively for 5–10 min (the exact time was noted and used in the calculation of the gel point), depending on the rate of polymerisation. 560 µL of this mixture was then transferred to the rheometer using an Eppendorf pipette. The gel point was calculated from the intersection of the G' and G" curves (determined graphically from recorded data) [29,30], by adding the stirring and transfer time. Details for the determination of the gel points can be found in Table S2 in the Supporting Information. An example of the graphical determination of the gel point is given in the Supporting Information (Figure S1). The measurements for the samples with 2 mol% of the different crosslinkers were repeated twice. The determined gel points deviated by approx. ±10%.

4.5. Electrochemical Impedance Spectroscopy (EIS)

Impedance spectra were recorded on a Gamry Potentiostat 400 (C3 Prozess und Analysetechnik GmbH, Munich, Germany). Sample preparation followed the above procedure for the preparation of crosslinked DADMAC hydrogels using 1 mol% of the desired crosslinker BIS, TAAB, TAPB, or TAMPB). After adding the KPS initiator and stirring intensively for 10 min, the mixtures were then transferred to standard electroporation cuvettes (Gap width = 4 mm), equipped with two opposing aluminium electrodes of 9 × 18 mm² (blocking electrodes). The solution was covered with a layer of silicon oil to exclude oxygen and prevent evaporation. Over a course of 16 days, the impedance Z was recorded every hour using a voltage of 141 mV. As shown in Figure S20, the phase angle $\varphi = 90°$ at an ac frequency of 100 kHz and with

$$Z = Z_{real} + iZ_{imag} = |Z|e^{i\varphi}$$

follows that $Z_{imag} \approx 0$ and $Z = Z_{real}$.

4.6. Swelling Experiments (Teabag Tests)

Approx. 100 mg of lyophilized hydrogel were weighed into polyester filter-bags and submerged in 0.5 L of double-distilled water at 22 °C. The mass increase was initially recorded every 15 min, later every 30 min to one hour, depending on the previous mass changes. After removing the teabags from the solutions, they were carefully stripped and hung up for 5 min to drain the nonabsorbed liquid. The reported maximum swelling ratios were determined by averaging the last three recorded values. The swelling experiments were each repeated once. It was found that the determined values varied by ±5% at low degrees of swelling. For high degrees of swelling (>200 g/g), deviations of up to ±14% were found.

4.7. NMR-Spectroscopy

¹H-NMR and ¹³C-NMR spectra were recorded on a Mercury 400 spectrometer (Varian, Palo Alto, CA, USA). Chemical shifts were calculated using the HDO signal at 4.64 ppm or the CDCL₃ signal at 7.26 ppm as a reference.

Supplementary Materials: The following are available online at https://www.mdpi.com/article/10.3390/gels8020100/s1, Figure S1: Example of the rheological determination of G' and G" of a DADMAC gelation mixture, Figure S2: ¹H-NMR-spectrum (400 MHz, D₂O) of Tetraallyl ammonium bromide (TAAB), Figure S3: ¹³C-NMR-spectrum (400 MHz, D₂O) of Tetraallyl ammonium bromide (TAAB), Figure S4: ¹³C-NMR-APT-spectrum (400 MHz, D₂O) of Tetraallyl ammonium bromide (TAAB), Figure S5: ATR-IR-spectrum of Tetraallyl ammonium bromide (TAAB), Figure S6: ¹H-NMR-spectrum (400 MHz, CDCl₃) of N,N'-diallyl piperazine (DAP), Figure S7: ¹H-NMR-spectrum (400 MHz, D₂O) of N,N,N',N'-tetraallyl piperazinium dibromide (TAPB), Figure S8: ¹³C-NMR-spectrum (400 MHz, D₂O) of N,N,N',N'-tetraallyl piperazinium dibromide (TAPB), Figure S9: ¹³C-NMR-APT-spectrum (400 MHz, D₂O) of N,N,N',N'-tetraallyl piperazinium dibromide (TAPB), Figure S10: ATR-IR-spectrum of N,N,N',N'-tetraallyl piperazinium dibromide (TAPB), Figure S11: ¹H-NMR-spectrum (400 MHz, CDCl₃) of N,N'-diallyl trimethylene dipiperidine (DAMP), Figure S12: ¹H-NMR-spectrum (400 MHz, D₂O) of N,N,N',N'-tetraallyl trimethylene dipiperidine dibromide(TAMPB), Figure S13: ¹³C-NMR-spectrum (400 MHz, D₂O) of N,N,N',N'-tetraallyl trimethylene dipiperidine dibromide(TAMPB), Figure S14: ¹³C-NMR-APT-spectrum (400 MHz, D₂O) of N,N,N',N'-tetraallyl trimethylene dipiperidine dibromide(TAMPB), Figure S15: ATR-IR-spectrum of N,N,N',N'-tetraallyl trimethylene dipiperidine dibromide(TAMPB), Figure S16: Copolymerisation diagram of poly(DADMAC-co- N,N'-Methylenebisacrylamide) calculated according to alfrey and price, Figure S17: ¹H-NMR-spectra (400 MHz, D₂O) pre-swollen in D₂O, Figure S18: ¹³C-NMR-spectra(400 MHz, D₂O) pre-swollen in D₂O, Figure S19: ¹H-NMR-spectrum (400 MHz, D₂O) of Poly-tetraallyl ammonium bromide (pTAAB), Figure S20: Exemplary bodeplot of a DADMAC gelation mixture with 1 mol % TAMPB after 10 h of polymerisation, Figure S21: Poly(DADMAC) crosslinked with 1%w.t. BIS and swollen in double distilled Water. Table S1: Details on the sample composition used to prepare crosslinked DADMAC hydrogels, Table S2: Details on the determination of the gelation points of crosslinked DADMAC hydrogels.

Author Contributions: All authors contributed to the study conception and design. Material preparation, data collection and analysis were performed by T.B.M.; Original draft preparation, T.B.M.; Review and editing, O.W. All authors have read and agreed to the published version of the manuscript.

Funding: The research project was carried out in the framework of the industrial collective research programme (grant no. ZF4140609). It was supported by the Federal Ministry for Economic Affairs and Energy (BMWi) through the AiF (German Federation of Industrial Research Associations eV) based on a decision taken by the German Bundestag.

Institutional Review Board Statement: Not applicable.

Informed Consent Statement: Not applicable.

Data Availability Statement: Data is available from the authors upon request.

Conflicts of Interest: The authors declare that they have no conflict of interest.

References

1. Ahmed, E.M. Hydrogel: Preparation, characterization, and applications: A review. *J. Adv. Res.* **2015**, *6*, 105–121. [CrossRef]
2. Zhu, Q.; Barney, C.W.; Erk, K.A. Effect of ionic crosslinking on the swelling and mechanical response of model superabsorbent polymer hydrogels for internally cured concrete. *Mater. Struct.* **2014**, *48*, 2261–2276. [CrossRef]
3. Dhara, D.; Chatterji, P.R. Swelling and deswelling pathways in non-ionic poly(N-isopropylacrylamide) hydrogels in presence of additives. *Polymer* **2000**, *41*, 6133–6143. [CrossRef]
4. Rivero, R.; Alustiza, F.; Capella, V.; Liaudat, C.; Rodriguez, N.; Bosch, P.; Barbero, C.; Rivarola, C. Physicochemical properties of ionic and non-ionic biocompatible hydrogels in water and cell culture conditions: Relation with type of morphologies of bovine fetal fibroblasts in contact with the surfaces. *Colloids Surf. B Biointerfaces* **2017**, *158*, 488–497. [CrossRef]
5. Buchanan, K.J.; Hird, B.; Letcher, T.M. Crosslinked poly(sodium acrylate) hydrogels. *Polym. Bull.* **1986**, *15*, 325–332. [CrossRef]
6. El-Sherbiny, I.M.; Yacoub, M.H. Hydrogel scaffolds for tissue engineering: Progress and challenges. *Glob. Cardiol. Sci. Pract.* **2013**, *2013*, 316–342. [CrossRef]
7. Yoshioka, H.; Mori, Y.; Shimizu, M. Separation and recovery of DNA fragments by electrophoresis through a thermoreversible hydrogel composed of poly(ethylene oxide) and poly(propylene oxide). *Anal. Biochem.* **2003**, *323*, 218–223. [CrossRef]
8. Akashi, M.; Saihata, S.; Yashima, E.; Sugita, S.; Marumo, K. Novel nonionic and cationic hydrogels prepared from N-vinylacetamide. *J. Polym. Sci. Part A Polym. Chem.* **1993**, *31*, 1153–1160. [CrossRef]
9. Campbell, R.L.; Seymour, J.L.; Stone, L.C.; Milligan, M.C. Clinical studies with disposable diapers containing absorbent gelling materials: Evaluation of effects on infant skin condition. *J. Am. Acad. Dermatol.* **1987**, *17*, 978–987. [CrossRef]
10. Barakat, M.A.; Sahiner, N. Cationic hydrogels for toxic arsenate removal from aqueous environment. *J. Environ. Manag.* **2008**, *88*, 955–961. [CrossRef] [PubMed]
11. Wong, H.S. Concrete with superabsorbent polymer. In *Eco-Efficient Repair and Rehabilitation of Concrete Infrastructures*; Woodhead Publishing: Cambridge, UK, 2018; pp. 467–499.
12. Jung, A.; Weichold, O. Preparation and characterisation of highly alkaline hydrogels for the re-alkalisation of carbonated cementitious materials. *Soft Matter* **2018**, *14*, 8105–8111. [CrossRef]
13. Jung, A.; Faulhaber, A.; Weichold, O. Alkaline hydrogels as ion-conducting coupling material for electrochemical chloride extraction. *Mater. Corros.* **2021**, *72*, 1448–1455. [CrossRef]
14. Yin, Y.-L.; Prud'homme, R.K.; Stanley, F. Relationship Between Poly(acrylic acid) Gel Structure and Synthesis. In *Polyelectrolyte Gels*; ACS Symposium Series; American Chemical Society Publication: Washington, DC, USA, 1992; pp. 91–113.
15. Senkal, B.F.; Erkal, D.; Yavuz, E. Removal of dyes from water by poly(vinyl pyrrolidone) hydrogel. *Polym. Adv. Technol.* **2006**, *17*, 924–927. [CrossRef]
16. Biçeak, N.; Koza, G. A Nonhydrolyzable-Water Soluble Crosslinker: Tetrallylpiperazinium Dichloride and its Copolymers with Acrylic Acid and Acrylamide. *J. Macromol. Sci. Part A* **1996**, *33*, 375–380. [CrossRef]
17. Ali, S.A.; Ahmed, S.Z.; Hamad, E.Z. Cyclopolymerization studies of diallyl- and tetraallylpiperazinium salts. *J. Appl. Polym. Sci.* **1996**, *61*, 1077–1085. [CrossRef]
18. Hirano, Y. Crosslinked Polymer, Method for Manufacturing It and Use Thereof. U.S. Patent 7,094,853B2, 7 March 2002.
19. Kabiri, K.; Zohuriaan-Mehr, M.J. Superabsorbent hydrogel composites. *Polym. Adv. Technol.* **2003**, *14*, 438–444. [CrossRef]
20. Kozempel, S.; Tauer, K.; Rother, G. Aqueous heterophase polymerization of styrene—a study by means of multi-angle laser light scattering. *Polymer* **2005**, *46*, 1169–1179. [CrossRef]
21. Alfrey, T.; Price, C.C. Relative reactivities in vinyl copolymerization. *J. Polym. Sci.* **1947**, *2*, 101–106. [CrossRef]
22. Shukla, A.; Srivastava, A.K. Free Radical Copolymerization of Acrylamide and Linalool with Functional Group as a Pendant. *High Perform. Polym.* **2016**, *15*, 243–257. [CrossRef]
23. Blicke, F.F.; Hotelling, E.B. Polycyclic Quaternary Ammonium Salts. III. *J. Am. Chem. Soc.* **2002**, *76*, 5099–5103. [CrossRef]
24. Lee, C.J.; Wu, H.; Hu, Y.; Young, M.; Wang, H.; Lynch, D.; Xu, F.; Cong, H.; Cheng, G. Ionic Conductivity of Polyelectrolyte Hydrogels. *ACS Appl. Mater. Interfaces* **2018**, *10*, 5845–5852. [CrossRef]
25. Hoti, G.; Caldera, F.; Cecone, C.; Rubin Pedrazzo, A.; Anceschi, A.; Appleton, S.L.; Khazaei Monfared, Y.; Trotta, F. Effect of the Cross-Linking Density on the Swelling and Rheological Behavior of Ester-Bridged beta-Cyclodextrin Nanosponges. *Materials* **2021**, *14*, 478. [CrossRef] [PubMed]
26. Zohuriaan-Mehr, M.J.; Kabiri, K. Superabsorbent polymer materials: A review. *Iran. Polym. J.* **2008**, *17*, 451–477.

27. Jung, A.; Endres, M.B.; Weichold, O. Influence of Environmental Factors on the Swelling Capacities of Superabsorbent Polymers Used in Concrete. *Polymers* **2020**, *12*, 2185. [CrossRef]
28. Butler, G.B.; Bunch, R.L. Preparation and Polymerization of Unsaturated Quaternary Ammonium Compounds. *J. Am. Chem. Soc.* **2002**, *71*, 3120–3122. [CrossRef]
29. Tcharkhtchi, A.; Nony, F.; Khelladi, S.; Fitoussi, J.; Farzaneh, S. Epoxy/amine reactive systems for composites materials and their thermomechanical properties. In *Advances in Composites Manufacturing and Process Design*; Woodhead Publishing: Cambridge, UK, 2015; pp. 269–296.
30. Kroutilová, I.; Matějka, L.; Sikora, A.; Souček, K.; Staš, L. Curing of epoxy systems at sub-glass transition temperature. *J. Appl. Polym. Sci.* **2006**, *99*, 3669–3676. [CrossRef]

Article

Non-Invasive Assessment of PVA-Borax Hydrogel Effectiveness in Removing Metal Corrosion Products on Stones by Portable NMR

Valeria Stagno [1,2,*], Alessandro Ciccola [3], Roberta Curini [3], Paolo Postorino [4], Gabriele Favero [5,*] and Silvia Capuani [2]

1. Earth Sciences Department, Sapienza University of Rome, Piazzale Aldo Moro 5, 00185 Rome, Italy
2. Physics Department, National Research Council—Institute for Complex Systems (CNR-ISC), Sapienza University of Rome, Piazzale Aldo Moro 5, 00185 Rome, Italy; silvia.capuani@isc.cnr.it
3. Department of Chemistry, Sapienza University of Rome, Piazzale Aldo Moro 5, 00185 Rome, Italy; alessandro.ciccola@uniroma1.it (A.C.); roberta.curini@uniroma1.it (R.C.)
4. Department of Physics, Sapienza University of Rome, Piazzale Aldo Moro 5, 00185 Rome, Italy; paolo.postorino@roma1.infn.it
5. Department of Environmental Biology, Sapienza University of Rome, Piazzale Aldo Moro 5, 00185 Rome, Italy
* Correspondence: valeria.stagno@uniroma1.it (V.S.); gabriele.favero@uniroma1.it (G.F.)

Citation: Stagno, V.; Ciccola, A.; Curini, R.; Postorino, P.; Favero, G.; Capuani, S. Non-Invasive Assessment of PVA-Borax Hydrogel Effectiveness in Removing Metal Corrosion Products on Stones by Portable NMR. *Gels* **2021**, *7*, 265. https://doi.org/10.3390/gels7040265

Academic Editor: Wei Ji

Received: 15 November 2021
Accepted: 13 December 2021
Published: 14 December 2021

Publisher's Note: MDPI stays neutral with regard to jurisdictional claims in published maps and institutional affiliations.

Copyright: © 2021 by the authors. Licensee MDPI, Basel, Switzerland. This article is an open access article distributed under the terms and conditions of the Creative Commons Attribution (CC BY) license (https://creativecommons.org/licenses/by/4.0/).

Abstract: The cleaning of buildings, statues, and artworks composed of stone materials from metal corrosion is an important topic in the cultural heritage field. In this work the cleaning effectiveness of a PVA-PEO-borax hydrogel in removing metal corrosion products from different porosity stones has been assessed by using a multidisciplinary and non-destructive approach based on relaxation times measurement by single-sided portable Nuclear Magnetic Resonance (NMR), Scanning Electron Microscopy—Energy Dispersive Spectroscopy (SEM-EDS), and Raman Spectroscopy. To this end, samples of two lithotypes, Travertine and Carrara marble, have been soiled by triggering acidic corrosion of some copper coins in contact with the stone surface. Then, a PVA-PEO-borax hydrogel was used to clean the stone surface. NMR data were collected in untreated, soiled with corrosion products, and hydrogel-cleaned samples. Raman spectroscopy was performed on PVA-PEO-borax hydrogel before and after cleaning of metal corrosion. Furthermore, the characterization of the dirty gel was obtained by SEM-EDS. The combination of NMR, SEM-EDS and Raman results suggests that the mechanism behind the hydrogel cleaning action is to trap heavy metal corrosion products, such as Cu^{2+} between adjacent boron ions cross-linked with PVA. Moreover, the PVA-PEO-borax hydrogel cleaning effectiveness depends on the stone porosity, being better in Carrara marble compared to Travertine.

Keywords: portable NMR; PVA-PEO-borax hydrogel; porous stones; coin corrosion products; raman spectroscopy; SEM-EDS

1. Introduction

The surface cleaning of works of art composed of stone materials is a critical concern for conservators and restorers. The cleaning process involves the removal of dirt, dust, pollutants, metal ions, or microorganisms. In particular, all these agents continuously endanger the life of stone artworks exposed to outdoor. When a stone building, statue or monument under atmospheric conditions is in contact with metallic parts, such as clamps, pivots, or plaques [1], corrosion of the latter may be induced and the corrosion products may hinder the correct readability of the artwork [2–4]. In particular, in the case of copper and its alloys, corrosion is a chemical attack mainly promoted by the affinity of metals and pollutants (i.e., sulfur, carbon dioxide, chlorides). This process leads to a corrosive layer called patina [5,6], which can be protective (noble patina) or unprotective (vile patina) [7–9], depending on the concentration of pollutants and acid rain. As a result, geographic location, precipitation, and pollution level all have an impact on patina composition and morphology [9,10]. So, the atmospheric exposure of copper produces the oxidation-reduction reactions leading to different corrosion products: copper(I) oxide which is red, copper(II) oxide which is black, black copper sulfide, various colored salts and nantokite, green-blue atacamite and clinoatacamite [6]. These corrosion products are responsible for the discoloration of the stone [1,11].

In the case of iron in contact with a stone artwork and in the presence of oxygen and water, it will corrode depending on the pH value of the surrounding environment [1,4,11]. Generally, corrosion is activated by acidic conditions, but it can also take place in an alkaline environment [11]. The result of iron corrosion will not be a patina adhered to the metal surface but a powdery rust layer produced by electrochemical processes [6]. So, the rust consists of the stratification of the oxides, usually green hydrated magnetite, black anhydrous magnetite and, only externally, the ferric hydroxides [6], which are responsible for cervices formation in the stone surrounding the metallic part [11]. Moreover, also when iron is covered by lead a damage to the stone should be expected because lead itself can be attacked [11]. Behind the surface alterations of stone artefacts induced by the corrosion of metallic components, other factors can cause the aesthetical modification of the artwork. Among these factors, there are old restorations and protecting interventions, deposition of particulate and pollutants, graffiti, and vandalism [12]. Specifically, because of the well-known interaction among SO_2, PM_{10} and rain pH, black crusts are the most common alteration for stone artworks in cities [9,13]. The black crusts composition reflects that one of the air in which the artwork is exposed and they lead to a mechanical, aesthetical and chemical damage [9].

In this scenario, the cleaning process and the choice of the cleaning substance are of fundamental importance. In the last years, gels or gel-like systems have been widely employed for the cleaning procedure of stone artworks and cultural heritage in general [1,14–17]. Gels or gel-like systems, as well as high-viscous-polymeric-dispersion [18], have shown great potential due to their high selectivity, low toxicity, and low environmental impact [16,18,19]. Among these, there are gel-systems based on polyvinyl alcohol (PVA). PVA is a water soluble and biocompatible polymer with good resistance to mechanical stress, capable to form hydrogen bonds, thanks to its hydroxyl groups, and ion complexes [19–21]. It has emerged as a potential adsorbent because of its high swelling capacity and its resistance to dissolution, mainly due to the formation of cross-links between network chains [22]. In particular, it can capture contaminants entrapping them between the fine pores of hydrogel developed via crosslinking networks [23]. The high-water-content and porous structure networks help to diffuse the solute with contaminants [24]. Moreover, it has been used together with both natural and synthetic compounds to produce different types of hydrogels, which may also be physically or chemically cross-linked. One of these, the PVA-borax hydrogel, obtained by cross-linking of PVA with borate ions [18,20,21,25], can be described as a viscoelastic dispersion with a dynamic network. In fact, the increase in borax concentration expands the system network due to electrostatic repulsions in the polymer chain [20,25,26]. When PVA-borax aqueous gel is applied on the surface of cultural heritage for cleaning purposes, this should be able to remove degradation products by capillarity absorption through their pores [19]. To this end, Riedo et al. [20] studied the effect of polyethylene oxide (PEO) addition to the PVA-borax hydrogel. The authors showed that PEO, which is a water-soluble and biocompatible thermoplastic polymer, would increase the pore size of the system in agreement with other studies [27,28]. Moreover, PEO seemed to improve the mechanical properties of the gel and the retention of its liquid phase [20,29].

PVA-PEO-borax gels, as well as all the gel-systems, can be easily removed in one step from a surface simply by peeling. This represents a great advantage for the conservation treatment of cultural heritage. However, the ease of gel removal, together with its capability of retention of the liquid phase and the absence of gel residues also depend on the characteristics of the surface to be cleaned [12,29].

In this work the cleaning effectiveness of a PVA-PEO-borax hydrogel in removing metal corrosion products from two different porosity stones has been assessed by using a multidisciplinary and non-destructive approach, combining the single-sided portable Nuclear Magnetic Resonance (NMR) to investigate stone samples cleaning, with Scanning Electron Microscopy—Energy Dispersive Spectroscopy (SEM-EDS) to characterize the gel composition after the stone cleaning process and Raman Spectroscopy measurements to study the gel before and after the cleaning procedure.

To this end, samples of two lithotypes, Travertine and Carrara marble, have been soiled by triggering the acidic corrosion of some copper coins in contact with the stone surface. Then, a PVA-PEO-borax hydrogel was used in order to clean the stone surface. NMR relaxation times were evaluated in untreated and treated samples. Pure gel and dirty gel obtained after sample cleaning were analyzed by Raman Spectroscopy. Moreover, the composition of the dirty gel was obtained by SEM-EDS. The novelty of this study is the use of portable NMR as a non-invasive and non-destructive tool for the monitoring of the gel cleaning procedure. Indeed, the NMR protocol that we developed can be employed for in situ analyses to evaluate the cleaning efficacy and action of different gels on different materials.

2. Results and Discussion
2.1. NMR Characterization of Travertine and Carrara marble

In a first phase, we characterized the untreated samples of Travertine and Carrara marble with the aim of differentiating the samples on the base of their porosity and morphology, by using NMR T_1 and T_2 parameters. In Figure 1 the T_1 and T_2 relaxation time distributions obtained for the three untreated samples are shown. All the samples are characterized by two T_1 and two T_2 components associated with two different pore size compartments. Regarding the T_1 relaxation time, Travertine 2 is characterized by the higher values (blue curve in Figure 1a), whereas Carrara marble the smaller ones (red curve in Figure 1a). On the other hand, Carrara marble is characterized by the highest T_2 values (red curve in Figure 1b).

In Figure 1, the different T_1 and T_2 mean values among the two Travertine samples and Carrara marble, can be explained on the basis of their different porous structure. Due to the inverse relationship between T_2 or T_1 and the surface-to-volume ratio (S/V) of the pores in a porous medium [30], our results can provide information about the different porosities of the three analyzed samples. The two T_1 and T_2 components detected for all three samples, suggest that two main different pore size compartments exist in the stones. In contrast to our previous work [12], we investigated samples at relative humidity (RH) value equal to 50 ± 3%. For this reason we did not detect the long T_2 component (around few or tens of ms) obtained in the study performed at RH = 94% [12]. Indeed, at RH = 50% less water molecules in vapor form wet the stone pores compared to those present at RH = 94%. About the two Travertine samples, they were cut from the same slab and parallel to the bedding planes. Despite having the same chemical composition, the two Travertines show a rather different pore structure inside the 2 mm layer studied by portable NMR. Indeed, Travertine 2 shows higher T_1 and T_2 than Travertine 1. This result indicates that Travertine 2 pores are characterized by a smaller S/V than those of Travertine 1. Concerning the Carrara marble sample, its T_1 is the smallest, whereas its T_2 is the highest (see red-line in Figure 1a,b). In the Carrara marble, a very few structure metal ions are present, which are responsible for the characteristic color of its veins [31]. Therefore, the shorter T_1 values of Carrara marble displayed in Figure 1a may be attributed to the effect of paramagnetic ions on T_1 [32–34].

On the other hand, the longest T_2 values of Carrara marble inform about its pores, which have the smallest S/V. These results suggest that Travertine has larger pores than Carrara marble, in agreement with the literature [31,35].

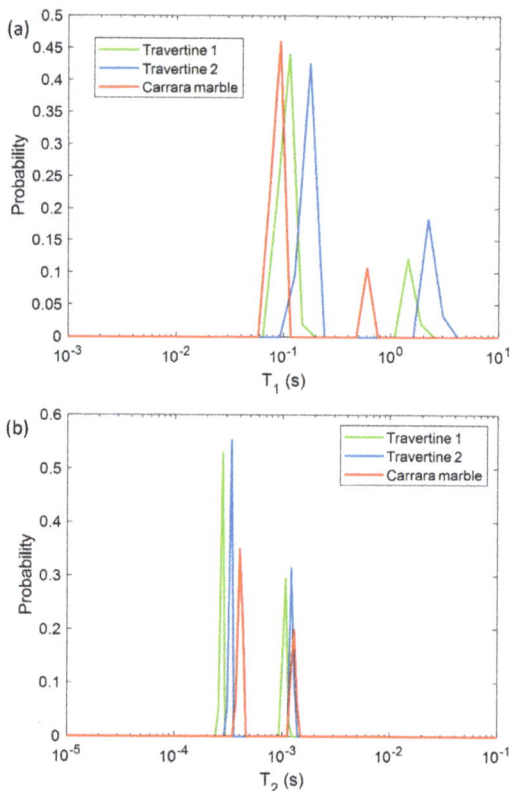

Figure 1. Longitudinal relaxation time T_1 (**a**) and transversal relaxation time T_2 (**b**) distribution for the three untreated samples analyzed in this study: two Travertines (green and blue lines), and one Carrara marble (red line). Travertine 1 refers to the untreated sample before being soiled with corrosion products from penny, whereas Travertine 2 and Carrara marble refer to the untreated samples before being soiled with corrosion products from euro cents.

2.2. NMR Monitoring of the PVA-Gel Cleaning of Metal Corrosion Products from Stones Surface

In the first step, we performed NMR measurements on the soiled samples in order to test if the T_1 and T_2 parameters were affected by the presence of metal corrosion products and if they could inform us about different degrees of soiling of the samples. Then, we repeated the same experiments on the stone surface after the hydrogel cleaning process, with the aim of highlighting the cleaning efficacy and detecting possible gel residues. To this end, in Figures 2 and 3 the T_1 and T_2 distributions before (black solid-line) and after the soiling process (green dashed-line), and after the cleaning process (pink dashed-line) for the two Travertines (Figures 2a,b and 3a,b) and the Carrara marble sample (Figures 2c and 3c) are displayed. Figures 2 and 3 show how the presence of metal corrosion products on the stones surface affects the NMR relaxation times (green-dashed lines). The longitudinal relaxation time T_1 seems to be strongly influenced by the heavy metal corrosion products and it significantly decreases in all the three stained samples compared to the untreated ones (see Figure 2). This result is in agreement with the literature [32–34] about the shortening effect of paramagnetic ions on the T_1. Conversely, T_2 is less affected by metal ions. The two main T_2 components of Travertine soiled with penny corrosion products are reduced compared to the untreated sample (see Figure 3a) confirming the already observed effect of paramagnetic ions. Travertine soiled with corrosion products from euro cents in Figure 3b, is characterized by T_2 that is not significantly affected by the presence of metal corrosion products. Furthermore, Carrara marble (Figure 3c) shows one short T_2 component, which is shortened because of the metal corrosion products, and one long T_2, which seems to increase. These different results can be explained on the basis of the different degree of corrosion of the coins used. Indeed, Travertine soiled with corrosion products from the penny has the highest variation in its relaxation times because of the greater degradation state of the already aged penny, which led to a greater deposition of metallic corrosion products (see Section 4.1).

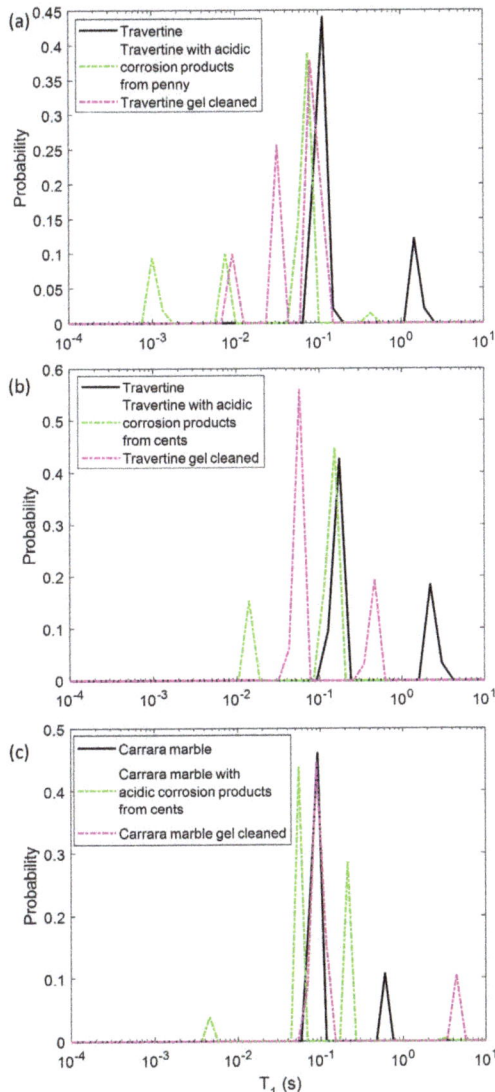

Figure 2. Longitudinal relaxation time T_1 distribution for (**a**) Travertine 1, (**b**) Travertine 2 and (**c**) Carrara marble before (solid-line) and after the soiling process (green dashed-line), and after the gel cleaning (pink dashed-line).

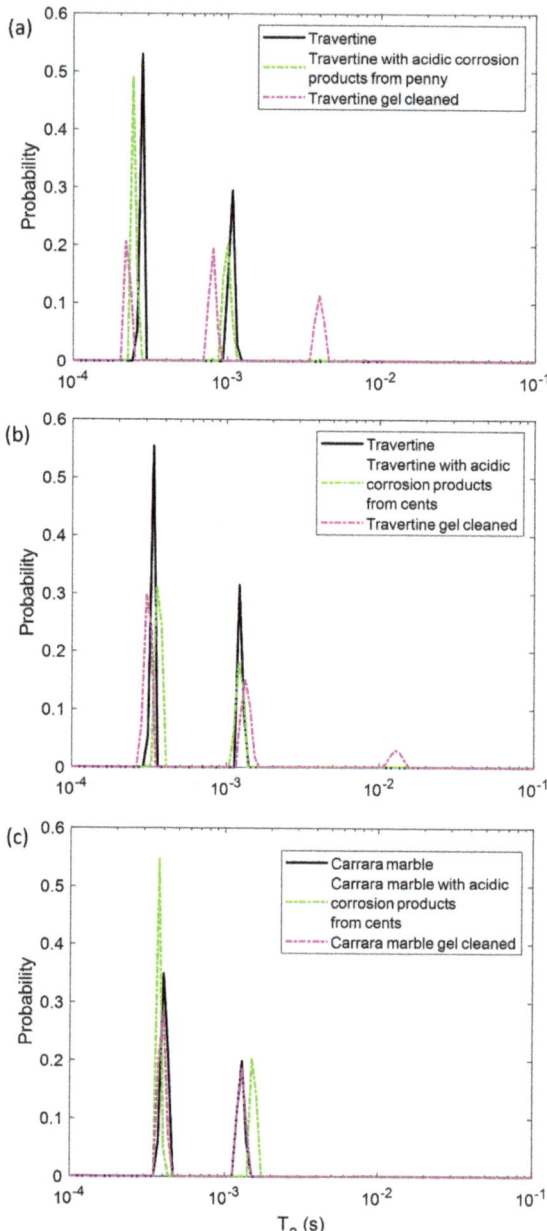

Figure 3. Transversal relaxation time T_2 distribution for (**a**) Travertine 1, (**b**) Travertine 2 and (**c**) Carrara marble before (solid-line) and after the soiling process (green dashed-line), and after the gel cleaning (pink dashed-line).

In principle, PVA-gel cleaning should result in a return of the NMR relaxation times to their values measured on the untreated samples. After the gel cleaning, the T_1 of the two Travertine surfaces are slightly increased compared to those ones measured before cleaning. Nevertheless, their T_1 did not return to its original value, acquired on the untreated surfaces.

This result may suggest that the PVA-based hydrogel did not completely remove the metal corrosion products from the stone. In addition, the fact that the T_1 did not return to the initial values may also suggest that the gel used has left residues in the pores, as observed in previous studies [12,29]. In particular, in Travertine soiled with corrosion products from the penny (Travertine 1) the T_1 and T_2 distributions have the same behavior with both T_1 and T_2 mean values that did not return to their original value after the cleaning process (Figure 3). While the T_1 components (Figure 2a) measured after the gel cleaning increased compared to those of the soiled surface, the T_2 components did not change (Figure 3a). Moreover, a third component of T_2 around 4 ms was detected after the gel cleaning, probably due to the gel residues (observable by the naked eye, see Section 4.1) or to the removal of dust from the stone surface. A similar consideration can be formulated for the second sample of Travertine (Travertine 2) soiled with corrosion products from euro cents (Figures 2b and 3b). Here, after the gel cleaning, both the T_2 components are close to the initial values measured on the untreated surface (black-solid line). Again, the third T_2 component around 13 ms may suggest gel residues or dust removal. However, because of the very small variation of the T_2 distribution in Travertine 2, our discussions must be considered as preliminary considerations, and further measurements will be performed in future work.

In Carrara marble, after the gel cleaning, both the T_1 and T_2 returned close to their original values measured on the untreated surface (see Figures 2c and 3c). This result indicates that the PVA-borax gel cleaned the Carrara marble surface without leaving residues. However, in Figure 2c the second T_1 component is higher than that one of the untreated sample. This effect can be ascribable to a deep dust removal probably present in the untreated samples. The absence of gel residues in Carrara marble can be explained by the lack of macropores, inside which the gel can penetrate and becomes difficult to remove by peeling, compared to Travertine.

2.3. Raman Spectroscopy of the Gel Layers Removed after Cleaning of the Stones Surface

The purpose of Raman analyses was to detect structural variations of the dirty gel (i.e., used for the stone cleaning from metal corrosion products) compared to the pure gel, likely induced by the formation of new bonds among metal corrosion products and the hydrogel polymers.

In Figure 4 the images of the pure gel and of the two layers of gel removed after cleaning of the Travertine and Carrara marble surface are displayed. Figure 4a shows the porous structure of the gel and a brighter appearance compared to the gel used to clean Travertine from penny corrosion (Figure 4b) and Carrara marble from euro cent corrosion (Figure 4c). This result suggests that the PVA-PEO-borax hydrogel interacted with the metal corrosion products.

In Figures 5 and 6 the Raman spectra acquired are displayed. The spectra acquired for the PVA-borax hydrogel show a great reproducibility (Figure 5), where all the features of the chemical matrices are observable and summed up in Table 1: the broad signal at 1125 cm^{-1} is attributable to B-O-C bond groups, while the intense signal at 1440 cm^{-1}, along with the lower intensity band at 1355 cm^{-1}, is related to the C-H bending modes in the PVA moieties. At higher wavenumbers, it is possible to observe a main signal at 2913 cm^{-1}, characteristic of C-H stretching along with the shoulders at around 2855 and 2935 cm^{-1}. These spectral features confirm the evidences reported in literature about PVA and PVA-borax hydrogels [36–39]. Regarding the hydrogel used for the cleaning of metal corrosion products (Figure 6), some aspects must be highlighted.

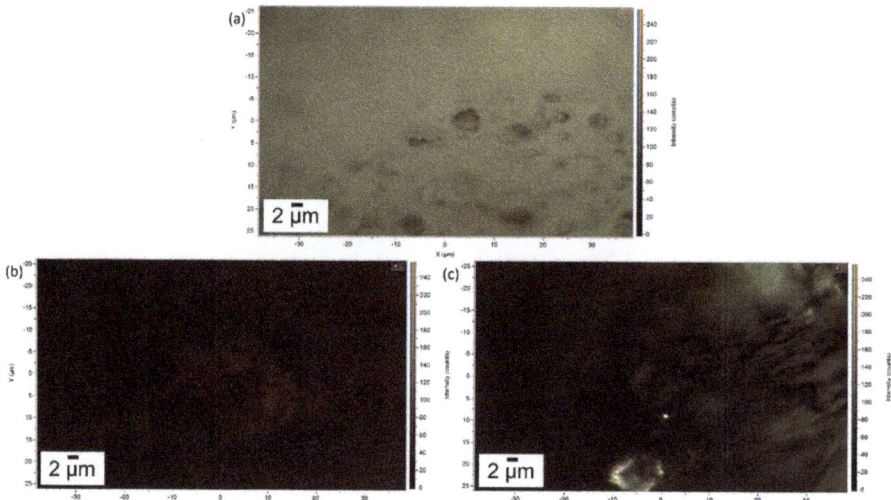

Figure 4. Image of (**a**) the pure gel, (**b**) the gel used to remove penny corrosion products from Travertine and (**c**) the gel used to clean Carrara marble from euro cents corrosion products. The scale bar is 2 µm and the image magnification is 100×.

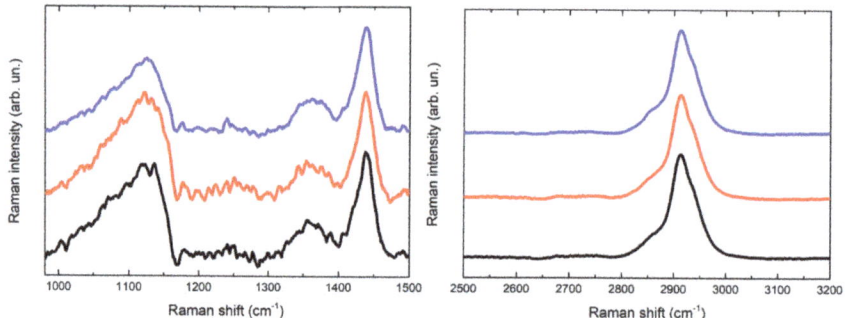

Figure 5. Comparison of three Raman spectra acquired for the reference of PVA-borax hydrogel.

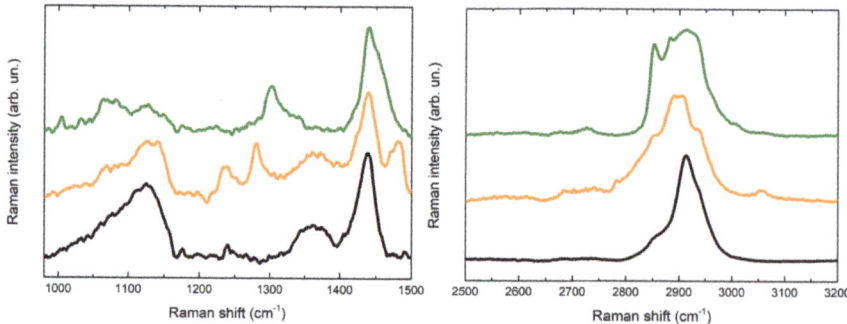

Figure 6. Comparison of Raman spectra acquired for the reference of PVA-borax hydrogel (black line), the cleaning gel for the cent stains on Carrara marble (orange line) and the cleaning gel for the penny stains on Travertine (green line).

Table 1. Tentative assignations of observed Raman peaks for PVA-borax hydrogel.

Wavenumber	Assignation
1125	B-O-C stretching
1355	C-H bending
1440	C-H bending
2855	C-H stretching
2913	C-H stretching
2935	C-H stretching

First, it is relevant to emphasize that no particles of corrosion products are visible on the hydrogel surface, while the whole matrix is homogeneously colored (Figure 4). For this reason, the spectra were acquired at different random points on the surface. Some differences in the spectra are observable in comparison to the PVA-borax reference spectra. In particular, in one of the spectra acquired from the hydrogel used for the cleaning of the penny corrosion, the band at 1125 cm^{-1} is split into two signals at 1072 and 1127 cm^{-1}, while the band at 1355 cm^{-1} disappears and it is replaced by another signal at 1304 cm^{-1}. A new band at around 1656 cm^{-1} appears, while, in the higher wavenumber range, defined peaks at 2852, 2882 and 3007 cm^{-1} overlap with the C-H stretching of the original hydrogel matrix. At two points, instead, the main signals of the gel are present, but two peaks at 972 and 1273 cm^{-1} also appear.

Analogously, in the Raman spectrum of the hydrogel used for the cleaning of the euro cent stains on Carrara marble, new signals at 1239, 1282 and 1481 cm^{-1} are visible, while in the C-H stretching range the signal broadens at lower wavenumbers, with the appearance of a shoulder at around 2854 cm^{-1} and a low intensity peak at 3055 cm^{-1}. Moreover, even if it is less evident than for the previous cleaning gel, a shoulder at around 1070 cm^{-1} is visible close to the B-O-C signal at 1125 cm^{-1}.

Considering the reproducibility of Raman spectra of the hydrogel matrix reference and the absence of visible particles attributable to corrosion products, it is possible to hypothesize that the spectral variations observable in the case of the hydrogel samples used for the cleaning could be attributed to the removal process, which should not have a mechanical origin but probably chemical, otherwise the corrosion products would be clearly visible as aggregates on the gel surface. Moreover, the shoulder close to the B-O-C signal at 1125 cm^{-1} would suggest that the boron atom could be involved in this process. In this regard, our result are in agreement with Saeed et al. [40] paper. The authors, using XRD and FTIR techniques, suggest that PVA-borax hydrogel entraps divalent transition metal ions such as Cu^{2+}, Fe^{2+} and Zn^{2+} that crosslink complexes of PVA-borax. Specifically, they report that the C-O stretching peak in the presence of Cu (II) and Zn (II) appears at, 1124.50 cm^{-1} and 1118.71 cm^{-1}, respectively, and the 1074 cm^{-1} peak was attributed to the B-O-C stretching frequency [40], when divalent transition metal ions are entrapped.

However, it is also important to mention that, in order to deepen this behavior, a higher number of Raman measurements is needed. Furthermore, new experiments with the related statistical analysis are expected to investigate the mechanism of removal of the corrosion products from the different matrices, involving the combination of heteronuclear high resolution NMR and Raman spectroscopy.

2.4. SEM-EDS of the Gel Layers Removed after Cleaning of the Stones Surface

The purpose of SEM-EDS analyses was to detect the presence of metals in the hydrogel layer that was peeled off from the soiled stone samples. Metals trapped in the gel network indicate that the PVA-PEO-borax hydrogel was able to remove coin corrosion products from the stone surface.

To this end, we obtained a high-resolution image of one dirty gel layer, shown in Figure 7. The heterogeneous porous structure of the PVA-PEO-borax hydrogel is visible in Figure 7. Here, many impurities in the form of crystals trapped into the hydrogel network can be observed. They confirmed the mechanical action of the gel towards particles and dust deposited on the Travertine surface before the soiling process, as suggested by the presence of Ca in the dirty hydrogel layer (see Figure 7), which comes from the Travertine structure made of $CaCO_3$. The spectrum of one inclusion revealed the presence of 1.2% of iron (see Figure 7, spectrum 1). Because the composition of the penny coin was copper and zinc, we can suggest that some particles of iron were already present on the stone surface before that the soiling process was obtained by the deposition of acidic corrosion products from the penny. Figure 7 also shows spectrum 2 acquired in a different point, where a 0.9% of copper was detected. The presence of copper is ascribable to the corrosion products of the penny.

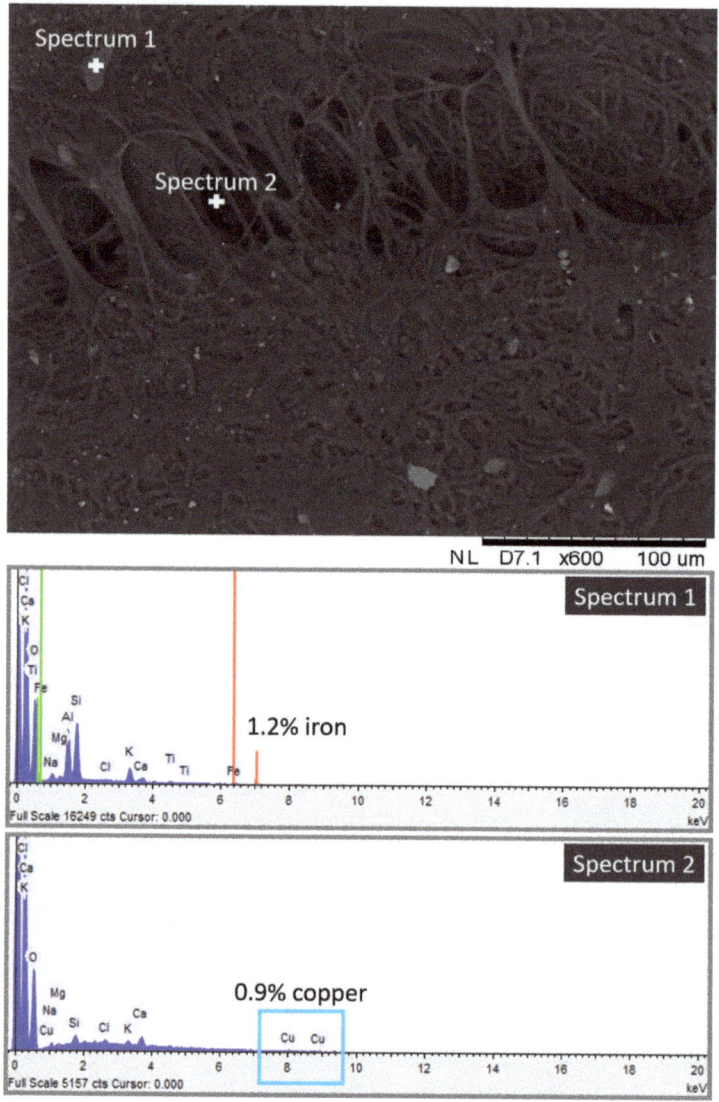

Figure 7. SEM-EDS spectrum and image of the dirty hydrogel layer removed after cleaning of the Travertine surface soiled with acidic corrosion products from penny.

SEM-EDS analysis suggested the presence of copper chemically bonded to the hydrogel structure and of iron and other inclusions not chemically bonded to the hydrogel. This indicates that the cleaning procedure of the PVA-PEO-borax hydrogel is both chemical and mechanical. Metal corrosion products are removed by a chemical action, whereas other impurities of the stone surface are removed by a mechanical action. Moreover, the detection of Ca in the hydrogel layer suggests that some particles of $CaCO_3$ from the Travertine structure were removed.

2.5. Limits of PVA-PEO-Borax Hydrogel to Clean Porous Materials

In this work, the formulation of PVA-PEO-borax in water solution developed by Riedo et al. [20] has been used for the hydrogel preparation. This has shown many weaknesses, in agreement with previous studies [20,21,29,41–43]. First of all, it does not seem suitable for the surface cleaning of macro-porous materials, such as the Travertine samples, because of the difficulty in the complete removal of the hydrogel layer, which leads to solid-like residues (see Section 4.1). Despite the fact that the addition of PEO increases the pore size and the liquid phase retention of the gel network, as well as improves its mechanical properties, it has been observed that the residues left on the stone surface may be due to the fact that PEO is the only polymer not chemically bonded to the network [20,21,29,42]. So, PEO seems the major responsible for residues together with the type and state of conservation of the sample analyzed.

NMR, Raman and SEM-EDS results are compatible with a scenario for which metal corrosion products are trapped between two adjacent boron ions as suggested by Saeed et al. [40]. Therefore, the removal efficacy of metals from the sample surface would increase with the increase of boron ions crosslinked to the gel network. This can be achieved thanks to the addition of organic solvents to the PVA-PEO-borax water solution. Indeed, boron ions are insoluble in organic solvents and prefer to be in a cross-linked state [41]. Moreover, it has been noticed that the presence of organic solvents increases the thermal stability of the gel network and the folding of the PVA chains increasing the cross-linking [41].

The PVA-borax hydrogel properties can also be improved by adding another polymer. Some studies [21,43] investigated the effect of agarose on the PVA-borax gel network. Agarose seems to increase the shape stability of the gel. The PVA-borax-agarose gel shows enhanced liquid phase retention and improved mechanical properties [21,43]. The PVA-B/AG gel is more suitable for the cleaning of porous materials, and if it is applied for an adequate contact time, no residues are left [21,43].

3. Conclusions

The study we presented here, provides new information for the possible use of PVA-borax hydrogel to clean monuments and marble statues from metal corrosion products. For this reason, we used a portable NMR instrument that allows in situ measurement to study the cleaning efficacy of PVA-borax hydrogel. We confirmed the great potential of single-sided portable NMR as a tool to evaluate the effectiveness of the cleaning procedure. Among the NMR parameters used to investigate the cleaning effectiveness with respect to corrosion products such as copper and iron alloys, as expected, the T_1 relaxation time has proved to be extremely sensitive (it strongly reduces its value in the presence of metallic corrosion products), and therefore we suggest it in a future monitoring protocol with portable NMR instrument.

The NMR, SEM-EDS and Raman data obtained in this work, integrate the data in the literature related to the characterization of PVA-borax hydrogels obtained with multimodal approaches of different microscopic and spectroscopic techniques [20,26,29,40,44]. The NMR study carried out on the surface of Travertine and Carrara marble samples up to 2 mm in deep has shown that is easier to remove PVA-borax hydrogel in Carrara marble compared to Travertine, as Carrara marble is characterized by pores on average smaller with less dispersion than Travertine porosity, which is much more heterogeneous and with the presence of macroscopic pores that prevent the perfect removal of the hydrogel. Therefore, the cleaning action depends on porosity features of stone.

Preliminary experiments of Raman spectroscopy and SEM-EDS performed on PVA-borax hydrogel before and after the cleaning suggest a chemical rather than a physical-mechanical action to remove metal corrosion products. However, for future work, we plan to improve the quality and increase the number of Raman experiments in order to obtain more information.

In conclusion, by combining the NMR, Raman and SEM-EDS results, this work suggests that the mechanism behind the hydrogel cleaning action is to trap metal corrosion products (such as copper or iron, as we used copper coin corrosion for this study) between adjacent boron ions crosslinked with PVA.

4. Materials and Methods

Two Travertine samples and one Carrara marble sample with size of $5 \times 5 \times 2$ cm^3 have been chosen to be studied in this work. Roman Travertine samples have been cut from the same slab and therefore have the same chemical composition characterized by 97–99% CaCO$_3$ [45,46]. The Travertines used in this work have been cut parallel to the bedding planes, which means they have larger and more interconnected pores in comparison to samples cut perpendicular to the bedding planes [12,47]. Their porosity is around 6% [48,49]. Intercrystalline and intergranular porosity are poorly sorted and range between 0.01 and 10 µm. Open porosity has a size of up to 100 µm and can achieve a centimetric-size [50]. These macropores are poorly connected [35].

Carrara marble, coming from Carrara region in the central part of Italy, is a calcitic marble mainly composed by calcium carbonate (>99%) but with also low amount of mineral impurities (clay, silt, sand, iron oxide) that produces its characteristic color [31,51–53]. Carrara marble is characterized by a total porosity from about 1.6% to 3.3% [31,49,52]. While Travertine has also macropores, Carrara marble shows only micropores and mesopores (0.001 µm < r < 10 µm) [31,50].

Moreover, two different types of coins have been used. One aged penny, which is the British decimal currency composed of copper and zinc, and several 1 euro cent coins made of copper-covered steel.

In order to trigger the acidic corrosion of both penny and euro cent, a solution 0.2 M of citric acid was used. The solution was characterized by pH of 2.5.

The PVA-PEO-borax hydrogel employed in this research was conceived and described in a previous study by Riedo et al. [20] and investigated by using portable NMR by Stagno et al. [12]. Its components are poly(vinylalcohol) (87–89% hydrolyzed, Mw 85,000–12,400, Sigma-Aldrich, Milano, Italy), poly(ethyleneoxide) (Mw. 37,000–4400, Sigma-Aldrich, Milano, Italy), and sodium tetraborate decahydrate (Sigma-Aldrich, Milano, Italy). In Figure 8, the chemical structure of PVA-borate and of PEO is shown.

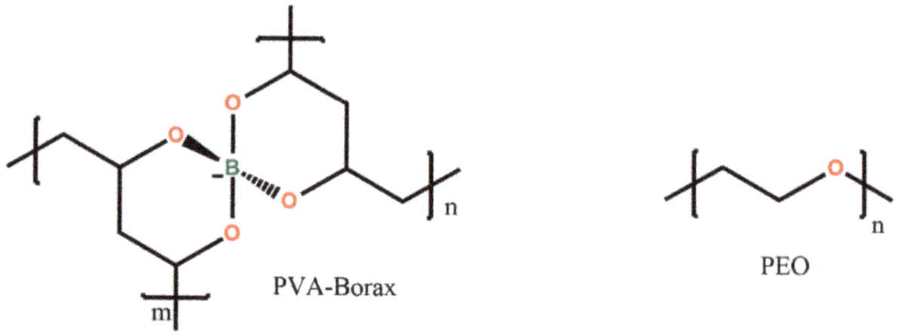

Figure 8. Chemical structure of the hydrogel compounds.

4.1. Sample Preparation

The two Travertine samples have been soiled in different way. On the surface of one Travertine (Travertine 1) an acidic corrosion of one aged penny was induced by using the solution of citric acid (Figure 9a,b), whereas on the surface of the second Travertine sample (Travertine 2) two coins of euro cents with citric acid solution (Figure 9c) have been left to react. The surface of Carrara marble was stained with the corrosion products coming from the reaction of four coins of euro cents with the citric acid solution (Figure 9d).

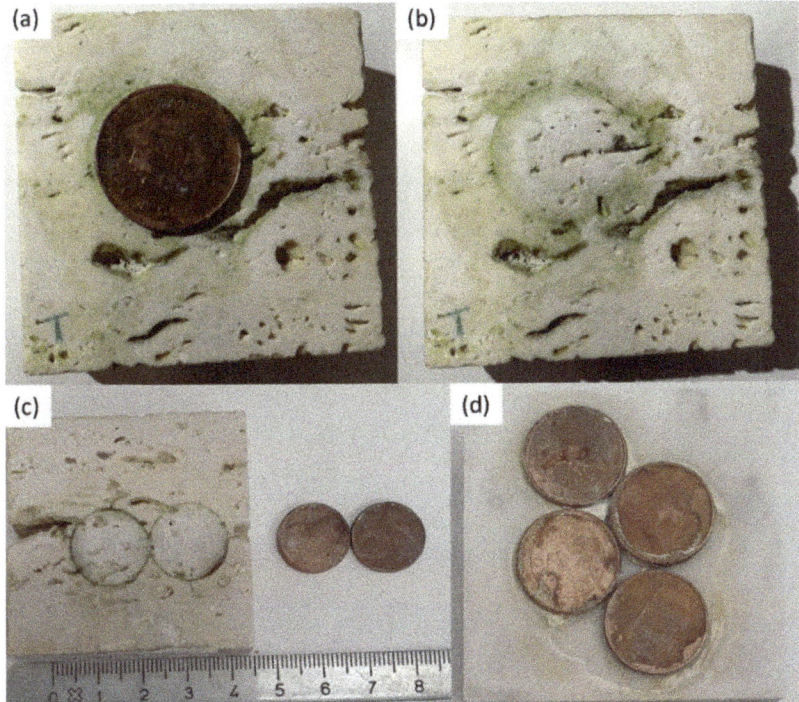

Figure 9. In (**a**,**b**) Travertine sample stained by corrosion products of one penny reacting with citric acid; the greenish-black stain on Travertine produced by reaction of two euro cents with citric acid in (**c**); in (**d**) Carrara marble surface with acidic corrosion of four euro cents with citric acid.

The coins and the reagent (i.e., citric acid) were left on the stones surface for one week. After that, the coins were removed from the stones surface and a greenish-black stain was observed (see Figure 9). Each stone surface was treated by the PVA-PEO-borax gel, which was applied in a thin layer and left on the stone surface for 30 min. After this time, the gel layer was peeled off (see Figure 10) and it was left drying and then preserved for Raman analysis. The cleaned stones surfaces (Figure 11) showed a different appearance. The two Travertine samples showed at the naked eye the presence of transparent gel residues in the pores (zoomed portions of Figure 11a,b). All the surfaces were left drying before performing NMR measurements.

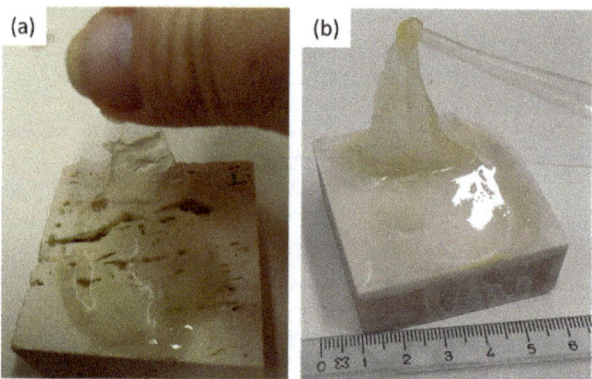

Figure 10. PVA-PEO-borax hydrogel peeling from (**a**) Travertine surface and (**b**) Carrara marble surface.

Figure 11. Stone surfaces after PVA-PEO-borax hydrogel cleaning of (**a**) Travertine 1 soiled with corrosion products from penny, (**b**) Travertine 2 soiled with corrosion products from euro cents, and (**c**) Carrara marble soiled with corrosion products from euro cents. In the zoomed portions transparent gel residues are observable in the pores.

4.2. Portable NMR Measurements

A Bruker minispec mq-ProFiler with a single-sided magnet that generates a static magnetic field of 0.35 T was used. It was equipped with a RF-probe for performing experiments with a 2 mm depth from the sample surface, characterized by a ^{1}H-resonance frequency of 17 MHz and dead time of 2 µs. The two Travertine samples and the Carrara marble sample have been characterized in their natural state by measuring the longitudinal (T_1) and transversal (T_2) relaxation time. T_1 and T_2 of the stones surfaces have been also monitored after the soiling process and after the gel cleaning process. All the NMR measurements have been performed in a controlled environment characterized by a temperature (T) of 25 ± 1 °C and a relative humidity (RH) around 50 ± 3%.

The longitudinal relaxation time (T_1) was acquired by using a Saturation-Recovery (SR) sequence with repetition time (TR) of 0.2 s, 65 steps from 0.05 ms to 8000 ms, increment factor of 1.2 and number of scans (NS) of 2048.

The transversal relaxation time (T_2) was acquired by using a Carr-Purcell-Meiboom-Gill (CPMG) sequence with TR of 1 s, echo time (TE) of 30 µs, 200 echoes, NS = 2048.

Each T_1 and T_2 measurement was repeated three times in order to test the reproducibility and minimize the error.

All data were elaborated by using the Inverse Laplace Transform (ILT) algorithm [54] in MATLAB 2021a to obtain the T_1 and T_2 distribution.

4.3. Raman Measurements

The gel layer used to clean each surface of the stones from the corrosion products was analyzed by using Raman spectroscopy. The instrumental setup is represented by a Horiba Jobin-Yvon HR Evolution micro-Raman spectrometer. This is equipped with a He-Ne laser (λ = 632 nm), coupled with a microscope with a set of interchangeable objectives. A 100× objective was used. Intensity of radiation has been set at 15 mW by neutral filters; the acquisition time has been varied between 10 and 15 s for each scan, and the number of acquired scans has been changed between 30 and 60 scans. Images of the pure gel and of the gel layers removed from Travertine and Carrara marble have been acquired with 100× magnification. Moreover, spectra were acquired in three or five different points of pure gel and of gel used to clean Travertine soiled with corrosion products from penny and Carrara marble soiled with corrosion products from euro cents.

4.4. SEM-EDS Measurements

The gel layer removed from the Travertine stone after cleaning of the penny corrosion products was analyzed by using a TM 3000 HITACHI scanning electron microscope (SEM) equipped with an EDX SWIFT ED 3000 probe. No coating treatment was applied to the sample. The accelerating voltage was 15 kV. From the sample, an image with a 600× magnification and an acquisition time of 10 min was acquired. Then, several points on the image were detected and the elemental composition in each point was obtained. The acquisition time of each spectrum was 3 min.

Author Contributions: Conceptualization, G.F. and S.C.; methodology, V.S., A.C. and S.C.; software, V.S.; validation, V.S., A.C., S.C., P.P. and R.C.; formal analysis, V.S., A.C. and S.C.; investigation, V.S., A.C., S.C.; resources, G.F., R.C., P.P. and S.C.; data curation, V.S. and A.C.; writing—original draft preparation, V.S. and S.C.; writing—review and editing, V.S., S.C., A.C. and G.F.; visualization, V.S. and A.C.; supervision, G.F., S.C. and P.P.; project administration, G.F. and S.C.; funding acquisition, G.F. All authors have read and agreed to the published version of the manuscript.

Funding: The APC was funded by CollectionCare project, which has received funding from the European Union's Horizon 2020 research and innovation programme under grant agreement No 814624 (https://www.collectioncare.eu/, accessed 12 December 2021). No other external funding was received for this study.

Data Availability Statement: The data presented in this study are available on request from the corresponding author.

Acknowledgments: G.F. thanks the CollectionCare project (European Union's Horizon 2020 research and innovation programme under grant agreement No 814624) for the financial support of the APC.

Conflicts of Interest: The authors declare no conflict of interest.

References

1. Sansonetti, A.; Bertasa, M.; Corti, C.; Rampazzi, L.; Monticelli, D.; Scalarone, D.; Sassella, A.; Canevali, C. Optimization of Copper Stain Removal from Marble through the Formation of Cu(II) Complexes in Agar Gels. *Gels* **2021**, *7*, 111. [CrossRef] [PubMed]
2. Canevali, C.; Fasoli, M.; Bertasa, M.; Botteon, A.; Colombo, A.; Di Tullio, V.; Capitani, D.; Proietti, N.; Scalarone, D.; Sansonetti, A. A multi-analytical approach for the study of copper stain removal by agar gels. *Microchem. J.* **2016**, *129*, 249–258. [CrossRef]
3. Young, M.E.; Urquhart, D.; Laing, R. Maintenance and repair issues for stone cleaned sandstone and granite building façades. *Build. Environ.* **2003**, *38*, 1125–1131. [CrossRef]
4. Macchia, A.; Sammartino, M.; Tabasso, M.L. A new method to remove copper corrosion stains from stone surfaces. *J. Archaeol. Sci.* **2011**, *38*, 1300–1307. [CrossRef]
5. FitzGerald, K.; Nairn, J.; Skennerton, G.; Atrens, A. Atmospheric corrosion of copper and the colour, structure and composition of natural patinas on copper. *Corros. Sci.* **2006**, *48*, 2480–2509. [CrossRef]
6. Albini, M.; Ridolfi, S.; Giuliani, C. Multi-spectroscopic approach for the non-invasive characterization of paintings on metal surfaces. *Front. Chem.* **2020**, *8*, 289. [CrossRef] [PubMed]
7. Chiavari, C.; Rahmouni, K.; Takenouti, H.; Joiret, S.; Vermaut, P.; Robbiola, L. Composition and electrochemical properties of natural patinas of outdoor bronze monuments. *Electrochim. Acta* **2007**, *52*, 7760–7769. [CrossRef]
8. Rahmouni, K.; Takenouti, H.; Hajjaji, N.; Srhiri, A.; Robbiola, L. Protection of ancient and historic bronzes by triazole derivatives. *Electrochim. Acta* **2009**, *54*, 5206–5215. [CrossRef]

9. Di Turo, F.; Proietti, C.; Screpanti, A.; Fornasier, M.F.; Cionni, I.; Favero, G.; De Marco, A. Impacts of air pollution on cultural heritage corrosion at European level: What has been achieved and what are the future scenarios. *Environ. Pollut.* **2016**, *218*, 586–594. [CrossRef]
10. Fonseca, I.; Picciochi, R.; Mendonça, M.; Ramos, A. The atmospheric corrosion of copper at two sites in Portugal: A comparative study. *Corros. Sci.* **2004**, *46*, 547–561. [CrossRef]
11. Stambolov, T.; Van Asperen De Boer, J.R.J. *The Deterioration and Conservation of Porous Building Materials in Monuments*; A Preliminary Review; The International Centre for the Study of the Preservation and Restoration of Cultural Property (ICCROM): Rome, Italy, 1967.
12. Stagno, V.; Genova, C.; Zoratto, N.; Favero, G.; Capuani, S. Single-Sided Portable NMR Investigation to Assess and Monitor Cleaning Action of PVA-Borax Hydrogel in Travertine and Lecce Stone. *Molecules* **2021**, *26*, 3697. [CrossRef]
13. Cachier, H.; Sarda-Estève, R.; Oikonomou, K.; Sciare, J.; Bonazza, A.; Sabbioni, C.; Greco, M.; Reyes, J.; Hermosin, B.; Saiz-Jimenez, C. Aerosol characterization and sources in different European urban atmospheres: Paris, Seville, Florence and Milan. In *Air Pollution and Cultural Heritage*; CRC Press: London, UK, 2004; pp. 3–14.
14. Baglioni, M.; Giorgi, R.; Berti, D.; Baglioni, P. Smart cleaning of cultural heritage: A new challenge for soft nanoscience. *Nanoscale* **2012**, *4*, 42–53. [CrossRef] [PubMed]
15. Bonelli, N.; Poggi, G.; Chelazzi, D.; Giorgi, R.; Baglioni, P. Poly(vinyl alcohol)/poly(vinyl pyrrolidone) hydrogels for the cleaning of art. *J. Colloid Interface Sci.* **2019**, *536*, 339–348. [CrossRef] [PubMed]
16. Baglioni, M.; Poggi, G.; Giorgi, R.; Rivella, P.; Ogura, T.; Baglioni, P. Selective removal of over-paintings from "Street Art" using an environmentally friendly nanostructured fluid loaded in highly retentive hydrogels. *J. Colloid Interface Sci.* **2021**, *595*, 187–201. [CrossRef] [PubMed]
17. Carretti, E.; Natali, I.; Matarrese, C.; Bracco, P.; Weiss, R.G.; Baglioni, P.; Salvini, A.; Dei, L. A new family of high viscosity polymeric dispersions for cleaning easel paintings. *J. Cult. Herit.* **2010**, *11*, 373–380. [CrossRef]
18. Chelazzi, D.; Fratini, E.; Giorgi, R.; Mastrangelo, R.; Rossi, M.; Baglioni, P. Gels for the Cleaning of Works of Art. In *Gels and Other Soft Amorphous Solids*; ACS Symposium Series; American Chemical Society (ACS): Washington, DC, USA, 2018; pp. 291–314.
19. Mazzuca, C.; Severini, L.; Domenici, F.; Toumia, Y.; Mazzotta, F.; Micheli, L.; Titubante, M.; Di Napoli, B.; Paradossi, G.; Palleschi, A. Polyvinyl alcohol based hydrogels as new tunable materials for application in the cultural heritage field. *Colloids Surf. B Biointerfaces* **2020**, *188*, 110777. [CrossRef]
20. Riedo, C.; Caldera, F.; Poli, T.; Chiantore, O. Poly(vinylalcohol)-borate hydrogels with improved features for the cleaning of cultural heritage surfaces. *Herit. Sci.* **2015**, *3*, 23. [CrossRef]
21. Al-Emam, E.; Soenen, H.; Caen, J.; Janssens, K. Characterization of polyvinyl alcohol-borax/agarose (PVA-B/AG) double network hydrogel utilized for the cleaning of works of art. *Herit. Sci.* **2020**, *8*, 1–14. [CrossRef]
22. Ahmed, E.M. Hydrogel: Preparation, characterization, and applications: A review. *J. Adv. Res.* **2015**, *6*, 105–121. [CrossRef]
23. Bao, S.; Wu, D.; Wang, Q.; Su, T. Functional Elastic Hydrogel as Recyclable Membrane for the Adsorption and Degradation of Methylene Blue. *PLoS ONE* **2014**, *9*, e88802. [CrossRef]
24. He, Y.; Wu, P.; Xiao, W.; Li, G.; Yi, J.; He, Y.; Chen, C.; Ding, P.; Duan, Y. Efficient removal of Pb(II) from aqueous solution by a novel ion imprinted magnetic biosorbent: Adsorption kinetics and mechanisms. *PLoS ONE* **2019**, *14*, e0213377. [CrossRef]
25. Lin, H.-L.; Liu, Y.-F.; Yu, T.L.; Liu, W.-H.; Rwei, S.-P. Light scattering and viscoelasticity study of poly(vinyl alcohol)–borax aqueous solutions and gels. *Polymer* **2005**, *46*, 5541–5549. [CrossRef]
26. Lawrence, M.B.; Desa, J.; Aswal, V.K. Reentrant behaviour in polyvinyl alcohol–borax hydrogels. *Mater. Res. Express* **2018**, *5*, 015315. [CrossRef]
27. Lian, Z.; Ye, L. Effect of PEO on the network structure of PVA hydrogels prepared by freezing/thawing method. *J. Appl. Polym. Sci.* **2013**, *128*, 3325–3329. [CrossRef]
28. Zhang, Y.; Ye, L. Improvement of Permeability of Poly(vinyl alcohol) Hydrogel by Using Poly(ethylene glycol) as Porogen. *Polym. Technol. Eng.* **2011**, *50*, 776–782. [CrossRef]
29. Riedo, C.; Rollo, G.; Chiantore, O.; Scalarone, D. Detection and Identification of Possible Gel Residues on the Surface of Paintings after Cleaning Treatments. *Heritage* **2021**, *4*, 19. [CrossRef]
30. Luo, Z.-X.; Paulsen, J.; Song, Y.-Q. Robust determination of surface relaxivity from nuclear magnetic resonance DT2 measurements. *J. Magn. Reson.* **2015**, *259*, 146–152. [CrossRef]
31. Cantisani, E.; Fratini, F.; Malesani, P.; Molli, G. Mineralogical and petrophysical characterisation of white Apuan marble. *Period. Mineral.* **2005**, *74*, 117–138.
32. Yilmaz, A.; Yurdakoc, M.; Isik, B. Influence of transition metal ions on NMR proton T1 relaxation times of serum, blood, and red cells. *Biol. Trace Elem. Res.* **1999**, *67*, 187–193. [CrossRef] [PubMed]
33. McDonald, P.; Korb, J.-P.; Mitchell, J.; Monteilhet, L. Surface relaxation and chemical exchange in hydrating cement pastes: A two-dimensional NMR relaxation study. *Phys. Rev. E* **2005**, *72*, 011409. [CrossRef]
34. Kleinberg, R.; Kenyon, W.; Mitra, P. Mechanism of NMR Relaxation of Fluids in Rock. *J. Magn. Reson. Ser. A* **1994**, *108*, 206–214. [CrossRef]
35. Benavente, D.; Martínez-Martínez, J.; Cueto, N.; Ordóñez, S.; Garcia-Del-Cura, M.A. Impact of salt and frost weathering on the physical and durability properties of travertines and carbonate tufas used as building material. *Environ. Earth Sci.* **2018**, *77*, 147. [CrossRef]

36. Demirel, G.B.; Çaykara, T.; Demiray, M.; Guru, M. Effect of Pore-Forming Agent Type on Swelling Properties of Macroporous Poly(N-[3-(dimethylaminopropyl)]-methacrylamide-co-acrylamide) Hydrogels. *J. Macromol. Sci. Part A* **2008**, *46*, 58–64. [CrossRef]
37. Badr, Y.A.; El-Kader, K.M.A.; Khafagy, R.M. Raman spectroscopic study of CdS, PVA composite films. *J. Appl. Polym. Sci.* **2004**, *92*, 1984–1992. [CrossRef]
38. Tang, Q.; Qian, Y.; Yang, D.; Qiu, X.; Qin, Y.; Zhou, M. Lignin-based nanoparticles: A review on their preparations and applications. *Polymers* **2020**, *2*, 2471. [CrossRef] [PubMed]
39. Shi, Y.; Xiong, D.; Li, J.; Wang, K.; Wang, N. In situ repair of graphene defects and enhancement of its reinforcement effect in polyvinyl alcohol hydrogels. *RSC Adv.* **2017**, *7*, 1045–1055. [CrossRef]
40. Saeed, R.; Masood, S.; Abdeen, Z. Ionic Interaction of Transition Metal Salts with Polyvinyl Alcohol-Borax- Ethyl Acetate Mixtures. *Int. J. Sci. Technol.* **2013**, *3*, 132–142.
41. Angelova, L.V.; Terech, P.; Natali, I.; Dei, L.; Carretti, E.; Weiss, R.G. Cosolvent Gel-like Materials from Partially Hydrolyzed Poly(vinyl acetate)s and Borax. *Langmuir* **2011**, *27*, 11671–11682. [CrossRef]
42. Angelova, L.V.; Berrie, B.H.; De Ghetaldi, K.; Kerr, A.; Weiss, R.G. Partially hydrolyzed poly(vinyl acetate)-borax-based gel-like materials for conservation of art: Characterization and applications. *Stud. Conserv.* **2014**, *60*, 227–244. [CrossRef]
43. Al-Emam, E.; Motawea, A.G.; Janssens, K.; Caen, J. Evaluation of polyvinyl alcohol–borax/agarose (PVA–B/AG) blend hydrogels for removal of deteriorated consolidants from ancient Egyptian wall paintings. *Herit. Sci.* **2019**, *7*, 22. [CrossRef]
44. Yang, C.-C.; Chiu, S.-J.; Lee, K.-T.; Chien, W.-C.; Lin, C.-T.; Huang, C.-A. Study of poly(vinyl alcohol)/titanium oxide composite polymer membranes and their application on alkaline direct alcohol fuel cell. *J. Power Sources* **2008**, *184*, 44–51. [CrossRef]
45. Giampaolo, C.; Aldega, L. Il travertino: La pietra di Roma. *Rend. Online Della Soc. Geol. Ital.* **2013**, *27*, 98–109. [CrossRef]
46. Mancini, A.; Frondini, F.; Capezzuoli, E.; Mejia, E.G.; Lezzi, G.; Matarazzi, D.; Brogi, A.; Swennen, R. Porosity, bulk density and $CaCO_3$ content of travertines. A new dataset from Rapolano, Canino and Tivoli travertines (Italy). *Data Brief* **2019**, *25*, 104158. [CrossRef] [PubMed]
47. Gökçe, M.V. The effects of bedding directions on abrasion resistance in travertine rocks. *Turk. J. Earth Sci.* **2015**, *24*, 196–207. [CrossRef]
48. Garcia-Del-Cura, M.A.; Benavente, D.; Martínez-Martínez, J.; Cueto, N. Sedimentary structures and physical properties of travertine and carbonate tufa building stone. *Constr. Build. Mater.* **2012**, *28*, 456–467. [CrossRef]
49. Alesiani, M.; Capuani, S.; Maraviglia, B. NMR applications to low porosity carbonate stones. *Magn. Reson. Imaging* **2003**, *21*, 799–804. [CrossRef]
50. Alesiani, M.; Capuani, S.; Curzi, F.; Mancini, L.; Maraviglia, B. Evaluation of stone pore size distribution by means of NMR. In Proceedings of the 9th International Congress on Deterioration and Conservation of Stone, Venice, Italy, 19–24 June 2000.
51. Attanasio, D.; Armiento, G.; Brilli, M.; Emanuele, M.C.; Platania, R.; Turi, B. Multi-method marble provenance determinations: The carrara marbles as a case study for the combined use of isotopic, electron spin resonance and petrographic data. *Archaeometry* **2000**, *42*, 257–272. [CrossRef]
52. Sassoni, E.; Franzoni, E. Influence of porosity on artificial deterioration of marble and limestone by heating. *Appl. Phys. A* **2014**, *115*, 809–816. [CrossRef]
53. Siegesmund, S.; Ruedrich, J.; Koch, A. Marble bowing: Comparative studies of three different public building facades. *Environ. Earth Sci.* **2008**, *56*, 473–494. [CrossRef]
54. Venkataramanan, L.; Song, Y.; Hurlimann, M.D. Solving Fredholm integrals of the first kind with tensor product structure in 2 and 2.5 dimensions. *IEEE Trans. Signal Process.* **2002**, *50*, 1017–1026. [CrossRef]

Article

New Hydrogel Network Based on Alginate and a Spiroaceta Copolymer

Alina Elena Sandu, Loredana Elena Nita *, Aurica P. Chiriac, Nita Tudorachi, Alina Gabriela Rusu and Daniela Pamfil

"PetruPoni" Institute of Macromolecular Chemistry, Grigore Ghica Voda Alley 41-A, RO-700487 Iasi, Romania; sandu.alina@icmpp.ro (A.E.S.); achiriac@icmpp.ro (A.P.C.); ntudor@icmpp.ro (N.T.); rusu.alina@icmpp.ro (A.G.R.); dpamfil@icmpp.ro (D.P.)
* Correspondence: lnazare@icmpp.ro

Abstract: This study reports a strategy for developing a biohybrid complex based on a natural/synthetic polymer conjugate as a gel-type structure. Coupling synthetic polymers with natural compounds represents an important approach to generating gels with superior properties and with potential for biomedical applications. The study presents the preparation of hybrid gels with tunable characteristics by using a spiroacetal polymer and alginate as co-partners in different ratios. The new network formation was tested, and the structure was confirmed by FTIR and SEM techniques. The physical properties of the new gels, namely their thermal stability and swelling behavior, were investigated. The study showed that the increase in alginate content caused a smooth increase in thermal stability due to the additional crosslinking bridges that appeared. Moreover, increasing the content of the synthetic polymer in the structure of the gel network ensures a slower release of carvacrol, the encapsulated bioactive compound.

Keywords: hybrid gel network; natural/synthetic polymer conjugate; alginate; poly(itaconic anhydride-co-3,9-divinyl-2,4,8,10-tetra-oxa-spiro[5.5]undecane); bio-applications

Citation: Sandu, A.E.; Nita, L.E.; Chiriac, A.P.; Tudorachi, N.; Rusu, A.G.; Pamfil, D. New Hydrogel Network Based on Alginate and a Spiroacetal Copolymer. *Gels* **2021**, *7*, 241. https://doi.org/10.3390/gels7040241

Academic Editor: Wei Ji

Received: 5 November 2021
Accepted: 23 November 2021
Published: 27 November 2021

Publisher's Note: MDPI stays neutral with regard to jurisdictional claims in published maps and institutional affiliations.

Copyright: © 2021 by the authors. Licensee MDPI, Basel, Switzerland. This article is an open access article distributed under the terms and conditions of the Creative Commons Attribution (CC BY) license (https://creativecommons.org/licenses/by/4.0/).

1. Introduction

Hydrogels are three-dimensional, cross-linked networks of polymers and their individual physical properties are of particular interest for use in drug delivery applications, including in pharmaceutical patches or Transdermal Therapeutical Systems (TTS). Some of the most important characteristics of the hydrogels refer to their soft consistency and elasticity, as well as compatibility to body tissues [1,2]. These properties endorse the hydrogels as very attractive structures for biomaterials uses [1,3–6]. Hydrogels based on combinations between natural and synthetic polymers offer significant advantages, e.g., tunable mechanical properties, increased water content, enhanced biocompatibility and appropriateness to body tissue, and possibility of attaching chemical clues for further superior interfacial interactions. The use of polysaccharides as a base for the three–dimensional network structure preparation—recommended due to their properties such as biocompatibility, obtainment from renewable sources, and possibilities of "green" procedures for their modifying—is of major interest [1,7]. Among these compounds, the alginate has an important place. Alginic acid sodium salt (Alg–Na) is a naturally occurring biopolymer, biodegradable, biocompatible, and non-inflammatory, successfully used in medical applications as a carrier for drug delivery [8–10]. In the form of physically and chemically cross-linked systems, the alginate is an attractive starting material for the construction of hydrogels with desired morphology, stiffness, and bioactivity. However, the short residence of alginate time, due to a fast degradation process and poor mechanical characteristics, strongly limit the possibility of broadening its range of biomedical applications. Several chemical transformations of native alginate have been designed to provide mechanically and chemically robust materials and expand its range of application [8,11–13]. Moreover, the combination of alginate with synthetic polymers for obtaining hydrogels is of interest because in this way, the resistance and reproducibility of the materials increase. There is a series of studies [14–17] concerning the structures with synergistic properties resulting from combining synthetic polymers with polysaccharides [18]. Our group reported the preparation of poly(itaconic anhydride-co-3, 9-divinyl-2,4,8,10-tetraoxaspiro (5.5) undecane) (PITAU), a synthetic copolymer with pendant functional groups. PITAU presents specific properties, such as the possibility to create networks, biodegradability and biocompatibility, binding properties, amphiphilicity, thermal stability, and also sensitivity to pH and temperature [19,20]. The versatility and untapped potential of these polymeric systems make them promising agents for pharmaceutical delivery systems or support for bioactive compounds, among other biomedical applications. Because of these special characteristics of the PITAU copolymers, and taking into account our previous studies [19,20], the possibility of grafting PITAU onto alginate, to obtain biocompatible gels with improved properties it was investigated in the present work. This study presents the preparation of bioconjugated gels based on alginate and poly(itaconic anhydride-co-3,9-divinyl-2,4,8,10-tetraoxaspiro[5.5]undecane). The new prepared structures were characterized from structural, morphological, and thermal behavior points of view.

2. Results and Discussion

The proposed illustration structure of the new synthesized gel based on alginate and PITAU is presented in Figure 1.

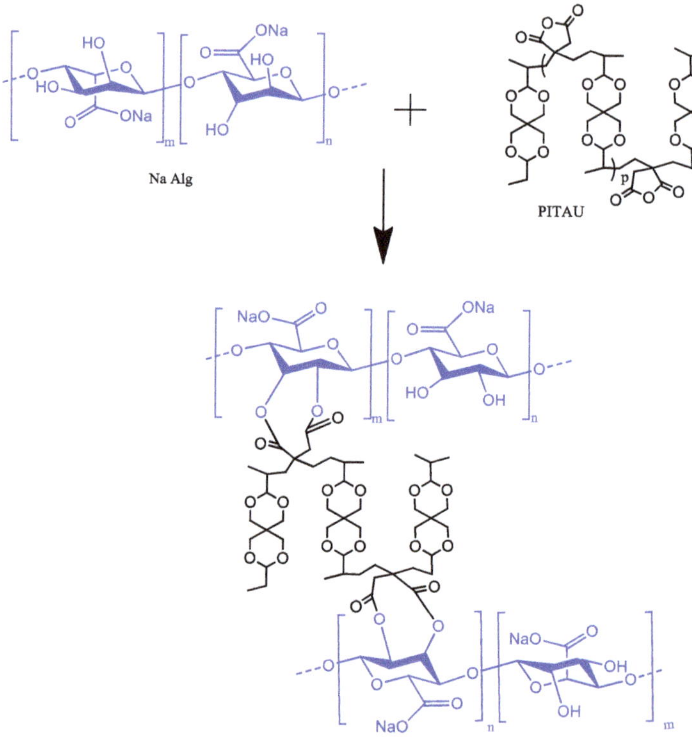

Figure 1. Schematized PITAU–alginate network.

2.1. FTIR Spectra

The FTIR spectra of the gel samples are presented in Figure 2. The chemical composition of PITAU–Alg gels was confirmed by FTIR spectroscopy. The characteristic peaks of the PITAU copolymer appeared at (1) 1780 cm^{-1} and 1856 cm^{-1} corresponding to C=O symmetric and asymmetric stretching of the five–member anhydride unit; (2) 1660 cm^{-1} from C=C stretching; and (3) 1400 cm^{-1} for =CH$_2$ in plane deformation, peaks evidenced as well by other authors [21]. The presence of a strong band around the 1000–1200 cm^{-1} region is attributed to ether C–O–C stretching from the spirochetal moieties. The presence of alginate is confirmed by characteristic alginate peaks registered at 2923 cm^{-1} and 2850 cm^{-1} that correspond to stretching vibrations of aliphatic C−H; at 1610 cm^{-1}, corresponding to the carboxylic groups C–O–O as a result of the asymmetric stretch; and the symmetric stretching at 1419 cm^{-1}. The band from 1024 cm^{-1} was attributed to the C–O stretching vibration, with contributions from C–C–H and C–O–H deformation data.

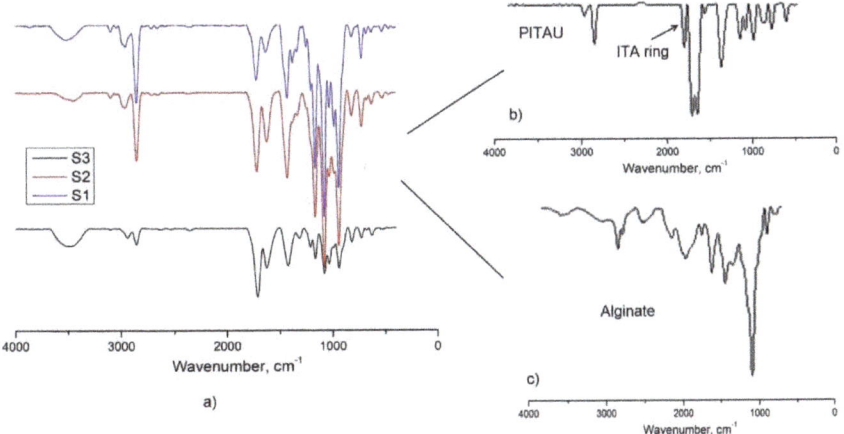

Figure 2. FTIR spectra of the (**a**) S1, S2, and S3 samples; (**b**) PITAU; (**c**) alginate.

The disappearance of the peaks corresponding to the vibrations of the anhydride ring from 1782 cm^{-1} and 1862 cm^{-1} (Figure 2) confirm the covalent bonds between alginate and PITAU governed by OH groups of the alginate, which open the ITA anhydride ring.

2.2. SEM Studies

The samples were analyzed using SEM microscopy to investigate the hydrogel microstructural architecture. The morphological analysis, illustrated in Figure 3, was performed in cross-section of the freeze–dried polymeric networks. According to the cross-sectional SEM pictures (Figure 3, S1–S3), the hydrogels present a continuous and porous configuration. As it is well known, the structural characteristics such as the porosity and topography governed by the chemical nature are very important in the performance of a gel. Moreover, a gel structure with a 3D network of interconnected channels and appropriate pores and pore size, allows a deeper penetration of liquids loaded with bioactive substances. The following images, representative of the studied gels, demonstrate the porous 3D architecture of the synthesized bioconjugate networks. SEM analyses confirm the important role of the amount of PITAU–Alg in generating the gel networks with superior properties. The pores of the gels have specific shapes and dimensions, in direct correlation with the PITAU–Alg ratio. Thus, the S3 sample, with higher amounts of alginate and smaller quantities of PITAU, leads to the formation of the irregular networks with larger pores. By increasing the PITAU amount (S2, S1), the gel networks are more ordered and present smaller pores. The morphological aspect of the bioconjugate matrices is correlated with the gel's capacity for swelling (Figure 8) and the ability to further incorporate, transport, and release a therapeutic agent.

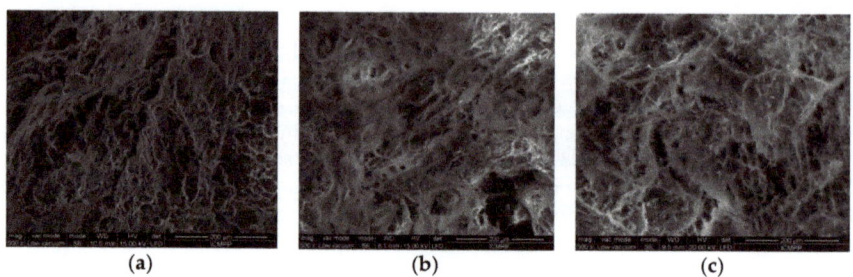

Figure 3. SEM microscopy of the freeze-dried gels: (**a**) S1, (**b**) S2, (**c**) S3.

2.3. Thermal Degradation

Figure 4 illustrates the thermogravimetric (TG) and derivative thermogravimetric (DTG) curves of the studied samples obtained with different gravimetric ratios between PITAU and alginate co-partners, and Table 1 presents the main thermal parameters of the samples. TG and DTG curves of the samples show similar shapes with different mass losses in the five stages of degradation. In the first stage, the mass losses of 4.15–6.52% were determined on the moisture removal, including the water adsorbed on the surface. The mass losses of 29.28–40.79% were recorded in the second stage of degradation with T_{onset} between 145–147 °C, due to the cleavage of hydroxyl, carboxyl, and carbonyl groups from samples and release of H_2O, CO_2, aldehydes, alcohols, and ketones [22]. The increase in the temperature over 220 °C generates two consecutive thermal processes (III, IV), which occur with low decomposition rates (3.5–4.0%/min) depending on the gravimetric ratio between PITAU and Alg in gels. The last thermal process with T_{onset} above 300 °C led to the structural units' decomposition of PITAU and the alginate glycosidic ring, with mass losses of about 16%, along with the release of CO_2 and high–molecular weight aliphatic derivatives, and the obtaining of Na_2CO_3 due to Na–alginate dehydration and decarbonylation. At 650 °C, Na_2CO_3 partially decomposes into CO_2 and Na_2O as residue. The increase in the Na–alginate content in the PITAU–Alg samples determined a slight increase in the thermal stability, which was attributed to the additional crosslinking bridges that appear in the new system. This observation is also supported by T20 and T40 values (temperatures with mass losses of 20% and 40%, respectively). These values are 181 °C and 227 °C in the case of S3, 177 °C and 206 °C for S2, and 172 °C and 203 °C for S1. Therefore, the thermal stability series grows in the following order: S3 > S2 > S1.

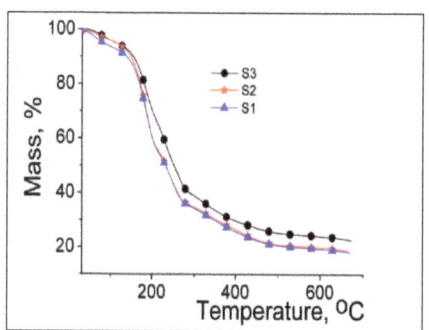
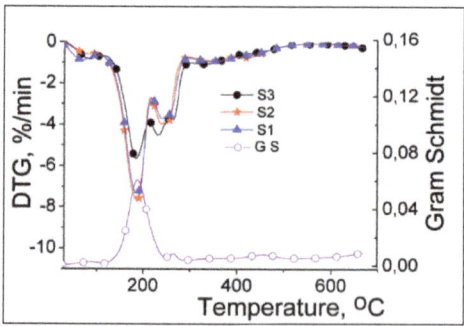

Figure 4. TG and DTG curves of PITAU–Alg gels.

Table 1. Thermal parameters.

Sample	Degradation Stage	T_{onset} °C	T_{peak} °C	W%	T_{20}, °C	T_{40}, °C	GS °C
S3	I	40	84	5.18	181	227	86
	II	145	185	29.28			190
	III	217	232	17.59			266
	IV	259	264	9.27			340
	V	309	325	16.13			460
	residue			22.55			
S2	I	42	86	4.15	177	206	
	II	147	193	40.79			
	III	224	244	10.39			
	IV	252	259	9.67			
	V	309	347	16.26			
	residue			18.74			
S1	I	34	71	6.52	172	203	
	II	144	189	39.56			
	III	224	243	11.28			
	IV	255	258	8.35			
	V	305	337	16.20			
	residue			18.09			

Heating rate—10 °C/min; T_{onset}—the temperature at which the thermal degradation starts; T_{peak}—the temperature at which the degradation rate is maximum; T_{20}, T_{40},—the temperatures corresponding to 20% and 40% mass losses, respectively; T_{GS}—the temperature at which the maximum amount of gases was released (determined from Gram–Schmidt curves using *Proteus* software); W—mass losses up to 680 °C.

The gases released were also examined by FTIR and mass spectrometry techniques concomitantly with the thermal decomposition of the S3 sample on the TG/FTIR/MS system, in the temperature range of 30–650 °C. Their 3D FT-IR spectrum is illustrated in Figure 5, and it can be noticed that the release of major gases during thermal decomposition takes place between 150–400 °C, according to the Gram–Schmidt and DTG curves. The main gases were identified based on the IR spectra and MS signals available in the literature and spectral libraries of the NIST [23]. From the 3D FT-IR spectrum, we extracted 2D spectra corresponding to the released gases at 195 °C and 426 °C (Figure 6). The absorption band from 3253 cm^{-1} was assigned to the MCT detector (ice band) of the TGA–IR external module cooled with liquid nitrogen [24]. The major gaseous degradation products from the studied sample were water, carbon dioxide, carboxylic derivatives, alcohols, saturated and unsaturated aliphatic hydrocarbons, cycloalkanes, ketones, aldehydes, anhydrides, and ethers. Thus, the main absorption bands located at 3853–3619, 1375–1304, and 1240 cm^{-1} can be assigned to water vapors and alcohols that can appear at the thermal degradation of the secondary hydroxyl or ester groups from alginate and PITAU units. The absorption bands from 3062–2847, 1523, 1462–1454, and 977–893 cm^{-1} are assigned to the vibration of CH, CH$_2$, and CH$_3$ groups located in the chemical structure of the saturated and unsaturated aliphatic hydrocarbons, cycloalkanes, ketones, and aldehydes. The higher signal present at 2355 cm^{-1} and the small signal at 677 cm^{-1} are attributed to carbon dioxide. The absorption bands between 1695 and 1648 cm^{-1} are assigned to the asymmetric carbonyl groups in acids, aldehydes, and ketones, and those from 1854–1793 and 1170–1086 cm^{-1} (νC=O vibrations) in ethers, anhydrides, and unsaturated aldehydes. These data are in agreement with the chemical structure of PITAU–Alg gels and correspond with gases that may result from their thermal degradation.

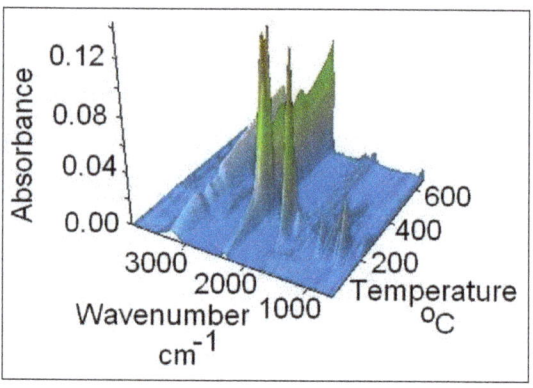

Figure 5. FTIR–3D spectrum of PITAU–Alg gel structure.

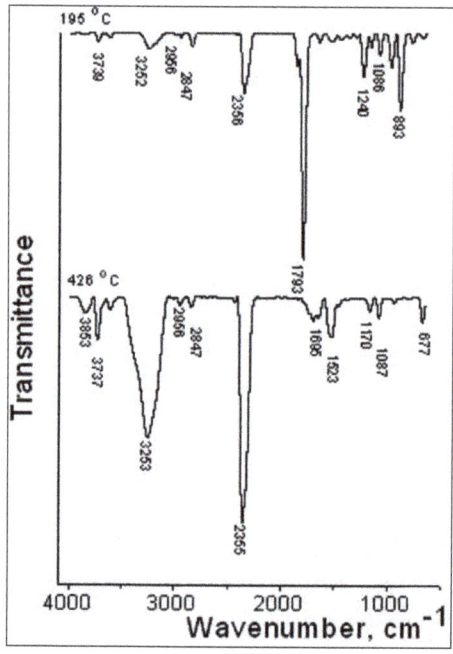

Figure 6. FTIR spectra of the evolved gases by thermal degradation of PITAU–Alg gel.

The data obtained from FTIR analysis of the evolved gases were also confirmed by MS spectrometry, and corresponding m/z signals are presented in Figure 6. With the increase in the temperature over 300 °C, a higher abundance of the gases developed, and the m/z ratio reached up to 120 for some high–molecular weight products. This confirms that up to 200 °C, there is a tearing of the thermal labile chemical bonds belonging to the functional groups (hydroxyl, carbonyl, carboxyl, and C–O links) from the structural units, and over 300 °C of the hydrocarbon bonds [25]. The main ionic fragments shown in Figure 7 are assigned as follows: HO^+ ($m/z = 17$), H_2O^+ ($m/z = 18$), CO_2^+ ($m/z = 44$). The ionic fragments of some saturated and unsaturated aliphatic derivatives are assigned as follows: CH_2^+ ($m/z = 14$), CH_3^+ ($m/z = 15$), ethane $C_2H_6^+$ ($m/z = 30$), cyclopropane $C_3H_6^+$ ($m/z = 42$), propane $C_3H_8^+$ ($m/z = 44$), cyclobutene $C_4H_6^+$ ($m/z = 54$), cyclobutene, 2–butene $C_4H_8^+$ ($m/z = 56$), butane $C_4H_{10}^+$ ($m/z = 58$), cyclohexane $C_6H_{12}^+$ ($m/z = 84$), carbonyl derivatives such as formaldehyde CH_2O^+ (m/z = 30), acetaldehyde $C_2H_4O^+$ (m/z = 44), acetone $C_3H_6O^+$ (m/z = 58), 2–cyclopenten–1–one $C_5H_6O^+$ (m/z = 82), itaconic anhydride $C_5H_4O_3$ (m/z = 112), alcohols as methanol CH_3OH+ (m/z = 32), ethanol $C_2H_5OH^+$ (m/z = 46), propanol $C_3H_8O^+$ (m/z = 60), acids as formic acid $CH_2O_2^+$ (m/z = 46), acetic acid $C_2H_4O_2^+$ (m/z = 60), propionic acid $C_3H_6O_2^+$ (m/z = 74), vinyl ethyl ether $C_4H_8O^+$ (m/z = 72), diethyl ether $C_4H_{10}O^+$ (m/z = 74).

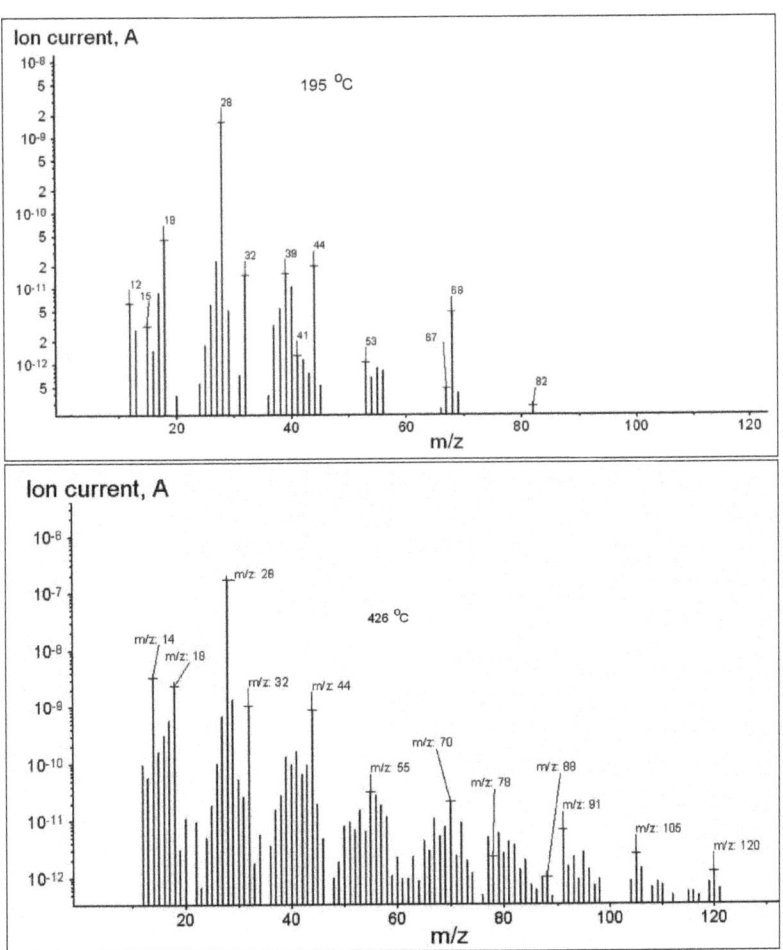

Figure 7. MS spectra of the evolved gases by thermal degradation of PITAU–Alg gel structures.

2.4. Swelling Study

One of the most important properties of the gels with pharmaceutical applicability is their capacity to swell when they come in contact with thermodynamically compatible solvents. In this case, the solvent molecules penetrate the polymeric network and determine the expanding of pores, which allow the incorporation of the drug or of other solvent molecules. As is well known, the swelling process is a consequence of the polymer–fluid interactive forces, which increase with the hydrophilic character of the macromolecules. The swelling degree of the studied gel samples as a function of time is illustrated in Figure 8. All the samples show a burst increase in swelling at the early stage, e.g., within the first 10 min. Then, the process is slowly continued up to 300 min.

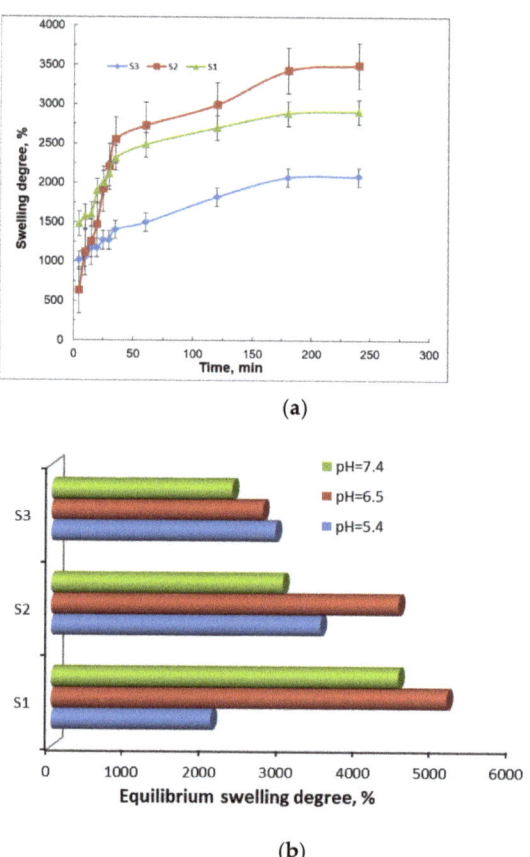

Figure 8. The swelling behavior of the gels as a function of time (**a**) and equilibrium swelling degree at different pH values (**b**).

The maximum degree of swelling corresponds to the network with the average amount of PITAU and alginate (S2), which is justified by a network structure with regulated pores (Figure 3b), due to PITAU copolymer presence, but at the same time owing to the alginate chains with relative mobility, which facilitates swelling. On the other hand, S1 samples have more possibilities for intense intramolecular bonds between alginate and PITAU governed by OH groups of the alginate, which open the ITA anhydride ring. The samples are pH–sensitive (Figure 8b), with the maximum swelling capacity registered at pH = 6.5.

The minimum swelling capacity is recorded for the sample with the minimum amount of PITAU and maximum amount of alginate (S3). In this case, the compact structure of the system induced a reduced swelling capacity. Subsequently, with the addition of synthetic polymer and the generation of an intermolecular network, the penetration of solvent molecules is easier, and consequently, the swelling capacity increased.

2.5. Release Study

The capacity of the PITAU–Alg network structure as a matrix was tested by the encapsulation and release of carvacrol. The carvacrol release profile presented in Figure 9 illustrates a burst effect highlighted in the first minutes of the samples' immersion in medium, while the equilibrium was reached after 4–11 h, depending on the PITAU–Alg hydrogel composition. Thus, the maximum amount of the drug released from the S2 hydrogel was reached after only 250 min, followed by the S1 hydrogel composition, with a more prolonged release time and a plateau reached after 540 min. The S3 hydrogel variant proceeds a fast release of bioactive compounds at the beginning, but then an extended release up to 720 min was observed. These observations correlate with the swelling study that attests a smaller degree of swelling in the case of the S3 sample and a higher degree for the S2 sample.

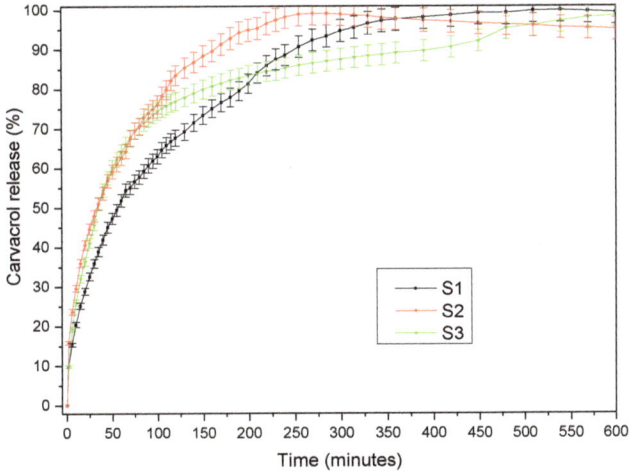

Figure 9. The release profile of carvacrol from PITAU–Alg hydrogel network.

The drug release kinetic parameters presented in Table 2 were calculated using the semi–empirical equation proposed by Korsmeyer and Peppas [26]. In the table above, the value of $n = 0.4931$ obtained for S1 indicates a Fickian diffusion mechanism of the drug from the sample, while the value of $n = 0.5581$ obtained for S3, which is between $0.5 < n < 1$, indicates an anomalous non–Fickian release behavior. The highest value of release rate constant k was obtained for S1, indicating the most accelerated drug release behavior, as observed in Figure 9, due to the chain disentanglement. These observations are also correlated with SEM images (Figure 3) which display a more ordered gel network with smaller pores in the S1 sample, in accord with the Fickian mechanism of release.

Table 2. Carvacrol release kinetic parameters.

Sample Name	n	R_n^2	k (min^{-n})	R_k^2
S1	0.4931	0.9976	0.0668	0.9971
S2	0.4055	0.9969	0.1209	0.9976
S3	0.5581	0.9982	0.0683	0.9989

n—release exponent; k—release rate constant; R_k^2 and R_n^2—correlation coefficients corresponding to the slope obtained for determination of n and k, respectively.

3. Materials and Methods

3.1. Materials

All reagents were of analytical purity and were used without further purification: alginic sodium salt (Alg) from brown algae was supplied by Acros Organics from Belgium, 3,9-divinyl-2,4,8,10-tetraoxaspiro[5.5] undecane (U) (purity 98%, Sigma-Aldrich, Hamburg, Germany), itaconic anhydride (ITA) (purity 98%, Aldrich), and 2,2′-Azobis (2-methylpropionitrile) (AIBN) (purity 98%, Sigma-Aldrich).

The solvents that we used 1,4 dioxane (D) (purity $\geq$ 99.0%) and diethyl ether (for precipitation)—were purchased from Sigma–Aldrich. The water used in the experiments was purified using an Ultra Clear TWF UV System.

3.2. Copolymer Synthesis

The PITAU preparation was described in detail before [20]. In brief, PITAU copolymers were synthesized through a radical process of polymerization with a total monomer concentration of about 20% (ratio between itaconic acid (ITA)/3, 9-divinyl-2,4,8,10-tetraoxaspiro (5.5) undecane (U) = 1.5/1), using AIBN as an initiator (0.9%) and 1,4-dioxane as a solvent. The reaction was conducted under a nitrogen atmosphere, in a constant temperature bath at 75 °C, with a stirring rate of 250 rpm, for about 17 h. The reaction mixture was further added dropwise into diethyl ether when the copolymer precipitated. The copolymer was further washed several times with diethyl ether and dried in a vacuum oven at room temperature with a 600 mm HG vacuum for 24 h.

3.3. Bioconjugate Samples Preparation

Three variants of bioconjugate structures based on PITAU and Alg were prepared as presented in Table 3. Precise amounts of copolymer (in dioxane solution) were mixed with specific amounts of alginate aqueous solution in order to have the following gravimetric ratios: PITAU:Alg = 3:1; 2:1, and 1.5:1. The gels formed rapidly within 20 min after mixing, and were left to mature for 24 h. Then, the gels were freeze–dried to remove the solvents and use the systems for further investigations.

Table 3. Codification and chemical composition of the studied samples.

Tested Samples	Gel Code	PITAU (g)	Alginate (g)
S1	PITAU_Alg_3_1	6.50	2.15
S2	PITAU_Alg_2_1	5.00	2.50
S3	PITAU_Alg_1.5_1	4.00	2.65

The recipe was calculated to obtain 100 mL gel.

3.4. Bioactive Compound Preparation

The carvacrol (Sigma Aldrich, Hamburg, Germany) was used as an antimicrobial compound. The ratio between the polymer and carvacrol was 2:1. The freeze–dried gels were immersed into a carvacrol solution and immersion was realized by diffusion mechanism. All the liquid was adsorbed into a polymer matrix.

3.5. Characterization of PITAU–Alg Gels

3.5.1. FT–IR Spectra

FT–IR spectra of the synthesized samples (S1, S2, S3) were recorded on a Vertex Bruker Spectrometer (Germany) in an absorption mode ranging from 400 to 4000 cm^{-1}. The samples were grounded with potassium bromide (KBr) powder and compressed into a disc for analysis. The spectra were generated at 4 cm^{-1} resolution with an average of 64 scans.

3.5.2. SEM Studies

The microstructural architecture of the samples was analyzed using SEM microscopy. The freeze-dried samples were examined by using a scanning electron microscope (Quanta 200). According to the cross–sectional SEM pictures, the hydrogels displayed a continuous and porous configuration acquired after freeze-drying.

3.5.3. Thermal Analysis

Thermogravimetric behavior of the samples was studied in dynamic conditions using a STA 449 F1 Jupiter apparatus (Netzsch, Germany) in a nitrogen atmosphere with a 10 °C min^{-1} heating rate, in the temperature range of 30–650 °C. Samples of 10–15 mg were placed in Al_2O_3 crucibles, and Al_2O_3 was used as reference material. The gases appearing by the thermal degradation of the samples were analyzed using an online connected spectrophotometer FT–IR (Vertex 70) equipped with an external module TGA–IR and Aëolos QMS 403C mass spectrometer. The acquisition of FT–IR spectra in 3D was executed with OPUS 6.5 software, with spectra recorded at 600–4000 cm^{-1} intervals, at a resolution of 4 cm^{-1}. The QMS 403C spectrometer worked at 10–5 mbar vacuum and electron impact ionization energy of 70 eV. The data acquisition was achieved in the range $m/z = 1$–200 with a measurement time of 0.5 s for one channel, resulting in a time/cycle of 100 s.

3.5.4. The Swelling Studies

The swelling degree of the gel samples was determined in PBS (phosphate-buffered saline) (0.01 M at pH = 5.4) and at 25 °C room temperature. The amount of the adsorbed solution was gravimetrically determined: the swollen gels were regularly taken out from the swelling medium, soaked on filter paper surface, weighed, and placed after every weighing in the same swelling medium. The measurements were continued until a constant weight was reached for each studied sample. All swelling experiments were performed in triplicate for S1, S2, and S3 samples. The degree of swelling (SD) was calculated by the following equation:

$$SD, \% = \frac{M(t) - M_0}{M_0} \times 100$$

where $M(t)$ is the weight of the swollen particles at time t, and M_0 is the weight of the sample before swelling (the weight of the dry sample).

The experiment was carried out in triplicate.

3.5.5. The Release Study

The in vitro release studies of carvacrol were carried out in a 708–DS Dissolution Apparatus coupled with a Cary 60 UV–VIS spectrophotometer (from Agilent Technologies, Germany) in 100 mL medium (85 mL 7.4 pH PBS/15 mL ethanol) at 25 ± 0.5 °C with a rotation speed of 100 rpm. Aliquots of the medium withdrawn at predetermined time intervals were analyzed with a λ_{max} value of 273 nm. The carvacrol concentrations were calculated based on the calibration curves determined at the same wavelengths.

The experiment was carried out in triplicate.

4. Conclusions

New gel network systems based on PITAU and alginate were developed in order to observe their characteristics, and to predict their potential applications in the biomedical and pharmaceutical fields.

According to SEM microscopy, the synthesized gels presented a continuous and porous morphology. SEM analyses also confirmed the important role of the ratio between PITAU and Alg co-partners in generating the bioconjugated structures with improved crosslinked networks. The also study underlined the dependence of the swelling properties of the synthesized networks on the ratio between PITAU and Alg co-partners. We found a better swelling capacity for the S2 sample (PITAU_Alg_2_1), justified by the larger pores generated by the intermolecular physical bonds between the macromolecular chains of the system. The study also demonstrated that the increase in the alginate content in the PITAU–Alg structures determined a smooth growth of the thermal stability, due to the additional crosslinking bridges that occur with increasing the amount of alginate. Moreover, the increase in the synthetic polymer content in the gel network structure ensured a slower release of carvacrol, the encapsulated bioactive compound.

Author Contributions: Conceptualization, A.E.S. and L.E.N.; methodology, A.E.S.; validation, L.E.N., A.P.C. and A.G.R.; formal analysis, A.E.S.; investigation, N.T. and D.P.; resources, L.E.N. and A.E.S.; writing—original draft preparation, L.E.N. and A.P.C.; supervision, L.E.N.; funding acquisition. All authors have read and agreed to the published version of the manuscript.

Funding: This work was financially supported by the grant of the Romanian National Authority for Scientific Research, CNCS–UEFISCDI, project number 339PED/2020, PN–III–P2–2.1–PED–2019–

2743 "New hybrid polymer/peptide hydrogels as innovative platforms designed for cell cultures applications", within PNCDI III.

Conflicts of Interest: The authors declare no conflict of interest.

References

1. Nita, L.E.; Chiriac, A.P.; Bercea, M.; Ghilan, A.; Rusu, A.G.; Dumitriu, R.P.; Mititelu–Tartau, L. Multifunctional hybrid 3D network based on hyaluronic acid and a copolymer containing pendant spiroacetal moieties. *Int. J. Biol. Macromol.* **2019**, *125*, 191–202. [CrossRef] [PubMed]
2. Fisher, O.Z.; Khademhosseini, A.; Langer, R.; Peppas, N.A. Bioinspired materials for controlling stem cell fate. *Acc. Chem. Res.* **2010**, *43*, 419–428. [CrossRef] [PubMed]
3. Fajardo, A.R.; Pereira, A.G.B.; Rubira, A.F.; Valente, A.J.M.; Muniza, E.C. Stimuli responsive polysaccharide–based hydrogels. In *Polysaccharide Hydrogels Characterization and Biomedical Applications*; Matricardi, P., Alhaique, F., Coviello, T., Eds.; Taylor & Francis Group: Abingdon-on-Thames, UK, 2016.
4. Costa, D.; Valente, A.J.M.; Miguel, M.G.; Queiroz, J. Gel network photodisruption: A new strategy for the codelivery of plasmid DNA and drugs. *Langmuir* **2011**, *27*, 13780–13789. [CrossRef] [PubMed]
5. Dugan, J.M.; Gough, J.E.; Eichhorn, S.J. Bacterial cellulose scaffolds and cellulose nanowhiskers for tissue engineering. *Nanomedicine* **2013**, *8*, 287–298. [CrossRef] [PubMed]
6. Durst, C.A.; Cuchiara, M.P.; Mansfield, E.G.; West, J.L.; Grande–Allen, K.J. Flexural characterization of cell encapsulated PEGDA hydrogels with applications for tissue engineered heart valves. *Acta Biomater.* **2011**, *7*, 2467–2476. [CrossRef] [PubMed]
7. Van Vlierberghe, S.; Dubruel, P.; Schacht, E. Biopolymerbased hydrogels as scaffolds for tissue engineering applications: A review. *Biomacromolecules* **2011**, *12*, 1387–1408. [CrossRef] [PubMed]
8. Diaconu, A.; Nita, L.E.; Chiriac, A.P.; Mititelu–Tartau, L.; Doroftei, F.; Vasile, C.; Pinteala, M. Self-linked polymer gels [based on hyaluronic acid and poly(itaconic anhydride-co-3,9-divinyl-2,4,8,10-tetraoxaspiro[5.5]undecane)] as potential drug delivery networks. In Proceedings of the 5th IEEE International Conference on E-Health and Bioengineering—EHB, Iasi, Romania, 19–21 November 2015.
9. Callesa, J.A.; Tártarac, L.I.; Lopez-Garcíad, A.; Dieboldd, Y.; Palmac, S.D.; Vallésa, E.M. Novel bioadhesive hyaluronan–itaconic acid crosslinked films for ocular therapy. *Int. J. Pharm.* **2013**, *455*, 48–56. [CrossRef] [PubMed]
10. Hoarea, T.R.; Kohaneb, D.S. Hydrogels in drug delivery: Progress and challenges. *Polymer* **2008**, *49*, 1993–2007. [CrossRef]
11. Borzacchiello, A.; Mayol, L.; Ramires, P.A.; Bartolo, C.; Pastorello, A.; Ambrosio, L.; Milella, E. Structural and rheological characterization of hyaluronic acid–based scaffolds for adipose tissue engineering. *Biomaterials* **2007**, *28*, 4399–4408. [CrossRef] [PubMed]
12. Borzacchiello, A.; Mayol, L.; Schiavinato, A.; Ambrosio, L. Effect of hyaluronic acid amide derivative on equine synovial fluid viscoelasticity. *J. Biomed. Mater. Res. A* **2010**, *92*, 1162–1170. [CrossRef] [PubMed]
13. Maltese, A.; Bucalo, C.; Maugeri, F.; Borzacchiello, A.; Mayol, L.; Nicolais, L.; Ambrosio, L. Novel polysaccharides based viscoelastic formulations for ophthalmic surgery: Rheological characterization. *Biomaterials* **2006**, *27*, 5134–5142. [CrossRef] [PubMed]
14. Imrea, B.; Garcíab, L.; Pugliad, D.; Vilaplana, F. Reactive compatibilization of plant polysaccharides and biobased polymers: Review on current strategies, expectations and reality. *Carbohydr. Polym.* **2019**, *209*, 20–37. [CrossRef] [PubMed]
15. Vasile, C.; Pamfil, D.; Stoleru, E.; Baican, M. New Developments in Medical Applications of Hybrid Hydrogels Containing Natural Polymers. *Molecules* **2020**, *25*, 1539. [CrossRef]
16. Rippe, M.; Cosenza, V.; Auzely–Velty, R. Design of Soft Nanocarriers Combining Hyaluronic Acid with Another Functional Polymer for Cancer Therapy and Other Biomedical Applications. *Pharmaceutics* **2019**, *11*, 338. [CrossRef] [PubMed]
17. Hu, Y.; Hu, S.; Zhang, S.; Dong, S.; Hu, J.; Kang, L.; Yang, X. A double-layer hydrogel based on alginate–carboxymethyl cellulose and synthetic polymer as sustained drug delivery system. *Sci. Rep.* **2021**, *11*, 9142. [CrossRef] [PubMed]
18. Diaconu, A.; Nita, L.E.; Bercea, M.; Chiriac, A.P.; Rusu, A.G.; Rusu, D. Hyaluronic acid gels with tunable properties by conjugating with a synthetic copolymer. *Biochem. Eng. J.* **2017**, *125*, 135–143. [CrossRef]
19. Chiriac, A.P.; Nita, L.E.; Diaconu, A.; Bercea, M.; Tudorachi, N.; Pamfil, D.; Mititelu-Tartau, L. Hybrid gels by conjugation of hyaluronic acid with poly(itaconic anhydride-co-3,9-divinyl-2,4,8,10-tetraoxaspiro[5.5]undecane) copolymers. *Int. J. Biol. Macromol.* **2017**, *98*, 407–418. [CrossRef] [PubMed]
20. Diaconu, A.; Chiriac, A.P.; Nita, L.E.; Tudorachi, N.; Neamtu, I.; Vasile, C.; Pinteala, M. Design and synthesis of a new polymer network containing pendant spiroacetal moieties. *Des. Monomers Polym.* **2015**, *18*, 780–788. [CrossRef]
21. Shang, S.; Huang, S.J.; Weiss, R.A. Synthesis and characterization of itaconic anhydride and stearyl methacrylate copolymers. *Polymer* **2009**, *50*, 3119–3127. [CrossRef]
22. Liua, Y.; Zhang, C.J.; Zhao, J.C.; Guo, Y.; Zhu, P.; Wang, D.Y. Bio-based barium alginate film: Preparation, flame retardancy and thermal degradation behavior. *Carbohydr. Polym.* **2016**, *139*, 106–114. [CrossRef] [PubMed]
23. Search for Species Data by Chemical Name. Available online: http://webbook.nist.gov/chemistry/name--ser.html (accessed on 1 November 2021).
24. TGA–IR User Manual, Bruker. Available online: https://www.bruker.com/en/products-and-solutions/infrared-and-raman/ft-ir-research-spectrometers/tg-ftir-thermogravimetric-analysis.html (accessed on 1 November 2021).

25. Nita, L.E.; Chiriac, A.P.; Rusu, A.G.; Bercea, M.; Diaconu, A.; Tudorachi, N. Interpenetrating polymer network systems based on poly(dimethylaminoethyl methacrylate) and a copolymer containing pendant spiroacetal moieties. *Mater. Sci. Eng.* **2018**, *87*, 22–31. [CrossRef] [PubMed]
26. Korsmeyer, R.W.; Lustig, S.R.; Peppas, N.A. Solute and penetrant diffusion in swellable polymers. I. Mathematical modeling. *J. Polym. Sci. Part B Polym. Phys.* **1986**, *24*, 395–408. [CrossRef]

MDPI
St. Alban-Anlage 66
4052 Basel
Switzerland
Tel. +41 61 683 77 34
Fax +41 61 302 89 18
www.mdpi.com

Gels Editorial Office
E-mail: gels@mdpi.com
www.mdpi.com/journal/gels

www.ingramcontent.com/pod-product-compliance
Lightning Source LLC
LaVergne TN
LVHW070657100526
838202LV00013B/984